Dennis Phillips

The Symmetric Eigenvalue Problem

THE SYMMETRIC EIGENVALUE PROBLEM

BERESFORD N. PARLETT
University of California
Berkeley, California

The greatest gift is the power to estimate correctly the value of things.

François de la Rochfoucauld (1613–1680)
Maxim number 244

Prentice-Hall, Inc., Englewood Cliffs, N. J. 07632

Library of Congress Cataloging in Publication Data

Parlett, Beresford N

The symmetric eigenvalue problem.

Bibliography: p.

Includes index.

1. Symmetric matrices. 2. Eigenvalues. I. Title.

QA188.P37 512.9'434 79-27221

ISBN 0-13-880047-2

Prentice-Hall Series in Computational Mathematics

Cleve Moler, Advisor

Editorial/production supervision and interior design by *Karen J. Clemments*
Cover design by *Edsal Enterprises*
Manufacturing buyer: *Gordon Osbourne and Joyce Levatino*

Printed in the United States of America

10 9 8 7 6 5 4 3 2 1

PRENTICE-HALL INTERNATIONAL, INC., *London*
PRENTICE-HALL OF AUSTRALIA PTY. LIMITED, *Sydney*
PRENTICE-HALL OF CANADA, LTD., *Toronto*
PRENTICE-HALL OF INDIA PRIVATE LIMITED, *New Delhi*
PRENTICE-HALL OF JAPAN, INC., *Tokyo*
PRENTICE-HALL OF SOUTHEAST ASIA PTE. LTD., *Singapore*
WHITEHALL BOOKS LIMITED, *Wellington, New Zealand*

To my parents,
Terèse and Norman

Contents

The more difficult material is indicated with a star both in the table of contents, and in the text.

||3|| COUNTING EIGENVALUES *38*

||4|| SIMPLE VECTOR ITERATIONS *58*

||5|| DEFLATION *81*

||6|| USEFUL ORTHOGONAL MATRICES (TOOLS OF THE TRADE) *87*

||7|| TRIDIAGONAL FORM *111*

Preface

Whenas in silks my Julia goes
Then, then methinks how sweetly flows
The liquefaction of her clothes.
Next when I cast mine eyes and see
That brave vibration, each way free,
Oh, how that glittering taketh me.

Robert Herrick (1591–1674)

The fact of harmony between Heaven and Earth and Man does not come
from a physical union, from a direct action,
it comes from a tuning on the same note producing vibrations in unison.

Tong Tshung-chu (second century B.C.)

$$\Delta\psi + \frac{8\pi^2 m}{h}(E - V)\psi = 0.$$

E. Schrödinger (1925)

Vibrations are everywhere and so too are the eigenvalues (or frequencies) associated with them. The concert-goer unconsciously analyzes the quivering of his eardrum, the spectroscopist identifies the constituents of a gas by looking at eigenvalues, and in California the State Building Department requires that the natural frequencies of many new buildings should lie outside the earthquake band. Indeed, as mathematical models invade more and more disciplines we can anticipate a demand for eigenvalue calculations in an ever richer variety of contexts. The reader who is not sure what an eigenvalue is should turn to Chapter 1 or some other work.

The interesting differences between various eigenvalue problems disappear when the formulation is made sufficiently abstract and the practical problems reduce to the single task of computing the eigenvalues of a square matrix with real or complex entries. Nevertheless there is one distinction worth maintaining: some matrices have real eigenvalues and others do not. The former usually come from so-called self-adjoint problems which are much nicer than the others. Apart from a handful of numerical analysts there appear to be few people interested in computations for both self-adjoint and nonself-adjoint problems. There are advantages in treating the two cases separately and this book is devoted to eigenvalue computations for real symmetric matrices. This may be the easier case but expectations are correspondingly higher.

For the newcomer to eigenvalue calculations I would like to give a brief summary of the situation. (A more detailed discussion of the contents of the book follows in the introduction.) Matrices are either small or large, as described in Chapter 2. For the small ones, good programs are now available in most scientific computing centers for virtually all the requests that users are likely to make. Furthermore, the understanding of the methods used by the programs is essentially complete. By dint of being worked and reworked the theory has become simple to the point of elegance. That story is told in Chapters 6 through 9.

Attention has now turned to large matrices. The tasks are harder, some good methods have been developed but the subject is far from tidy. There is as yet no general consensus on the right techniques for each task and there are financial incentives, in addition to the intellectual ones, for making more progress. In 1978 I was told that the computation of 30 eigenvalue/eigenvector pairs of a certain matrix of order 12,000 required $12,000 of computer time (excluding the cost of program development). The last five chapters present the tools which any large scale eigenvalue prospector should have at hand.

Any author would like to be both brief and intelligible; of these two virtues, the latter depends the more strongly on the background and fortitude of the reader. The level of difficulty of this book goes up and down according to the topic but please, gentle reader, do not expect to go through the proofs as you would a novel. The transition from one line to the next sometimes requires the proper marshalling of facts presented earlier and only by engaging in that irksome activity can the material be made your own. The exercises at the end of each section are there to reinforce this exhortation.

I hope there is something of interest in each chapter, even for the expert. Classical topics have been reworked and a fair proportion of the results, though not always new, are either little known or else unavailable. One strong incentive for writing this book was the realization that my friend and colleague W. Kahan would never narrow down his interests sufficiently to publish his own very useful insights.

The most difficult part of the enterprise was trying to impose some sort of order on the mass of worthwhile information. After many rearrangements the final list of contents seemed to emerge of its own accord. The book is intended to be a place of reference for the most important material as of 1978. As such it aims to be a sequel to Chapter 5 of J. H. Wilkinson's 1965 masterpiece "The Algebraic Eigenvalue Problem".

Selections from the material before Chapter 10 are in use as part of a senior level course at Berkeley and some of the later material was covered in a graduate seminar. There is a noticeable increase in difficulty after Chapter 9. Despite the mathematical setting of the discussions they are actually intended to enlighten anyone with a need to compute eigenvalues. Nothing would please me more than to learn that some consumers of methods, who make no claim to be numerical analysts, have read parts of the book with profit, or even a little pleasure.

At this point I would like to acknowledge some of my debts. My thesis adviser, the late George Forsythe, made numerical analysis seem both useful and interesting to all his students. Jim Wilkinson led many numerical analysts out of the wasteland of unenlightening analyses and showed us how to think right about matrix computations. Vel Kahan has been a patient and helpful guide for my study of symmetric matrices, and many other topics as well. The persevering reader will learn the magnitude of his contribution, but those who know him will appreciate that with any closer cooperation this book would never have seen the light of day. Naturally I have learnt a lot from regular contacts with Gene Golub, Cleve Moler, Chris Paige and Pete (G. W.) Stewart, and I have been helped by my present and former students. The manuscript was read by Joel Franklin, Gene Isaacson, Cleve Moler and Gil Strang. I am grateful for the blemishes they caught, and the encouragement they gave.

Next I wish to thank Dick Lau and the late Leila Bram, of the Office of Naval Research, whose generous support enabled me to write the book. Ruth (bionic fingers) Suzuki once again lived up to her exalted reputation for technical typing.

Beresford N. Parlett

Introduction

Principal notations are indicated inside the front cover.

At many places in the book, reference is made to more or less well known facts from matrix theory. For completeness these results had to be present but for brevity I have omitted proofs and elementary definitions. These omissions may help the reader to see the subject as a whole, an outcome that does not automatically follow a course in linear algebra. Not all the facts in Chapter 1 are elementary.

Chapter 2 is what the reader should know about the computer's influence on the eigenvalue problem. To my mind it is the need to harmonize conflicting requirements that makes the concoction of algorithms a fascinating task. At the other extreme my heart always sinks when the subject of roundoff error is mentioned. The sometimes bizarre effects of fixed precision arithmetic should be at the heart of work in matrix computations. Yet formal treatment of roundoff, though it seems to be necessary, rarely enlightens me. Fortunately there is Wilkinson's excellent little book, *Rounding Errors in Algebraic Processes*, which supplies a thorough and readable treatment of the fundamentals. Thus I felt free to concentrate on those topics which I consider essential, such as the much maligned phenomenon of cancellation. In addition I point to some of the excellent programs that are now available.

The book gets underway in Chapter 3 which centers on the remarkable fact that the number of negative pivots in the standard Gaussian elimination process applied to $\mathbf{A} - \xi\mathbf{I}$ equals the number of $\mathbf{A}$'s eigenvalues less than ξ.

The power method and inverse iteration are very simple processes but the ideas behind them are quite far reaching. It is customary to analyze the methods by means of eigenvector expansions but a little plane trigonometry yields results which are sharper and simpler than the usual formulations. In the same vein I have tried to find the most elegant proofs of the well-known error bounds which also seem to belong to Chapter 4.

The Rayleigh quotient iteration is perhaps the best known version of inverse iteration with variable shifts. Its rapid convergence, when it does converge to an eigenvalue/eigenvector pair, is well known. Kahan's discovery that the iteration does in fact converge from almost all starting vectors rounds out the theory very nicely but is little known. The proof (Sec. 4-9) is relatively difficult, and should be omitted from an undergraduate course.

The important point about "deflating" a known eigenpair from a matrix is that the safety and desirability of the operation depend on how it is done. It seemed useful to put the various techniques side by side for easy comparison.

Chapter 6 prepares the way for discussing explicit similarity transformations, but in fact orthogonal matrices play an important role in many other matrix calculations.

It does not seem to be appreciated very widely that any orthogonal matrix can be written as the product of reflector matrices. Thus the class of reflections is rich enough for all occasions and yet each member is characterized by a single vector which serves to describe its mirror. Pride of place must go to these tools despite the fact that they were generally ignored until Householder advocated their use in 1958.

Plane rotations made a comeback into favor in the late 1960's because, when proper scaling is used, they are no less efficient than reflections and are particularly suited for use with sparse matrices. In addition to the material mentioned above, Chapter 6 presents Wilkinson's simple demonstration of the very small effect that roundoff has on a sequence of explicit orthogonal similarity transformations. Lastly the QR factorization is introduced as simply a matrix description of the Gram–Schmidt orthonormalizing process. The actual way that the factors Q and R should be computed is a different question and the answer depends on the circumstances. The effect of roundoff on orthogonalization is treated in the final section.

Tridiagonal matrices have been singled out for special consideration ever since 1954 when Wallace Givens suggested reducing small full matrices to this form as an intermediate stage in computing the eigenvalues of the original matrix. Chapter 7 begins with the useful theorem which says to what extent the tridiagonal form is determined by the original matrix. Next comes a neglected but valuable characterization of the off-diagonal elements in the tridiagonal form. After this preparation the techniques of reduction are developed together with the analysis, due to Wilkinson, that an apparent instability in the reduction is not to be feared.

For later use I found it necessary to include one section which presents the interesting relationships governing elements in the eigenvectors of tridiagonal matrices, another section on the difficult question of when an off-diagonal element is small enough to be neglected, and a brief account of how to recover the tridiagonal matrix from its eigenvalues together with some extra information. These later sections are not standard material.

The champion technique for diagonalizing small tridiagonal matrices is the QR algorithm (the latest version is called the QL algorithm, just to add to the confusion). It is rather satisfying to the numerical analyst that the best method has turned out not to be some obvious technique like the power method, or the characteristic polynomial, or some form of Newton's method, but a sequence of sophisticated similarity transformations. Understanding of the algorithm has increased gradually and we now have an elegant explanation of why it always works. The theory is presented first, followed by the equally interesting schemes for implementing the process. For the sake of efficiency, the QL algorithm is usually applied to matrices of narrow bandwidth.

Over a hundred years older than QR is the Jacobi method for diagonalizing a symmetric matrix. The idea behind the method is simple and Chapter 9 covers the salient features. I must confess that I cannot see the niche where Jacobi methods will fit in the computing scene of the future. These doubts are made explicit in the final section.

With Chapter 10 the focus of the book shifts to large problems, although the chapter itself is pure matrix theory. It brings together results, old and new, which estimate eigenvalues of a given matrix in terms of eigenvalues of submatrices or low rank modifications. Up till now these results have been scattered through the literature. Here they are brought together and given in complete detail. This makes the chapter highly nutritious but rather indigestible at first encounter. (It came from notes used by Kahan in various lectures.)

Chapter 11 is a rather thorough description of a useful tool called the Rayleigh–Ritz procedure, but confined to matrices instead of differential operators. Section 11-4 should make clear in what senses the Ritz approximations are optimal and in what senses they are not. I often got confused on this point before I came to write the book.

The next chapter introduces a special sequence of subspaces, called Krylov spaces for no sufficient reason, and applies the Rayleigh–Ritz procedure to them. I believe that the attractive approximation theory in this chapter will be new to most readers. It establishes the great potential of the Lanczos algorithm for finding eigenvalues.

Chapter 13 is concerned with the widespread fear that roundoff will prevent us from reaping the harvest promised by Chapter 12. In his unpublished 1971 thesis, Paige showed that the fear is not warranted. In giving a proof of the main theorem I have tried to strip away the minutiae which encrusted Paige's penetrating, pioneering analysis of the simple Lanczos algorithm. Next comes selective orthogonalization, a modification guided by Paige's theory, which subdues quite economically the much feared loss of orthogonality among the Lanczos vectors. The method was announced in 1978, and is still ripe for further development. The chapter ends with a discussion of block Lanczos schemes which are likely to be needed for coping with the biggest matrices.

Chapter 14 treats the highly polished and effective methods based on block inverse iteration. These were refined by physicists, engineers, and numerical analysts from the 1960's onwards while the Lanczos algorithm lay under a cloud. I doubt that they will maintain their superiority.

The last chapter turns to the generalized problem and begins with some basic properties which are not universally appreciated. Indeed, I have seen books which get them wrong. The next topic, a key issue, is whether the given problem should be reduced to the standard form either explicitly or implicitly or not at all. The rest of the chapter, though long, is simply concerned with how the ideas and methods in the rest of the book can be extended to the general problem.

Both appendices have proved useful to me; the first describes elementary matrices, and the second lists the most useful properties of Chebyshev polynomials.

From time to time there occur applications in which special conditions permit the use of economical techniques which are, in general, unstable or inapplicable. The power to recognise these situations is one of the benefits of understanding a whole corpus of numerical methods.

For large matrices A the choice of method depends strongly on how many of the following operations are feasible:

1. Multiply any conformable vector x by A,
2. split A into $\mathsf{H} + \mathsf{V}$ where V is small and the triangular factors of H can be computed readily,

3. triangular factorization of A,
4. triangular factorization of A $-$ σI for many values of σ.

The more than can be done with A, the more powerful the methods which can be invoked to compute its eigenvalues. Early drafts of this book presented the methods according to these operations and, although that arrangement has been abandoned, this scheme provides a useful structure for thinking about possible algorithms.

Sections 15-9-1 and 15-13 extend this discussion to the generalized eigenvalue problem.

1

Basic Facts About Self-Adjoint Matrices

1-1 INTRODUCTION

This chapter offers a quick tour of those parts of linear algebra which are relevant to a discussion of numerical methods for approximating eigenvalues and eigenvectors of symmetric matrices. The notational conventions used throughout the book are introduced along the way, but they are also collected inside the covers for those readers who want to skip ahead. The list of facts may be of use to readers who wish to gauge their mastery of linear algebra by actually proving the assertions made on the tour. Those who would like a more leisurely trip should consult the annotated bibliography for a suitable text.

Before beginning, it is worth recalling that notation assists us by hiding information. Any particular notation is successful in so far as it displays only what is necessary. Matrix notation suppresses reference to individual elements of vectors and matrices as much as possible. The result is confusion before this language is mastered, and satisfaction thereafter.

1-2 EUCLIDEAN SPACE

This book is not concerned with the abstract notion of a vector space as any set of objects in which certain axioms hold; instead it goes straight to the coordinate representation and takes $\mathcal{R}^n$ as the vector space of all

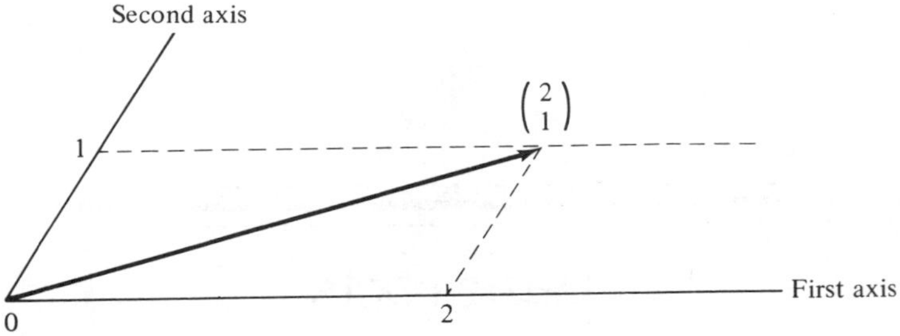

Figure 1-2-1 Axes

n-dimensional column vectors with real components. However linear algebra embraces complex vectors as easily as it does real and for this chapter, and this chapter only, each number (or scalar[†]) is complex unless the contrary is stated. Throughout the book scalars are usually denoted by lower case Greek letters (α, β), vectors by lower case Roman letters (x, q_1). The space of all complex n-dimensional vectors is denoted by $\mathcal{C}^n$.

To aid the imagination a vector x in $\mathcal{R}^2$ can be thought of as a line segment, or arrow, directed from the chosen origin o to a point in the plane with coordinates $\mathsf{x} = \begin{bmatrix} \xi_1 \\ \xi_2 \end{bmatrix}$. (See Fig. 1-2-1.)

The proper setting for problems involving real symmetric matrices is not just $\mathcal{R}^n$ but n-dimensional Euclidean space, called $\mathcal{E}^n$, which is $\mathcal{R}^n$ embellished with some extra structure. What distinguishes $\mathcal{E}^n$ from $\mathcal{R}^n$ or $\mathcal{C}^n$ is the idea that any pair of vectors x and y make a certain **angle**. More precisely, the extra structure enjoyed by $\mathcal{E}^n$ is an **inner product function** which assigns to each pair x (with components $\xi_1, \ldots, \xi_n$) and y (with components $\eta_1, \ldots, \eta_n$) a number written as (x, y) and defined by

$$(\mathsf{x}, \mathsf{y}) \equiv \sum_{i=1}^{n} \bar{\eta}_i \xi_i. \tag{1-2-1}$$

Throughout the book $\equiv$ denotes a definition and $\bar{\alpha}$ denotes the complex conjugate of α. Other more complicated definitions of the inner product function are appropriate in certain applications, such as the analysis of the stability of buildings, and are discussed later. However (1-2-1) is the simplest and distinguishes Euclidean space from other inner product spaces. In some contexts, such as the solution of most systems of linear equations, it is not appropriate to introduce angles and $\mathcal{R}^n$ is the proper setting.

[†]A quantity having magnitude but no direction (*Concise Oxford Dictionary*).

Apart from the next section the whole book is focused on real Euclidean space $\mathcal{E}^n$. Strictly speaking $\mathcal{E}^2$, for example, is not the set of points in the plane (or arrow tips) because it makes no sense to speak of the angle between two points. Pictorially $\mathcal{E}^2$ is the set of all arrows emanating from some chosen origin o. The axes, to which all the other vectors are referred, are taken **perpendicular** (or **orthogonal**) to each other. The powerful notions of orthogonality, angle, and length are all grounded in the inner product and are defined in the next paragraph.

The Euclidean **length**, or **norm**, of x is defined by

$$\|\mathsf{x}\| \equiv \sqrt{(\mathsf{x}, \mathsf{x})}\,. \tag{1-2-2}$$

There are many other norms which can be imposed on $\mathcal{C}^n$ or $\mathcal{R}^n$ but (1-2-2) is the natural norm for $\mathcal{E}^n$. The famous **Cauchy-Schwarz inequality**, namely

$$|(\mathsf{x}, \mathsf{y})| \leqslant \|\mathsf{x}\| \cdot \|\mathsf{y}\|, \text{ for all } \mathsf{x}, \mathsf{y} \text{ in } \mathcal{E}^n, \tag{1-2-3}$$

justifies the definition of angle given in (1-2-4) below. The angle in radians between x and y, written $\angle(\mathsf{x}, \mathsf{y})$, is the real number θ satisfying $0 \leqslant \theta \leqslant \pi$ and

$$\cos\theta = \frac{\mathrm{Re}(\mathsf{x}, \mathsf{y})}{\|\mathsf{x}\| \cdot \|\mathsf{y}\|}. \tag{1-2-4}$$

Usually it is not just x which is of interest but the line determined by x, namely the collection of all scalar multiples of x, including $-$x. We call this Span(x). The (acute) angle ϕ between the line on x and the line on y is the real number ϕ satisfying $0 \leqslant \phi \leqslant \pi/2$ and

$$\cos\phi = \frac{|(\mathsf{x}, \mathsf{y})|}{\|\mathsf{x}\| \cdot \|\mathsf{y}\|}. \tag{1-2-5}$$

Fig. 1-2-2 illustrates the difference between θ and ϕ. In particular, two vectors x and y are orthogonal if (x, y) = 0. A vector is normalized, or a unit vector, if $\|\mathsf{x}\| = 1$.

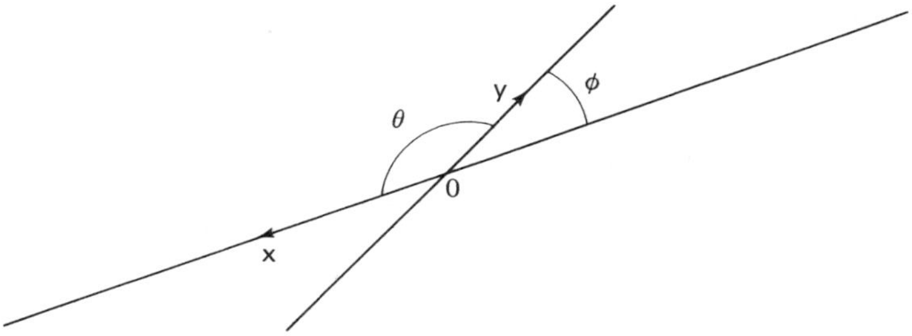

Figure 1-2-2 Angles between rays and lines in real $\mathcal{E}^2$

Matrices, always denoted by capital letters, are usually engaged in multiplying vectors. If F is m by n and w lies in $\mathcal{C}^n$ then the product Fw lies in $\mathcal{C}^m$. It is useful to imagine F multiplying all the vectors in $\mathcal{C}^n$ and in this way it transforms $\mathcal{C}^n$ into $\mathcal{C}^m$. The transformation is linear, i.e., $\mathsf{F}(\alpha\mathsf{x} + \beta\mathsf{y}) = \alpha\mathsf{F}\mathsf{x} + \beta\mathsf{F}\mathsf{y}$, and, in fact, any linear transformation of $\mathcal{C}^n$ into $\mathcal{C}^m$ can be represented by some m by n matrix F.

When F is square and invertible it can be regarded as simply effecting a change of basis in $\mathcal{C}^n$ instead of a transformation of vectors in $\mathcal{C}^n$. Thus Fw is either the image of w under F or, equally well, the new coordinates for the same old vector that used to have coordinates w. This flexibility is helpful to those who are used to it but confusing to the beginner.

The notation y* for the row vector $(\bar{y}_1, \ldots, \bar{y}_n)$ is widespread, yet fundamentally the conjugate transpose operation * is yet another consequence of the Euclidean inner product structure. Let F denote an m by n matrix. Its **adjoint** matrix F* is n by m and is defined abstractly by the property that, for all x in $\mathcal{E}^n$ and u in $\mathcal{E}^m$, F*u is the only vector satisfying

$$(\mathsf{x}, \mathsf{F}^*\mathsf{u}) = (\mathsf{F}\mathsf{x}, \mathsf{u}). \tag{1-2-6}$$

It can then be shown that F* is the familiar **conjugate transpose** of F (Ex. 1-2-3). For example, if $i^2 = -1$, α and β real, then

$$\begin{bmatrix} i & \alpha + i\beta \\ 0 & 1 \end{bmatrix}^* = \begin{bmatrix} -i & 0 \\ \alpha - i\beta & 1 \end{bmatrix}.$$

Whenever the product FG is defined then

$$(\mathsf{FG})^* = \mathsf{G}^*\mathsf{F}^*. \tag{1-2-7}$$

In particular $(\mathsf{Fw})^* = \mathsf{w}^*\mathsf{F}^*$. Even when F is real the same symbol F*, rather than F^H or F^T or F^t, can and will be used for the **transpose** of F. Furthermore, when F is square and invertible then

$$(\mathsf{F}^*)^{-1} = (\mathsf{F}^{-1})^* \tag{1-2-8}$$

and each will be written simply as F^{-*}.

An m by n matrix P, with $m \geqslant n$, is **orthonormal** if its columns are orthonormal, that is if

$$\mathsf{P}^*\mathsf{P} = \mathsf{I}_n \quad (= 1 \text{ for brevity}). \tag{1-2-9}$$

Square orthonormal matrices are called **unitary** (when complex) and **orthogonal** (when real). It is tempting to replace these two less-than-apt adjectives by the single natural one, **orthonormal**. The great importance of these matrices is that inner product formula (1-2-1) is preserved under their action, i.e.,

$$(\mathsf{Px}, \mathsf{Py}) = \mathsf{y}^*\mathsf{P}^*\mathsf{Px} = \mathsf{y}^*\mathsf{x} = (\mathsf{x}, \mathsf{y}). \tag{1-2-10}$$

When an orthogonal change of basis occurs in $\mathcal{E}^n$ then the coordinates of all vectors are multiplied by some orthonormal P but, mercifully, the values of norms and angles stay the same. In fact, the set of orthogonal transformations, together with simple translations, are the familiar rigid body motions of Euclidean geometry.

Exercises on Sec. 1-2

1-2-1. Derive the Cauchy-Schwarz inequality by noting that $(\mathsf{x} + e^{i\phi}\mu\mathsf{y})^*(\mathsf{x} + e^{i\phi}\mu\mathsf{y}) = \|\mathsf{x} + e^{i\phi}\mu\mathsf{y}\|^2 \geqslant 0$ and making the right choice for ϕ and μ.

1-2-2. Given complex x and y is it possible to choose a real θ such that $\exp(i\theta)\,\mathsf{x}$ and y are orthogonal? Consider (1-2-2).

1-2-3. By using coordinate vectors e_i for x and y show that (1-2-6) implies $(\mathsf{F}^*)_{ij} = (\bar{\mathsf{F}})_{ji}$.

1-2-4. Derive (1-2-7) from (1-2-6).

1-2-5. Prove that (1-2-10) holds if, and only if, (1-2-9) holds.

1-3 EIGENVALUES

The notions of eigenvalue and eigenvector do not depend on length, angle, or inner product and so we forsake $\mathcal{E}^n$ for this section in favor of $\mathcal{C}^n$. Of central importance in the study of any n by n matrix B are those special vectors in $\mathcal{C}^n$ whose directions are not changed (except possibly for sign) when multiplied by B. Any such vector z must satisfy

$$\mathsf{B}\mathsf{z} = \mathsf{z}\lambda = \lambda\mathsf{z}, \qquad \mathsf{z} \neq \mathsf{o} \tag{1-3-1}$$

for some scalar λ, called an **eigenvalue**† of B. Each nonzero multiple of z is an eigenvector‡, and λ and z belong to (or are associated with) each other. By convention o is never an eigenvector.

> FACT 1-1 Let $\mathsf{C} = \mathsf{F}\mathsf{B}\mathsf{F}^{-1}$. If $\{\lambda, \mathsf{z}\}$ is an eigenpair of B then $\{\lambda, \mathsf{F}\mathsf{z}\}$ is an eigenpair of C.

The mapping $\mathsf{B} \longrightarrow \mathsf{F}\mathsf{B}\mathsf{F}^{-1}$ is a **similarity transformation** (more briefly, a similarity) of B. The algebraic view is that similarity is an

---

†In German the word eigen means characteristic or special.

‡If we adopted the usual symbol for an eigenvector, namely x, the matrix of eigenvectors $\mathsf{X} \equiv (\mathsf{x}_1, \ldots, \mathsf{x}_n)$ would then be denoted by a symmetric letter thus violating a useful convention which is introduced in the next section.

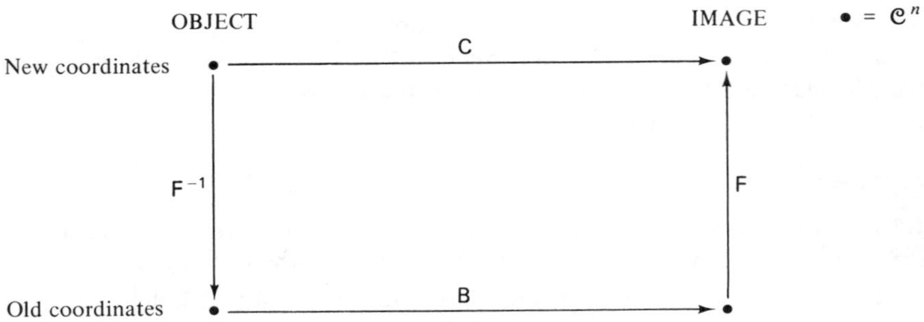

Figure 1-3-1 The similarity transformation

equivalence relation on the set of n by n matrices which preserves eigenvalues and changes eigenvectors in a simple way. The geometric interpretation of Fact 1-1 (and the bane of new students) is that B and C each represent the same (abstract) linear transformation on $\mathcal{C}^n$ while F defines a change of basis in $\mathcal{C}^n$ (old coordinates w, new coordinates Fw). This relationship is shown in Fig. 1-3-1; there are two ways of going from the top left to the top right corner.

The **characteristic polynomial**, χ, is defined by

$$\chi(\zeta) \equiv \chi_{\mathsf{B}}(\zeta) \equiv \det[\zeta \mathsf{I} - \mathsf{B}]. \tag{1-3-2}$$

By the theory of linear equations (1-3-1) has a nonzero solution z if, and only if $\chi(\lambda) = 0$. So B can have at most n eigenvalues. The set of these values, in the complex plane, constitutes B's **spectrum**.

1-4 SELF-ADJOINT MATRICES

We now go back to n by n matrices pre-multiplying vectors in $\mathcal{E}^n$ and focus on those matrices which satisfy

$$\mathsf{M}^* = \mathsf{M}. \tag{1-4-1}$$

Note that (1-4-1) is not preserved under arbitrary change of basis in $\mathcal{C}^n$ or $\mathcal{R}^n$. In more general contexts than $\mathcal{E}^n$ linear operators which satisfy (1-4-1) are called **self-adjoint** and then there is no need to make a distinction between the real and complex case. Unfortunately in n dimensions such matrices are called **Hermitian**, when complex, and **symmetric**, when real. In this chapter we will persevere with the adjective self-adjoint; afterward we shall throw generality to the winds and concentrate on real symmetric matrices. Unless the contrary is stated, letters which are **symmetric about a vertical axis**, namely A, H, . . . , Y, represent self-adjoint matrices.

FACT 1-2 All eigenvalues of self-adjoint matrices are real.

As a result of Fact 1-2, we may label eigenvalues in increasing order,

$$\lambda_1 \leqslant \lambda_2 \leqslant \cdots \leqslant \lambda_n \tag{1-4-2}$$

and $\lambda_j[\mathsf{M}]$ denotes the j-th (smallest) eigenvalue of M. Any normalized eigenvector belonging to λ_i is denoted by $\mathbf{z}_i$; $\mathsf{A}\mathbf{z}_i = \mathbf{z}_i\lambda_i$, $i = 1, \ldots, n$. For large values of n it is sometimes convenient to label the largest eigenvalue without reference to dimension. Hence, we also order the eigenvalues by

$$\lambda_{-n} \leqslant \cdots \leqslant \lambda_{-2} \leqslant \lambda_{-1}. \tag{1-4-3}$$

Thus, λ_{-1} is always the largest eigenvalue (algebraically).

FACT 1-3 If $\lambda_k \neq \lambda_j$ then $(\mathbf{z}_j, \mathbf{z}_k) = \mathbf{z}_k^*\mathbf{z}_j = 0$.

A few words must be said about multiple eigenvalues. It is tempting to interpret Fact 1-3 as saying that distinct eigenvectors of self-adjoint matrices are orthogonal. Consideration of the identity matrix I shows that the proper formulation is that eigenvectors of A may always be **chosen** to be pairwise orthogonal. Multiple eigenvalues furnish a wide choice of associated eigenvectors.

If λ is an eigenvalue of A, i.e., if $\lambda = \lambda_j[\mathsf{A}]$, then $\mathfrak{N}_\lambda$, the null space of $\mathsf{A} - \lambda$, i.e., the set of all $\mathbf{x}$ such that $(\mathsf{A} - \lambda)\mathbf{x} = \mathbf{o}$, is sometimes called the eigenspace belonging to λ. The only vector in $\mathfrak{N}_\lambda$ which is not an eigenvector is $\mathbf{o}$. The multiplicity of λ is the dimension of $\mathfrak{N}_\lambda$. (For nonsymmetric matrices the notion of eigenvalue multiplicity is more complicated.) Eigenspaces are the simplest invariant subspaces (discussed at the end of this section) and one consequence of the next fact is that all invariant subspaces are spanned by eigenvectors.

FACT 1-4 (The spectral theorem) Any A is similar to a diagonal matrix Λ via an orthonormal similarity. In symbols

$$\mathsf{A} = \mathsf{Z}\Lambda\mathsf{Z}^* = \sum_{i=1}^{n} \lambda_i \mathbf{z}_i \mathbf{z}_i^*,$$

$$\mathsf{I} = \mathsf{Z}\mathsf{Z}^* = \sum_{i=1}^{n} \mathbf{z}_i \mathbf{z}_i^*.$$

$\mathsf{Z} = (\mathbf{z}_1, \ldots, \mathbf{z}_n)$ is a matrix of orthonormalized eigenvectors of A.

Definition. A matrix E is a **projector** if $\mathsf{E}^2 = \mathsf{E}$. It is an **orthogonal projector** (in contrast to an oblique projector) if $\mathsf{Ey} = \mathsf{o}$ for all y orthogonal to E's range (also called span E). The condition for this is simply

$$\mathsf{E}^* = \mathsf{E}. \tag{1-4-4}$$

For any square matrix B the **spectral projector** (sometimes called idempotent) E_λ for an eigenvalue λ satisfies

$$\mathsf{BE}_\lambda = \mathsf{E}_\lambda\mathsf{B} = \lambda\mathsf{E}_\lambda. \tag{1-4-5}$$

To achieve a **unique** decomposition in the presence of multiple eigenvalues one uses the eigenspaces $\mathcal{N}_\lambda$, or equivalently the spectral projectors H_λ defined by

$$\mathsf{H}_\lambda\mathsf{x} = \begin{cases} \mathsf{x}, & \text{if } \mathsf{x} \in \mathcal{N}_\lambda, \\ \mathsf{o}, & \text{if } \mathsf{x} \in \mathcal{N}_\mu,\ \mu \neq \lambda. \end{cases} \tag{1-4-6}$$

Now the spectral theorem can be written unambiguously as

$$\mathsf{A} = \sum \lambda_j \mathsf{H}_j, \qquad \mathsf{I} = \sum \mathsf{H}_j \tag{1-4-7}$$

where the sums are over A's spectrum. Because A is self-adjoint its spectral projectors are also orthogonal projectors which is why we used a symmetric letter H in (1-4-6) and (1-4-7).

EXAMPLE 1-4-1

$$\mathsf{A} = \begin{bmatrix} 2 & 1 & 1 & 0 \\ 1 & 2 & 0 & 1 \\ 1 & 0 & 2 & 1 \\ 0 & 1 & 1 & 2 \end{bmatrix}; \qquad \begin{aligned} \alpha_1 &= 0 \\ \alpha_2 &= \alpha_3 = 2 \\ \alpha_4 &= 4 \end{aligned}$$

$$\mathsf{z}_1 = \frac{1}{2}\begin{bmatrix} 1 \\ -1 \\ -1 \\ 1 \end{bmatrix}, \qquad \mathsf{H}_1 = \mathsf{z}_1\mathsf{z}_1^* = \frac{1}{4}\begin{bmatrix} 1 & -1 & -1 & 1 \\ -1 & 1 & 1 & -1 \\ -1 & 1 & 1 & -1 \\ 1 & -1 & -1 & 1 \end{bmatrix},$$

$$\mathsf{z}_4 = \frac{1}{2}\begin{bmatrix} 1 \\ 1 \\ 1 \\ 1 \end{bmatrix}, \qquad \mathsf{H}_4 = \mathsf{z}_4\mathsf{z}_4^* = \frac{1}{4}\begin{bmatrix} 1 & 1 & 1 & 1 \\ 1 & 1 & 1 & 1 \\ 1 & 1 & 1 & 1 \\ 1 & 1 & 1 & 1 \end{bmatrix},$$

z_2, z_3 are not unique. One suitable pair is

$$\tfrac{1}{2}(1, -1, 1, -1)^*, \qquad \tfrac{1}{2}(1, 1, -1, -1)^*;$$

another pair is

$$(1/\sqrt{2})(1, 0, 0, -1)^*, \qquad (1/\sqrt{2})(0, 1, -1, 0)^*.$$

However,

$$H_2 = z_2z_2^* + z_3z_3^* = \frac{1}{2}\begin{bmatrix} 1 & 0 & 0 & -1 \\ 0 & 1 & -1 & 0 \\ 0 & -1 & 1 & 0 \\ -1 & 0 & 0 & 1 \end{bmatrix}$$

is unique and

$$A = 0 \cdot H_1 + 2 \cdot H_2 + 4 \cdot H_4, \qquad I = H_1 + H_2 + H_4.$$

1-4-1 INVARIANT SUBSPACES

An important consequence of Fact 1-4 is that any subspace $\mathcal{S}$ of $\mathcal{E}^n$ which is invariant under A, i.e., $A\mathcal{S} \subseteq \mathcal{S}$, is just the span (or direct sum) of some eigenvectors. Associated with each invariant $\mathcal{S}$ is the linear operator $A|_{\mathcal{S}}$, the **restriction** of A to $\mathcal{S}$, whose action is the same as A but whose domain is $\mathcal{S}$. Strictly speaking any function is defined by its action together with its domain, and so any change in domain produces a new operator which must receive its own name. It can be verified that $A|_{\mathcal{S}}$ is self-adjoint and its eigenvalues and eigenvectors are the appropriate subset of those of A.

1-4-2 HERMITIAN MATRICES

We close this section on a practical note.

> FACT 1-5 Hermitian matrices H may be replaced by real symmetric matrices $\hat{H}$ of twice the order for the purpose of computing eigensystems. See Ex. 1-4-6.

Fact 1-5 is useful in two situations; not only when programs for complex Hermitian matrices are unavailable but also when they are available and yet the complex arithmetic operations are implemented inefficiently. The local computer center experts should be consulted on this point. When complex arithmetic is implemented well those codes which treat the original Hermitian H should be twice as fast as the programs for real symmetric matrices working on $\hat{H}$.

Exercises on Sec. 1-4

1-4-1. Find a 2 by 2 complex symmetric matrix (not Hermitian) with complex eigenvalues.

1-4-2. Prove Fact 1-2 by considering $z_i^* A z_i$.

1-4-3. Prove Fact 1-3 by considering $z_i^* A z_j$.

1-4-4. Verify Fact 1-4 for $A = \begin{bmatrix} 0 & 1 \\ 1 & 0 \end{bmatrix}$. Compute $E_1 = z_1 z_1^*$ and $E_2 = z_2 z_2^*$.

1-4-5. Show that rank(E_j) = dimension($\mathcal{N}_j$) = m_j. Note that rank(E_j) = trace(E_j). Use the fact that $E_j^2 = E_j$ and trace(BC) = trace(CB). Trace(F) $\equiv \sum_{j=1}^{n} f_{jj}$.

1-4-6. Let $H = M + iS$ be Hermitian ($i^2 = -1$) and let $H(u + iv) = (u + iv)\lambda$, λ real. Verify that $\hat{H} = \begin{bmatrix} M & -S \\ S & M \end{bmatrix}$ is symmetric and has eigenvectors $\begin{bmatrix} u \\ v \end{bmatrix}$ and $\begin{bmatrix} -v \\ u \end{bmatrix}$ belonging to λ.

1-4-7. Use Fact 1-3 to show that a spectral projector is orthogonal when A is self-adjoint.

1-4-8. Show that $H_\lambda = \sum z_i z_i^*$ where $\{z_i\}$ is any orthonormal basis of $\mathcal{N}_\lambda$.

1-4-9. Show that if $\mathcal{S}$ is a subspace invariant under A then any eigenvalue of $A|_\mathcal{S}$ is an eigenvalue of A. Let $\{z_1, \ldots, z_n\}$ be an eigenvector basis for $\mathcal{E}^n$. Is it always true that some subset of $\{z_1, \ldots, z_n\}$ is a basis for $\mathcal{S}$?

1-5 QUADRATIC FORMS

Self-adjoint matrices arise naturally in the study of pure quadratic forms (or functions), not mixed with lower degree terms, typically,

$$\psi(x) \equiv x^*Ax = \sum_{i=1}^{n} \sum_{j=1}^{n} a_{ij}\bar{x}_i x_j. \tag{1-5-1}$$

These forms frequently represent some form of energy of a system, be it atom or skyscraper. Linear invertible changes of variable, say $x \longrightarrow y = F^{-1}x$, force a change in the form, i.e.,

$$\psi(x) = \hat{\psi}(y) \equiv y^*\hat{A}y, \qquad \text{for all } x, \tag{1-5-2}$$

if, and only if,

$$\hat{A} = F^*AF. \tag{1-5-3}$$

The mapping $A \longrightarrow F^*AF$ is a **congruence transformation** of A; we say $\hat{A}$ is **congruent** to A. These transformations preserve self-adjointness but do not, in general, preserve eigenvalues. Nevertheless congruencies do in some sense preserve the signs ($\pm$) of eigenvalues. That is the gist of the next fact.

> FACT 1-6 (Sylvester's Inertia Theorem.) Each A is congruent to a matrix diag($I_\pi, -I_\nu, O_\zeta$), where the number triple (π, ν, ζ) depends only on A and is called A's **inertia**. Moreover π, ν, ζ are the number of positive, negative, and zero eigenvalues of A.

In addition,

$$\begin{aligned} \pi + \nu + \zeta &= n, \\ \pi + \nu &= \text{rank}(\mathsf{A}), \qquad (1\text{-}5\text{-}4) \\ \pi - \nu &= \text{signature}(\mathsf{A}). \end{aligned}$$

Fact 1-6 shows that two self-adjoint matrices are congruent if, and only if, they have the same inertia.

Among quadratic forms are those which are strictly positive for all nonzero vectors: $\psi(\mathsf{x}) > 0$ if $\mathsf{x} \neq \mathsf{o}$. Such forms, and the self-adjoint matrices associated with them, are positive definite. [The word definite serves to distinguish these matrices from those which are merely positive, that is, matrices B satisfying $b_{ij} > 0$ for all i, j.] If $\psi(\mathsf{x}) \geqslant 0$ for all $\mathsf{x} \neq \mathsf{o}$ then it, and its matrix, are **positive semi-definite**. If $\psi(\mathsf{x})$ takes on both positive and negative values it is indefinite.

FACT 1-7 The following statements are equivalent:
1. A is positive definite,
2. A's inertia is $(n, 0, 0)$,
3. A's eigenvalues are positive.
4. A has a unique Choleski factor C which is upper triangular with positive diagonal and satisfies $\mathsf{A} = \mathsf{C}^*\mathsf{C}$ (see Chap. 3 on triangular factorization).

In general it requires less work to check 4 than any of the other three conditions.

The pure quadratic form ψ is homogeneous of degree 2: $\psi(\alpha\mathsf{x}) = \alpha^2\psi(\mathsf{x})$. Consequently there is no loss of information in restricting attention to its values on the unit sphere in $\mathcal{E}^n$. The new function is Rayleigh's quotient ρ and is usually defined by

$$\rho(\mathsf{u}) \equiv \rho(\mathsf{u}; \mathsf{A}) \equiv \frac{\mathsf{u}^*\mathsf{A}\mathsf{u}}{\mathsf{u}^*\mathsf{u}}, \qquad \mathsf{u} \neq \mathsf{o}. \qquad (1\text{-}5\text{-}5)$$

FACT 1-8 The Rayleigh quotient enjoys the following basic properties:

Homogeneity: $\rho(\alpha\mathsf{u}) = \rho(\mathsf{u}), \alpha \neq 0$ (degree 0).

Boundedness: $\rho(\mathsf{u})$ ranges over the interval $[\lambda_1, \lambda_{-1}]$ as u ranges over all nonzero n-vectors.

Stationarity: $\rho(\mathsf{u})$ is stationary (i.e., the gradient of ρ is o^*) at, and only at, the eigenvectors of A.

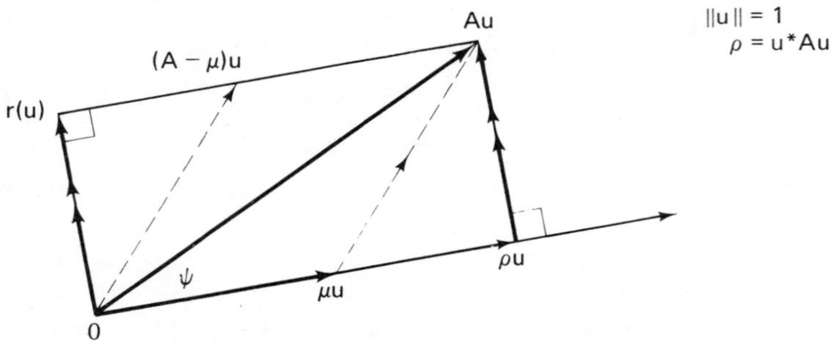

Figure 1-5-1 Geometric meaning of the residual

Rayleigh's quotient plays a big role in the computation of eigenvalues and eigenvectors and is discussed in more detail in Chap. 4. One basic property is worth mentioning now. For any $\mathsf{u} \neq \mathsf{o}$ define the special residual vector $\mathsf{r}(\mathsf{u})$ by

$$\mathsf{r}(\mathsf{u}) = [\mathsf{A} - \rho(\mathsf{u})]\mathsf{u}. \tag{1-5-6}$$

Figure 1-5-1 illustrates the fact that

$$\mathsf{Au} = \rho(\mathsf{u})\mathsf{u} + \mathsf{r}(\mathsf{u}) \tag{1-5-7}$$

is an orthogonal decomposition of Au. In other words,

> FACT 1-9 The minimal residual property. For each u in $\mathcal{E}^n$
>
> $$\|[\mathsf{A} - \rho(\mathsf{u})]\mathsf{u}\| \leqslant \|[\mathsf{A} - \mu]\mathsf{u}\| \text{ for all } \mu \text{ in } \mathcal{C}.$$

Exercises on Sec. 1-5

1-5-1. Express $\mathsf{x^*Ax}$ as a sum of squares.

$$\mathsf{A} = \begin{bmatrix} 10 & 1 & -14 \\ 1 & 17 & -4 \\ -14 & -4 & 20 \end{bmatrix}$$

1-5-2. Find the inertia of

$$\begin{bmatrix} 2 & 1 & 0 \\ 1 & 1 & 1 \\ 0 & 1 & -1 \end{bmatrix}$$

1-5-3. Show that the matrix $\mathsf{B^*B}$ is positive semi-definite for any m by n matrix B. When is it positive definite?

1-5-4. Show that the eigenvalues of a positive definite matrix are positive. Is the converse true?

1-5-5. Show that the triangular factorization

$$\mathsf{A} = \mathsf{L}\,\Delta\,\mathsf{L}^*$$

is unique when it exists. Here L is lower triangular with 1's on the diagonal and Δ is diagonal.

1-5-6. Establish Fact 1-8. For the third part try $\mathsf{u} = \mathsf{z} + \epsilon\mathsf{w}$ with $\mathsf{w}^*\mathsf{z} = 0$, $\|\mathsf{w}\| = 1$, $\mathsf{Az} = \mathsf{z}\lambda$ and let $\epsilon \longrightarrow 0$.

1-5-7. For the matrix A given in the example of Sec. 1-4 and $\mathsf{u} = (1, 1, 1, 0)^*/\sqrt{3}$, evaluate $\rho(u)$ and compute $\|\mathsf{r}(\mathsf{u})\|$.

1-5-8. Show that if $\mathsf{Az} = \mathsf{z}\lambda$ then $\rho(\mathsf{z}; \mathsf{A}) = \lambda$.

1-6 MATRIX NORMS

There is a special matrix norm associated with the Euclidean vector norm, namely

$$\|\mathsf{B}\| \equiv \max_{\mathsf{u}\neq 0} \frac{\|\mathsf{Bu}\|}{\|\mathsf{u}\|} = \sqrt{(\lambda_{-1}[\mathsf{B}^*\mathsf{B}])}. \tag{1-6-1}$$

This is called the **spectral norm** or the **bound norm**. It is the smallest norm which satisfies the useful inequality

$$\|\mathsf{Bu}\| \leqslant \mathrm{norm}(\mathsf{B})\|\mathsf{u}\|, \qquad \text{for all } \mathsf{u} \in \mathcal{E}^n. \tag{1-6-2}$$

Unfortunately it is expensive to compute. Another norm which satisfies (1-6-2) and is a simple function of the matrix elements is the **Frobenius** or **Schur** or **Hilbert-Schmidt** norm,

$$\|\mathsf{B}\|_F \equiv \sqrt{[\mathrm{trace}(\mathsf{B}^*\mathsf{B})]} = \left[\sum_i^n \sum_j^n |b_{ij}|^2\right]^{\frac{1}{2}}. \tag{1-6-3}$$

For any m by n matrix B, with $m \geqslant n$,

$$\|\mathsf{B}\| \leqslant \|\mathsf{B}\|_F \leqslant \sqrt{\mathrm{rank}(\mathsf{B})}\,\|\mathsf{B}\| \leqslant \sqrt{n}\,\|\mathsf{B}\|. \tag{1-6-4}$$

When $\mathsf{B} = \mathsf{A} = \mathsf{A}^*$ then

$$\|\mathsf{A}\| = \max\{|\lambda_1|, |\lambda_n|\} \equiv \text{spectral radius of } \mathsf{A}, \tag{1-6-5}$$

$$\|\mathsf{A}\|_F = \left[\sum_{i=1}^n \lambda_i^2\right]^{\frac{1}{2}}. \tag{1-6-6}$$

The next fact shows that a sequence of unitary transformations cannot change either norm.

FACT 1-10 (Unitary Invariance of the Norms.)

$$\|JBG\| = \|B\|, \qquad \|JBG\|_F = \|B\|_F$$

for all m by n B if, and only if, J and G are orthonormal, i.e.,

$$J^*J = JJ^* = I_m, \qquad G^*G = GG^* = I_n.$$

Each norm finds a natural place in the following useful inequalities.

FACT 1-11 (Eigenvalues are perfectly conditioned.)

$$\max_j |\lambda_j[A] - \lambda_j[M]| \leq \|A - M\|,$$

$$\sum_{j=1}^{n} (\lambda_j[A] - \lambda_j[M])^2 \leq \|A - M\|_F^2$$

The latter inequality is called the **Wielandt-Hoffman inequality**. An elementary but long proof can be found in [Wilkinson, 1965, chap. 3].

The first inequality is proved in Chap. 10. It is often interpreted as saying that each eigenvalue of a self-adjoint matrix is **perfectly conditioned**, that is, the (absolute) change in an eigenvalue is not more than the (absolute) change in the matrix. In other words, the problem of determining eigenvalues of self-adjoint matrices is always well posed; the solution is well determined by the data. This is not the case for some nonsymmetric matrices.

For eigenvectors the situation is more delicate. Let $Az = z\alpha$, $Ms = s\mu$. If μ is separated by a gap γ from A's eigenvalues other than α then

FACT 1-12 $|\sin \angle(z, s)| \leq \|A - M\|/\gamma$

This fact, and some extensions of it, are established in Chap. 11.

What is remarkable about these bounds is that they are not asymptotic; there is no requirement that the "perturbation" $M - A$ be small. On the other hand, without a gap eigenvectors can be very sensitive functions of the data. Suppose that A's elements are functions of a parameter t. If $A(t_0)$ has a multiple eigenvalue then there is no guarantee that the normalized eigenvectors vary continuously in a neighborhood of t_0 as revealed in the following example constructed by Givens.

Matrix: $\mathsf{A}(t) = \begin{bmatrix} 1 + t\cos(2/t) & t\sin(2/t) \\ t\sin(2/t) & 1 - t\cos(2/t) \end{bmatrix}$

Spectrum: $\{1 + t, 1 - t\}$

Eigenvectors: $\begin{bmatrix} \cos(1/t) \\ \sin(1/t) \end{bmatrix}, \begin{bmatrix} \sin(1/t) \\ -\cos(1/t) \end{bmatrix}$

As $t \longrightarrow 0$ the eigenvectors become dense in the unit disc in the plane and when $t = 0$, $\mathsf{A}(0) = \mathsf{I}$, the eigenvectors do fill out the unit disc, (Ex. 1-6-8).

The discontinuity at $t = 0$ in the formula for the eigenvectors must not be construed as signalling a pathological situation. It merely signals that two distinct eigenlines have joined together to become an eigenplane in which no single pair of vectors is distinguished from any other pair.

These remarks suggest that the proper object to compute when a few eigenvalues are tightly clustered is **any** orthonormal basis for the invariant subspace belonging to the whole cluster rather than separate eigenvectors for each eigenvalue. But where do you draw the line between merely close and tightly clustered eigenvalues?

Exercises on Sec. 1-6

1-6-1. Prove that $\|\mathsf{A}\| = \|\mathsf{A}\|_F$ if, and only if, $\mathsf{A} = \mathsf{uu}^*$ for some u.

1-6-2. Prove (1-6-5), using Fact 1-4.

1-6-3. Show that $\|\mathsf{B}\|_F$ is the Euclidean norm of B in the n^2-dimensional space of n by n matrices.

1-6-4. Prove (1-6-1) using Fact 1-4.

1-6-5. Prove Fact 1-10.

1-6-6. Prove that for each j there is a k such that $|\lambda_j[\mathsf{A}] - \lambda_k[\mathsf{M}]| \leqslant \|\mathsf{A} - \mathsf{M}\|$ by writing $\mathsf{A} - \mathsf{M} = \mathsf{H}$, $\mathsf{M} = \mathsf{P}\Lambda\mathsf{P}^*$ and using the fact that determinant $(\mathsf{M} + \mathsf{H} - \alpha)$ vanishes when $\alpha = \lambda_j[\mathsf{A}]$. Fact 1-6 must also be used. This result is a special case of the Bauer-Fike theorem: Given $\mathsf{B} = \mathsf{G}^{-1}\Delta\mathsf{G}$, $\Delta = \mathrm{diag}(\delta_1, \ldots, \delta_n)$, then to each δ_j there is an eigenvalue μ of $\mathsf{B} + \mathsf{C}$ such that $|\mu - \delta_j| \leqslant \|\mathsf{G}\|\|\mathsf{G}^{-1}\|\|\mathsf{C}\|$. The extra fact in the self-adjoint case is that the correspondence between μ and δ_j is one to one.

1-6-7. Verify Fact 1-12 when $\mathsf{A} = \begin{bmatrix} 2 & 1 \\ 1 & 2 \end{bmatrix}$, $\mathsf{M} = \begin{bmatrix} 0 & 1 \\ 1 & 0 \end{bmatrix}$ and $\mathsf{M} = \begin{bmatrix} 1 & 0 \\ 0 & 1 \end{bmatrix}$.

1-6-8. Compute one eigenvector of the matrix $\mathsf{A}(t)$ (from the given formula) when $t = 10^{-3}$ and $t = 10^{-4}$. Through what angle does the eigenvector change? A pocket calculator may help here.

1-6-9. Define $\|\mathsf{B}\|_k^2 \equiv \sum_{i=1}^{k} \lambda_{-i}[\mathsf{B}^*\mathsf{B}]$. Show that this is a unitarily invariant matrix norm.

1-7 THE GENERALIZED EIGENVALUE PROBLEM

In many branches of science the problems involve two quadratic forms. In Mechanics, for example, $u(t)$ may represent a state of some system, $\dot{u}$ its time derivative, $\dot{u}^*M\dot{u}$ its kinetic energy, and u^*Au its potential energy. Physical principles dictate that, in the absence of external forces, the actual state u will minimize the ratio of these energies. Consequently a key function in problems with two quadratic forms, A and M, is the **Rayleigh quotient**, defined for all $u \neq 0$ by

$$\rho(u) = \rho(u; A, M) = \frac{u^*Au}{u^*Mu} \qquad \text{(M positive definite).} \tag{1-7-1}$$

It turns out that ρ is stationary at z if, and only if,

$$(A - \lambda M)z = o \tag{1-7-2}$$

for some scalar λ. We say that (λ, z) is an eigenpair of the pair, or **pencil**, (A, M). Fact 1-8 continues to hold for $\rho(u; A, M)$ provided that either A or M, or some $\alpha A + \mu M$, is positive definite. We shall assume that M is positive definite.

The proper setting for this problem is not $\mathcal{E}^n$ but the inner product space $\mathcal{M}^n$ consisting of $\mathcal{C}^n$ (or $\mathcal{R}^n$) furnished with the inner product

$$(x, y) = y^*M^{-1}x \quad \text{or} \quad (x, y) = y^*Mx. \tag{1-7-3}$$

The fundamental notions of length and angle defined by (1-2-2), (1-2-4), (1-2-5) take on new values when the inner product given in (1-7-3) replaces the familiar Euclidean version, but the spectral theorem and most of Sec. 1-4 extend to pencils (A, M) with positive definite M.

Abstractly, this problem is indistinguishable from the standard eigenvalue problem (one inner product is as good as another) but, in practice, the presence of M complicates the task and increases the cost. Chapter 15 explores the problem in greater detail.

2

Tasks, Obstacles, and Aids

2-1 WHAT IS SMALL? WHAT IS LARGE?

In 1954 came the invention of the programming language Fortran and the ensuing ability of the programmer to access any element of a matrix A as easily as any other by simply writing $A(I, J)$. Each computer system can accommodate square matrices up to a certain order as conventional two-dimensional arrays. These are the storable, or stored matrices. Naturally the precise upper limit on the order of such matrices varies from system to system but the meaning of the adjective **small** is always that **all matrix elements are equally and rapidly accessible.** On most scientific computers a full 50 by 50 matrix is small.

Now that satisfactory methods are available for most eigenvalue tasks for small symmetric matrices attention has turned to harder problems. We say that a matrix is large if only part of it can be held at one time in high-speed storage. However, the order of the matrix is too crude a measure of storage demand. A 400 by 400 symmetric matrix with a half-bandwidth of 30 (i.e., $a_{ij} = 0$ if $|i - j| > 30$) is small for many computing systems whereas a 200 by 200 symmetric matrix with a random pattern of zero elements is large. The two phrases *conventionally storable* and *not conventionally storable* would be more precise but we prefer the simpler words **small** and **large**.

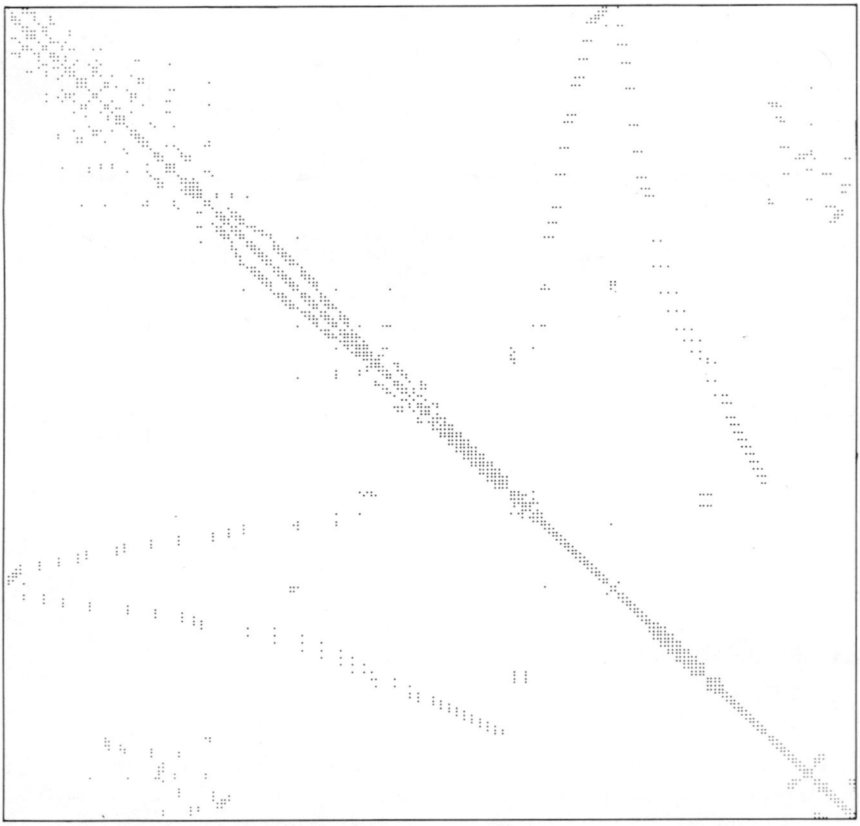

Figure 2-1-1 Portrait of a sparse matrix

It happens that most large matrices are also sparse (most of their elements are zero) but, as indicated above, what matters is how easily the pattern of zero elements can be exploited. A property such as narrow bandwidth† is too valuable to be lightly sacrificed.

A key factor in handling large sparse matrices is the way in which their nonzero elements are to be represented. For example, one obvious possibility is to create a list of the nonzero values and attach to each value, a_{ij}, five integers,

$$m, \; i, \; j, \; k, \; l$$

where m is the position of this element in the list, k points to the position of the next nonzero element in row i, and l points to the next nonzero

---

†The half-bandwidth of A is defined as $\max|i - j|$ over all i, j such that $a_{ij} \neq 0$.

element in column j. There are better schemes but we shall not pursue this matter any further here although it is an important facet of the efficiency with which sparse matrices can be handled. A good introduction to this subject is given in [Jennings, 1977]. See also [Bunch and Rose, 1976]. A sparse matrix is shown in Fig. 2-1-1.

2-2 TASKS

Mathematically this book is concerned with a single problem—the computation of eigenvalues and eigenvectors of symmetric matrices. However, as soon as we get down to details and confront questions of efficiency the underlying unity gives way to a mosaic of different tasks. To a certain extent the best method for one task turns out to be inadequate for another and this fact leads to an unpleasant profusion of programs.

Before listing some typical requests in order to give the flavor of the subject we must emphasize a feature of today's tasks which sets them apart from those of 1950.

> The results of many, but not all, matrix calculations are of no intrinsic interest. They are merely intermediate quantities needed in a larger computation.

1. Find the smallest eigenvalue, $\lambda_1[\mathsf{F}^*\mathsf{F}]$, where $\mathsf{F} = (\mathsf{f}_1, \ldots, \mathsf{f}_k)$, $\|\mathsf{f}_i\| = 1$, $i = 1, \ldots, k$, correct to 2 significant figures. The number $\sqrt{\lambda_1}$ is a simple measure of linear independence among the f_i.
2. Find all the eigenvalues of a small matrix A correct to working precision, but no eigenvectors. Such calculations occur, for example, in the course of computing a few eigenvalues of large matrices. In some analyses of electrical networks all the λ_i's themselves are wanted [Cullum, 1979].
3. Find the p smallest (leftmost) eigenvalues of large A and their matching eigenvectors correct to 4 significant decimals. Such tasks ($p = 30$, $n = 6000$) occur in nuclear shell model calculations.
4. Find the p smallest (leftmost) eigenvalues, λ_i, and their eigenvectors, z_i, for the pencil (A, M), i.e., satisfying $(\mathsf{A} - \lambda_i\mathsf{M})z_i = 0$, $i = 1, \ldots, p$. In the analysis of bridges, buildings, and vehicles suffering small displacements from equilibrium each z_i represents the shape of a

normal mode of vibration and $\sqrt{\lambda_i}$ represents the corresponding frequency of that mode. Moreover, if the vector $\mathsf{u}(t)$ represents the displacement of the structure at time t then $\mathsf{u^*Au}$ gives the potential (or strain) energy while $\dot{\mathsf{u}}\mathsf{M}\dot{\mathsf{u}}$ gives the kinetic energy of the system. Here $\dot{\mathsf{u}}$ is the derivative of u.

In practice A and M are themselves approximations and the higher eigenpairs have no physical meaning.

5. Find all the eigenvalues in a given interval and their matching eigenvectors. Such tasks occur in modelling the observable tides in the Atlantic Ocean [Cline, Golub, and Platzman, 1976] and also in testing whether structures are likely to withstand local earthquakes.
6. To discover significant variables in statistical analysis it is useful to find a few of the largest eigenpairs of a correlation matrix $\mathsf{FF^*}/(m-1)$ where F is n by m and $1 < m \ll n$.

2-3 CONFLICTING REQUIREMENTS

One of the attractions in the study of numerical methods is to see how well various methods satisfy the following mutually incompatible desiderata: reliability, appropriate accuracy, swift execution, low storage demands, and short programs. For any particular computing task one or more of these requirements may be in abeyance. The dramatic developments in computer technology keep upsetting the balance between these five goals and, no doubt, will continue to do so.

2-3-1 RELIABILITY

As computing tasks get harder there is a rise in the possibility that a method will sometimes fail. Failure is no disgrace; indeed, it is the only reasonable response to an illegitimate request: e.g., for the real square root of a negative number or the Choleski factorization of an indefinite matrix. The great sin is for a method to lie, to deliver results which appear to be reasonable but which are utterly wrong.

A numerical method, or a program, can be regarded as a function mapping input to output. As such the user is concerned with its domain, i.e., the set of matrices for which the method works. Sometimes this domain is unknown and then failure can be passed off as the discovery that a particular set of data (a certain matrix, perhaps) is not in the domain of the method. Reliability also concerns the way in which failure is signalled to the rest of the computation: message, abort, or something in between.

Of course, the user wants to have methods with broad and cleanly described domains. Whether the user can be satisfied depends on the task, the power of the method, and the completeness of its analysis.

2-3-2 ACCURACY

It seems reasonable, at first glance, to seek methods which are capable of giving results to any desired precision. We might hope that results with low accuracy would cost less than those with high accuracy. In practice, however, numerical methods do not work that way.

First, such flexibility may not be possible. The eigenvalues and eigenvectors of a matrix are strongly interconnected. It is quite plausible that acceptance of a 2-decimal approximation to λ_{max} may impair, or even prevent, the calculation of a 2-decimal approximation to λ_{min}, and so on.

Second, the number representation in most computers is quite rigid; arithmetic is performed in floating point mode with a fixed precision which is, in most cases, equivalent to 7, 14, or 16 decimals. Operations with twice the working precision are occasionally available.

The eigenvalues of symmetric A are sturdy functions of A's elements and so are eigenvectors which belong to isolated eigenvalues. For small A the major source of error is roundoff and the methods which will be described usually give the exact results for a matrix $\mathsf{A} + \mathsf{H}$ for some unknown H. In all cases we expect that

$$\|\mathsf{H}\| < n\epsilon\|\mathsf{A}\|$$

where ϵ is the relative precision of the arithmetic. This is a strong, simple statement but cannot be backed up by a proof. To be rigorous one must replace n by a fussy, uninformative function of n which depends on the details of the particular method. This function reflects the difficulty of the analysis quite as much as it reflects the real possibilities for error growth.

In any case, by Fact 1-11, the error in a computed eigenvalue is bounded by $\|\mathsf{H}\|$ and for most purposes this is far more accuracy than is needed. Two points must be made about this apparently excessive accuracy. The marginal savings reaped by accepting less accurate results are negligible, and second, the cost of small matrix computations has decreased steadily. All the eigenvalues of a 50 by 50 matrix can be computed for less than a 1970 dollar.

The situation is rather different for very large matrices ($n > 1000$). Roundoff error does not dominate the accuracy of the results and the iterative methods in current use sometimes converge quite slowly. There is a high premium on those methods which are capable of delivering the desired accuracy and no more!

2-3-3 *SWIFT EXECUTION*

This is the traditional criterion for comparing two methods each of which is reliable. For historical reasons the standard measure of execution time is the number of multiplications required. In complex computing systems, with so-called multiprocessing or time sharing, and optimizing compilers, the multiplication count had become much less relevant by 1975 than it was in 1960. However, it still has some value and gross differences in operation count (n^4 versus n^3 operations) will be reflected, in subdued form, in execution costs.

As computer technology advanced, the time required to perform one multiplication dropped (from 10^{-3} seconds in 1955 to 10^{-6} seconds in 1970) and so the duration of small matrix computations seems to have fallen below the threshold at which it is worth attention. Although this is the case for large scientific computers the picture may change significantly as more and more small computations are performed on mini-processors, desk top computers, and even on hand held calculators. The day may come when a fetch takes longer than a multiplication.

2-3-4 *LOW STORAGE REQUIREMENTS*

As storage devices have developed this criterion has suffered violent changes in status. Initially, around 1950, it was of paramount importance. With the advent of large (32,000 cell) ferrite core storage devices it seemed to vanish (circa 1960) and then, with the rise of large matrix problems and small computers, it has reappeared in full strength.

For small matrices the issue is the number of n by n arrays which are needed, and whether the original A must be saved. For large matrices the issue is the number of n-vectors which are needed in the fast storage device.

2-3-5 *SHORT PROGRAM*

This property of an algorithm is, in one way, subsumed under the storage needs. Nevertheless, it appears separately in two contexts. As reliable program libraries began to appear in the 1970s, containing more and more programs, the proliferation problem forced itself on the attention of the makers of these libraries. What a relief if one program could be made to carry out all eigenvalue tasks for small symmetric matrices even if it were not optimal for any of them! This thorny question is far from resolved and falls in the domain of mathematical software rather than numerical analysis.

The other environment where program length is paramount is the hand held calculator whose programs must fit into the 100 or 200 steps allotted to them. The demand for small matrix manipulation in desk top computers is imminent.

On the other hand, in many scientific computing laboratories program length is of no consequence at all.

2-4 FINITE PRECISION ARITHMETIC

Consider the representation of $10^{10}\pi = 10^{11} \times 0.314159265\ldots$ in a number system which permits only 5 decimal digits in the fractional part of its numbers. Many digital computers simply drop the extra digits to produce the chopped version,

	absolute error	relative error
$(10^{10}\pi)_{\text{chopped}} = 0.31415 \times 10^{11}$	0.93×10^{6}	0.29×10^{-4}

A more accurate, but more expensive approximation is the rounded version,

	absolute error	relative error
$(10^{10}\pi)_{\text{rounded}} = 0.31416 \times 10^{11}$	-0.73×10^{5}	-0.23×10^{-5}

Uniqueness in the representation of these computer numbers is obtained by normalizing them: The fractional part is less than one and has a nonzero leading digit.

What can be said in general about the size of the inevitable roundoff errors in such a system? A moment's reflection shows that an upper bound on the absolute error must involve the absolute value of the number whereas an upper bound on the relative error for normalized numbers does not. Consequently it is usually simpler to discuss relative error. For example, on the 5-digit system used above the relative error in chopping is always less than 10^{-4} (in absolute value). This bound is halved if results are rounded. (The fact that most computers work with numbers represented in bases 2 or 16 rather than 10 is irrelevant to the points being made here.)

Consider next the basic arithmetic operations $+$, $-$, $\times$, $/$ which act on normalized numbers held in the machine. To be specific we may suppose that the numbers are written in floating point (i.e., scientific)

decimal notation with unlimited exponents but only 5 fractional digits. Even if the exact result of an addition or multiplication were obtained by the arithmetic unit the computer must round or chop when storing it. Thus the best that can be hoped for is that the relative error in each arithmetic operation is bounded by 10^{-4}. This number, 10^{-4}, or its analogue for each computer, is called the **unit roundoff** or the relative precision of the computer and denoted by ϵ.

The system which we are discussing is called **floating point arithmetic** because the proper exponents of the results are produced automatically by the arithmetic unit. (In the 1950s some computers left this chore to the programmer and worked in fixed point mode.) A useful notational device is that the stored result of any calculation C is written *fl*(C). This is to be read as "the floating point result of C". The unit roundoff ϵ for any computer can now be characterized very neatly by

$$\epsilon = \text{minimum } \xi \text{ such that } fl(1 + \xi) > 1.$$

In order to avoid distracting details we shall use the following simple model.

> Each basic arithmetic operation is done exactly in the arithmetic unit. The only error occurs when the result is stored.

The actual behavior of most digital computers does not comform to the model but their departures from it are not significant for the analysis of the susceptibility of the methods described in this book to the hazards of roundoff error.

The model leads to some very simple relations on which error analyses are based. Let α and β be any normalized floating point numbers and let $\square$ stand for any of $+, -, \times, /$, then

$$fl(\alpha\square\beta) = (\alpha\square\beta)(1 + \rho), \qquad (\beta \neq 0 \text{ for } /), \tag{2-4-1}$$

where ρ, the relative error, is a complicated function of α, β and $\square$ and the precision of the arithmetic. Nevertheless, ρ satisfies

$$|\rho| < \epsilon \tag{2-4-2}$$

where ϵ is independent of α, β, and $\square$ and is the unit roundoff of the machine. The thrust of error analysis is to majorize ρ by ϵ and thus obtain simple but pessimistic error bounds which are as independent of the data as possible.

It should be pointed out that absolute and relative errors are not rivals. In more complicated calculations the absolute errors in the parts are needed for the relative error of the whole. The fundamental relation (2-4-1) can be interpreted in various ways, all valid and each useful on the appropriate occasion. For example, close to α and β are some real numbers $\bar{\alpha} \equiv \alpha(1 + \rho)$, and $\bar{\beta} \equiv \beta(1 + \rho)$, not usually computer numbers, which are related to α and β as follows:

$$
\begin{aligned}
fl(\alpha \pm \beta) &= \bar{\alpha} \pm \bar{\beta} \\
fl(\alpha\beta) &= \begin{cases} \bar{\alpha}\beta, \text{ and} \\ \alpha\bar{\beta}, \text{ and} \\ \left[\alpha(1+\rho)^{1/2}\right]\left[\beta(1+\rho)^{1/2}\right] \end{cases} \\
fl(\alpha/\beta) &= \begin{cases} \bar{\alpha}/\beta, \text{ and} \\ \alpha/\bar{\beta} \end{cases}
\end{aligned}
\tag{2-4-3}
$$

In each case the computed result is regarded as the exact operation with slightly perturbed data. This interpretation can be very useful and goes under the name of a **backward** analysis. It aims to put roundoff error on the same footing as uncertainty in the data. Wilkinson has used this approach to give intelligible yet rigorous error analyses of the majority of methods used for small matrix calculations. It is necessary to look at some earlier attempts to analyze the effects of roundoff in order to appreciate the great simplification he brought to the subject.

The more conventional approach which simply bounds the error in a final or intermediate result is called **forward** analysis. The two approaches are not rivals and success usually comes through an adroit use of both techniques.

A thorough and elementary presentation of error analysis is given in [Wilkinson, 1964] and so we hope that the reader will forgive us if we give an informal discussion of a few important issues.

2-5 CANCELLATION

Genuine subtraction produces the difference of two numbers with the same sign, say plus. There is a widespread misconception that subtraction of numbers which are very close is an inherently inaccurate process in which the relative error is greatly enhanced. This is known as **catastrophic cancellation**†.

†The reader will find no discussion of such catastrophes in the elegant theories of R. Thom. [See Zeeman, 1976.]

As with most misconceptions there is a grain of truth in the saying, but the simple fact is that the error in the subtraction of normalized floating point numbers with the same exponent is zero. Thus

$$fl(0.31416 \times 10^{11} - 0.31415 \times 10^{11}) = 0.10000 \times 10^{7}$$

What about close numbers with adjacent exponents? According to our model the exact result will be formed and will require at most 5 digits (probably fewer) for its representation. Thus, e.g.,

$$fl(0.10012 \times 10^{-8} - 0.99987 \times 10^{-9}) = 0.13300 \times 10^{-11}$$

and again there is no error at all.

At this point we must make a brief digression to mention that the most significant feature which prevents a number of computers (including the Control Data machines) from adhering to our simple model of arithmetic is the lack of a **guard digit**. A guard digit is an extra digit on the low order end of the arithmetic register whose purpose is to catch the low order digit which otherwise would be pushed out of existence when the decimal points are aligned. Without the guard digit the final 7 in Example 2-5-1 is lost and the relative error is **enormous**.

EXAMPLE 2-5-1

$$\begin{array}{lll} \alpha = & 0.10012 & \times 10^{-8} \\ \beta = & 0.09998[7] & \times 10^{-8} \\ \hline & 0.00014 & \times 10^{-8} = 0.14 \times 10^{-11} \text{ (normalized)} \end{array}$$

Absolute error: -0.70×10^{-13},

Relative error: $-0.53 \times 10^{-1} \approx -500\ \epsilon$!!

For machines without a guard digit, we have the anomalous result that the relative error in the subtraction of very close numbers is usually 0 but can be as great as 9 ($\approx 10^5\epsilon$!) when the exponents happen to differ. Thus the bound in (2-4-2) fails completely for such machines in these special cases. Nevertheless, it is still the case that

$$fl(\alpha - \beta) = \bar{\alpha} - \beta \qquad (|\alpha| \geqslant |\beta|)$$

where

$$\bar{\alpha} = \alpha(1 + \rho), \qquad |\rho| < \epsilon.$$

Thus $\bar{\alpha} = 0.100127 \times 10^{-8}$ in the example above.

Let us return to our model in which the subtraction of close numbers is always exact. The numbers α and β will often, but not always, be the result of previous calculations and thus have an inherent uncertainty. It is the relative **uncertainty** in the difference which is large when cancellation occurs, not the relative error in subtraction. this distinction is **not** academic; sometimes the close numbers are exact and their difference has no error. The phenomenon of cancellation is not limited to computer arithmetic and can be expressed formally as follows. Let $0 < \beta < \alpha$, $\beta \approx \alpha$, and let η_α and η_β be the relative uncertainty in α and β, respectively; then

$$\alpha(1 + \eta_\alpha) - \beta(1 + \eta_\beta) = \alpha - \beta + (\alpha\eta_\alpha - \beta\eta_\beta)$$
$$= (\alpha - \beta)(1 + \rho),$$

where

$$\rho = (\alpha\eta_\alpha - \beta\eta_\beta)/\ (\alpha - \beta)$$

and

$$|\rho| < \left(\frac{\alpha}{\alpha - \beta}\right)(|\eta_\alpha| + |\eta_\beta|).$$

In exact arithmetic the magnification factor $(\frac{\alpha}{\alpha - \beta})$ can be arbitrarily large but in our decimal system it is bounded by $10/\epsilon$.

There is more to be said. If α and β were formed by rounding previous calculations then information, in the form of low order digits, was discarded in previous storage operations. It is this lost information we mourn when cancellation occurs. Those discarded digits seemed negligible beside α and β. The subtraction is not be blame; it merely signals the loss.† The following examples illustrate these remarks.

EXAMPLE 2-5-2

In our simple model of arithmetic the relative error in $fl(\alpha + \beta)$ is bounded by ϵ, the unit roundoff. However, the relative error in $fl(\alpha + \beta + \gamma)$ can be as large as 1. Take $\alpha = 1$, $\beta = 10^{17}$, and $\gamma = -10^{17}$. The first addition $fl(\alpha + \beta)$ produces β (α is annihilated) and the second $fl(\beta + \gamma)$ produces complete cancellation. This example also shows how the associative law of addition fails. The calculation $fl(\alpha + fl(\beta + \gamma))$ produces the correct value.

---

†The ancient Spartans, so I am told, used to execute messengers who had the misfortune to bring bad news.

EXAMPLE 2-5-3 (See Sec. 6-9 for the remedy.)

Find the unit vector x orthogonal to y in the plane span (y, z).

$$\mathbf{y} = \begin{bmatrix} 0.16087 \times 10^0 \\ -0.11852 \times 10^0 \\ 0.98216 \times 10^{-1} \end{bmatrix}, \qquad \mathbf{z} = \begin{bmatrix} -0.50069 \times 10^{-1} \\ 0.36889 \times 10^{-1} \\ -0.30569 \times 10^{-1} \end{bmatrix}$$

$$\theta = \frac{\mathbf{z}^*\mathbf{y}}{\mathbf{y}^*\mathbf{y}} = -0.31123 \text{ (in 5-decimal chopped arithmetic)}$$

$$\mathbf{w} = \theta\mathbf{y} = \begin{bmatrix} -0.50067[5701] \times 10^{-1} \\ 0.36886[9796] \times 10^{-1} \\ -0.30567[7657] \times 10^{-1} \end{bmatrix}$$

$$\tilde{\mathbf{x}} = \mathbf{z} - \mathbf{w} = \begin{bmatrix} -0.50069 \times 10^{-1} \\ 0.36889 \times 10^{-1} \\ -0.30569 \times 10^{-1} \end{bmatrix} - \begin{bmatrix} -0.50067 \times 10^{-1} \\ 0.36886 \times 10^{-1} \\ -0.30567 \times 10^{-1} \end{bmatrix}$$

$$= \begin{bmatrix} -2.0000 \times 10^{-6} \\ 3.0000 \times 10^{-6} \\ -2.0000 \times 10^{-6} \end{bmatrix}$$

$$\mathbf{x} = \frac{\tilde{\mathbf{x}}}{\|\tilde{\mathbf{x}}\|} \qquad \text{Exact Solution}$$

$$\begin{bmatrix} -0.48507 \\ 0.72760 \\ -0.48507 \end{bmatrix} \qquad \begin{bmatrix} 0.62307 \\ 0.77789 \\ -0.81837 \times 10^{-1} \end{bmatrix}$$

$$\cos\phi = \frac{x^*y}{\|x\|\|y\|} = -0.95177; \qquad \phi = 2.8298 \text{ rad} \approx 162°!$$

We conclude this section with three pieces of numerical wisdom.

1. When an algorithm does turn out to be unreliable it is not because millions of tiny roundoff errors gradually build up enough to contaminate the results; rather it is because the rounding of a few numbers (perhaps only one) discards crucial information. Error analysis aims to detect such sensitive places.
2. Severe cancellation is not always a bad thing. It depends on the role of the difference in the rest of the computation. A backward error analysis often resolves the problem.

3. Subtractions which may provoke severe cancellation can sometimes be transformed algebraically into products or quotients, e.g.,

$$|\alpha| - \sqrt{\alpha^2 - \beta^2} = \beta^2 / \left(|\alpha| + \sqrt{\alpha^2 - \beta^2}\right)$$

$$\alpha^2 - \beta^2 = (\alpha - \beta)(\alpha + \beta)$$

2-6 INNER PRODUCT ANALYSIS

2-6-1 NUMERICAL EXAMPLE

We begin with a detailed account of the computation of $\mathbf{y}^*\mathbf{z}$ for the vectors of Example 2-5-3 given above. When these details are mastered the general pattern becomes clear. In this section y_i denotes the ith element of the vector $\mathbf{y}$.

The product y_1z_1 is $-0.805460003 \times 10^{-2}$ and the stored result $(-0.80546 \times 10^{-2})$ can be written $\bar{y}_1z_1$ where $\bar{y}_1 = y_1(0.80546/0.805460003) = 0.1608699992$. Similarly the product y_2z_2 is $-0.437208428 \times 10^{-2}$ and the stored result $(-0.43720 \times 10^{-2})$ can be written $\bar{y}_2z_2$ where $\bar{y}_2 = y_2(0.43720/0.437208428) = -0.1185177153$. The sum $s_2 = \bar{y}_1z_1 + \bar{y}_2z_2$ is $-0.1242660 \times 10^{-1}$ and the stored result $(-0.12426 \times 10^{-1})$ can be written $\bar{\bar{y}}_1z_1 + \bar{\bar{y}}_2z_2$ where

$$\bar{\bar{y}}_1 = 0.1608622315 = \bar{y}_1(0.12426/0.1242660),$$

$$\bar{\bar{y}}_2 = -0.1185119926 = \bar{y}_2(0.12426/0.1242660).$$

The final product y_3z_3 is $-0.3002364904 \times 10^{-2}$ and the stored result $(-0.30023 \times 10^{-2})$ can be written $\bar{y}_3z_3$ where $\bar{y}_3 = y_3(0.30023/0.3002364904) = 0.9821387679 \times 10^{-1}$. The final sum $s_3 = \bar{s}_2 + \bar{y}_3z_3$ is $-0.1542830 \times 10^{-1}$ and the stored inner product $(-0.15428 \times 10^{-1})$ can be written $\tilde{\mathbf{y}}^*\mathbf{z}$ where

$$\tilde{y}_1 = 0.1608591034 = \bar{\bar{y}}_1\phi, \qquad \textit{whereas } y_1 = 0.16087$$

$$\tilde{y}_2 = -0.1185096878 = \bar{\bar{y}}_2\phi, \qquad \textit{whereas } y_2 = -0.11852$$

$$\tilde{y}_3 = 0.9821196701 \times 10^{-1} = \bar{\bar{y}}_3\phi, \qquad \textit{whereas } y_3 = 0.98216 \times 10^{-1}$$

$$\phi = 0.15428/0.1542830 \approx 1 - 1.944 \times 10^{-5}.$$

The true inner product $\mathbf{y}^*\mathbf{z}$ is $-0.154294921 \times 10^{-1}$. In this case there is no cancellation in the summation and the relative error in the computed value is less than $\frac{1}{2}10^{-4}$. The backward error analysis has exhibited a vector $\tilde{\mathbf{y}}$, close to $\mathbf{y}$, such that

$$fl(\mathbf{y}^*\mathbf{z}) = \tilde{\mathbf{y}}^*\mathbf{z}.$$

A different, less systematic, rearrangement of the errors in the additions could produce another vector $\hat{\mathbf{y}}$, even closer to $\mathbf{y}$ than $\tilde{\mathbf{y}}$, such that for it too

$$fl(\mathbf{y}^*\mathbf{z}) = \hat{\mathbf{y}}^*\mathbf{z}.$$

2-6-2 THE GENERAL CASE

We now turn to a formal backward error analysis of $fl(\mathbf{x}^*\mathbf{y})$. Computer addition is not associative and so the order in which the summation is organized affects the value. The results are a little bit simpler when $\Sigma x_i y_i$ is calculated with i running from n down to 1. To be definite we shall obtain a vector $\tilde{x}$ such that

$$fl(\mathbf{x}^*\mathbf{y}) = \tilde{\mathbf{x}}^*\mathbf{y}$$

but, of course, the error could be shared between $\mathbf{x}$ and $\mathbf{y}$.

Formal error analyses are rather dull but if you have never studied one we urge you to work through this one line by line.

2-6-3 THE ALGORITHM

$s_n = 0,\qquad$ for $j = n, n-1, \ldots, 2, 1$

$$p_j = fl(x_j y_j), \tag{2-6-1}$$

$$s_{j-1} = fl(p_j + s_j). \tag{2-6-2}$$

The only properties required of the arithmetic system are

$$fl(\alpha\beta) = (1 - \mu)\alpha\beta \qquad \text{where } |\mu| < \epsilon, \tag{2-6-3}$$

$$fl(\alpha + \beta) = (1 - \rho)\alpha + (1 - \sigma)\beta \qquad \text{where } |\rho| < \epsilon, |\sigma| < \epsilon. \tag{2-6-4}$$

In our simple model $\rho = \sigma$ but no great advantage accrues from that property in this application. Moreover, most computers (with or without a guard digit) satisfy (2-6-4).

2-6-4 NOTATION

$x_j^{(k)}$ denotes the result of multiplying x_j by k factors of the form $(1 - \tau)$ where $|\tau| < \epsilon$. Thus

$$|x_j^{(k)} - x_j|/|x_j| < (1 + \epsilon)^k - 1 \approx k\epsilon \quad (\text{if } k\epsilon < 0.01) \tag{2-6-5}$$

(Hold tight and watch the superscripts.)

$$
\begin{aligned}
p_n &= (1-\mu_n)x_n y_n \\
&\equiv x_n^{(1)}y_n, \qquad \text{by (2-6-1) and (2-6-3)} \\
s_{n-1} &= p_n, \qquad \text{since } s_n = 0, \\
p_{n-1} &= (1-\mu_{n-1})x_{n-1}y_{n-1} \equiv x_{n-1}^{(1)}y_{n-1}, \qquad \text{defining } x_{n-1}^{(1)} \\
s_{n-2} &= (1-\rho_{n-1})x_{n-1}^{(1)}y_{n-1} + (1-\sigma_{n-1})x_n^{(1)}y_n, \quad \text{by (2-6-2) and (2-6-4)}, \\
&\equiv x_{n-1}^{(2)}y_{n-1} + x_n^{(2)}y_n, \qquad \text{defining } x_{n-1}^{(2)} \text{ and } x_n^{(2)}, \\
p_{n-2} &= (1-\mu_{n-2})x_{n-2}\,y_{n-2} = x_{n-2}^{(1)}y_{n-2}, \\
s_{n-3} &= (1-\rho_{n-2})x_{n-2}^{(1)}y_{n-2} + (1-\sigma_{n-2})s_{n-2}, \quad \text{by (2-6-2) and (2-6-4)} \\
&\equiv x_{n-2}^{(2)}y_{n-2} + x_{n-1}^{(3)}y_{n-1} + x_n^{(3)}y_n, \qquad \text{defining the } x_i^{(k)}, \\
p_{n-3} &= (1-\mu_{n-3})x_{n-3}y_{n-3} \equiv x_{n-3}^{(1)}y_{n-3}, \qquad \text{defining } x_{n-3}^{(1)} \\
s_{n-4} &= (1-\rho_{n-3})x_{n-3}^{(1)}y_{n-3} + (1-\sigma_{n-3})s_{n-3}, \\
&\equiv x_{n-3}^{(2)}y_{n-3} + x_{n-2}^{(3)}y_{n-2} + x_{n-1}^{(4)}y_{n-1} + x_n^{(4)}y_n, \\
&\qquad \cdots \\
s_{j-1} &= x_j^{(2)}y_j + x_{j+1}^{(3)}y_{j+1} + \cdots + x_n^{(n-j+1)}y_n, \qquad \text{(proof by induction)} \\
&\cdots \\
p_1 &= (1-\mu_1)x_1y_1 \equiv x_1^{(1)}y_1, \\
s_0 &= (1-\rho_1)x_1^{(1)}y_1 + (1-\sigma_1)s_1, \\
&\equiv x_1^{(2)}y_1 + x_2^{(3)}y_2 + \cdots + x_{n-1}^{(n)}y_{n-1} + x_n^{(n)}y_n, \\
&\equiv \tilde{\mathbf{x}}^*\mathbf{y}, \qquad \text{defining } \tilde{\mathbf{x}}.
\end{aligned}
$$

Moreover,

$$|\tilde{x}_i - x_i|/|x_i| < (1+\epsilon)^{i+1} - 1 \approx (i+1)\epsilon \qquad (\text{if } n\epsilon < 0.01).$$

This completes the backward analysis but some further comments are in order.

1. No mention has been made of the error in s_0. In general no bound can be placed on the relative error because three or more additions are involved. For the absolute error

$$|s_0 - \mathbf{x}^*\mathbf{y}| < |(\tilde{\mathbf{x}} - \mathbf{x})^*\mathbf{y}| < \left[(1+\epsilon)^n - 1\right]\|\mathbf{x}\|\|\mathbf{y}\| \approx n\epsilon\|\mathbf{x}\|\|\mathbf{y}\|.$$

The factor n reflects the generality of the result rather than the behavior of the error.

2. If there is uncertainty in the elements of x (or y) of at least n units in the last place then this uncertainty dominates the effects of roundoff and we say that the computed value is as good as the data warrants.
3. $\tilde{\mathbf{x}}$ is close to x not just in norm but element by element.
4. When the elements of x and y are known to decrease in absolute value ($|x_i| > |x_{i+1}|$) it is worthwhile to sum from n down to 1.

2-6-6 MATRIX VECTOR PRODUCTS

It is a corollary of the inner product analysis that

$$fl(\mathbf{Ay}) = \tilde{\mathbf{A}}\mathbf{y}$$

where $\tilde{a}_{ij} = a_{ij}$ if $a_{ij} = 0$ and otherwise

$$|\tilde{a}_{ij} - a_{ij}|/|a_{ij}| < (1 + \epsilon)^n - 1.$$

If the indices are taken in decreasing order then

$$|\tilde{a}_{ij} - a_{ij}|/|a_{ij}| < (1 + \epsilon)^{j+1} - 1.$$

If the elements of A are uncertain by n units in the last place and $n\epsilon < 0.01$ then the computed product is as good as the data warrants.

2-6-7 DOUBLE PRECISION ACCUMULATION OF INNER PRODUCTS

It is possible to form the products $p_i = fl(x_i y_i)$ exactly so that $\mu_i = 0$ in the foregoing analysis. On the IBM 360 and 370 machines the cost of obtaining the long product of two short "words" (numbers with approximately 6 decimal digits) is no greater than obtaining the short (truncated) product. On CDC machines two separate operations are required to get the exact product of two single precision (14 decimal) numbers and so the cost is doubled.

The extra accuracy is worthless unless these products are summed in double precision. This can be done and the result is often, but not always, the exact inner product. Nevertheless, the result must be stored and so one single precision rounding error will be made. Note that this calculation requires no extra storage; the double length work is done in the arithmetic registers.

What are the rewards of this extra accuracy?

1. Often, but not always, $fl(\mathbf{x}^*\mathbf{y}) = \mathbf{x}^*\mathbf{y}(1 - \rho)$, $|\rho| < \epsilon$.
2. Always $fl(\mathbf{x}^*\mathbf{y}) = \tilde{\mathbf{x}}^*\mathbf{y}$ with $|\tilde{x}_i - x_i| < \epsilon|x_i|$, provided $n\epsilon < 0.01$.
3. Always $|fl(\mathbf{x}^*\mathbf{y}) - \mathbf{x}^*\mathbf{y}| < n\epsilon^2\|\mathbf{x}\|\ \|\mathbf{y}\| < \epsilon\|\mathbf{x}\|\ \|\mathbf{y}\|$, provided $n\epsilon < 0.01$.

In other words the factor n is removed from the error bounds and this helps in the construction of simple, even elegant, error analyses. However, the absolute error itself is not reduced by a factor n because the single precision result is already more accurate than our worst case analysis indicates. The upshot is that double precision accumulation of inner products is usually practiced only when the extra cost is small, e.g., less than 5%.

2-7 CAN SMALL EIGENVALUES BE FOUND WITH LOW RELATIVE ERROR?

The answer to the question is that we have no right to expect such accuracy in general, but if both the matrix itself is right and the numerical method is right then all eigenvalues can be obtained with low relative error.

All good numerical methods (for small matrices) produce numbers which are exact eigenvalues of a matrix $\mathsf{A} + \mathsf{W}$ close to the original matrix A. This means that $\|\mathsf{W}\|$ is small (like roundoff) compared to $\|\mathsf{A}\|$. Fact 1-11 in Chap. 1 ensures that the absolute error in each eigenvalue is bounded by $\|\mathsf{W}\|$. This result is the best that can be hoped for, in general, as the case $\mathsf{W} = \epsilon\mathsf{I}$ reveals. This is bad news for the accurate computation of the smallest eigenvalue (when they are tiny compared to $\|\mathsf{A}\|$). Consider

$$\mathsf{A}_1 \approx \begin{bmatrix} 1 + 2\gamma & 1 \\ 1 & 1 - 2\gamma \end{bmatrix}$$

when $\gamma^2 \approx \epsilon$, the unit roundoff. Add $3\gamma^2\mathsf{I}$ to A_1 and $\lambda_1[A]$ changes from $-2\gamma^2$ to γ^2. We say that λ_1 is very poorly determined by A_1 and no numerical method can determine λ_1 accurately without using extra precision. Small relative changes to any element of A_1 provoke large relative changes in λ_1. So much for the general case.

Now consider

$$\mathsf{A}_2 \approx \begin{bmatrix} 2 & \gamma^2 \\ \gamma^2 & -2\gamma^2 \end{bmatrix}$$

which has approximately the same spectrum as A_1. Note that if $3\gamma^2\mathsf{I}$ is added to A_2 a huge relative change is made in the (2, 2) element, precisely the same as the relative change in $\lambda_1[\mathsf{A}_2]$. What is special about A_2 is that small relative perturbations of the elements lead to small relative changes in both the eigenvalues. In fact, A_2 is an extreme example of a graded matrix. A less bizarre specimen is

$$A_3 = \begin{bmatrix} 1 & 10 & 10 & 0 & 0 \\ 10 & 10 & 10^2 & 10^2 & 0 \\ 10 & 10^2 & 10^2 & 10^3 & 10^3 \\ 0 & 10^2 & 10^3 & 10^3 & 10^4 \\ 0 & 0 & 10^3 & 10^4 & 10^4 \end{bmatrix}$$

Such a matrix can arise when a generalized eigenvalue problem is reduced to standard form $(A - \lambda I)z = o$.

It is legitimate to ask of a method that it find small eigenvalues accurately whenever this is possible. The QL and QR algorithms described in Chap. 8 can do so.

2-8 AVAILABLE PROGRAMS

There are now available a few high-quality packages of programs which compute eigenvalues and/or eigenvectors of small matrices of various types including real symmetric matrices, either full or tridiagonal ($a_{ij} = 0$, $|i - j| > 1$). These programs can be invoked in the same way as the elementary functions such as SIN and EXP. The number of input parameters has inevitably increased with the complexity of the task but the user needs very little knowledge of numerical methods, mainly because of the excellent documentation.

These programs are based on [Wilkinson and Reinsch, 1971],† which is a collection of procedures written in the language Algol 60 together with pertinent comments on method and the all important computational details. This collection was put together by J. H. Wilkinson and C. Reinsch, the acknowledged leaders in the field of matrix computations, but it represents a remarkably cooperative effort by many others working on these problems. What is more, all of the programs had been published previously and almost all had been modified and improved in the light of usage. (That fact is quite puzzling; it is rare indeed for mathematicians to publish regular improvements on their proofs, the aim is to get them right the first time!) The difficulties in turning a good method into a good program are not easy to identify and everybody involved was surprised at the length of time required to reach the proper level of reliability and efficiency.

In the United States the best eigenvalue package is a systematized collection of Fortran IV subroutines called EISPACK. It is well docu-

†Hereafter called the Handbook.

mented, independently certified, and disseminated throughout the country. It was developed (during 1973–1974) at Argonne National Laboratory under the NATS project (National Activity to Test Software) in collaboration with teams from Stanford University and the University of Texas at Austin.

It is worth mentioning that the NATS project was set up simply to explore the problems involved in turning good algorithms into useful program packages. At first glance it appears almost trivial to translate the much studied, highly polished Algol programs in the Handbook into a Fortran package. Nevertheless, the task proved surprisingly difficult, partly because no software enterprise of this quality and scale had been undertaken before. Incidentally, minor flaws in the algorithms in the Handbook were uncovered and various improvements were introduced (for example, the way scaling problems are handled).

Some snapshots of the state of the art (1974) are given in the next section. The final release of EISPACK was issued in 1977. The associated guide is published as Lecture Notes in Computer Science, Volume 51 (Springer-Verlag, 1977), authored by B. S. Garbow et al.

Requests for information and for the EISPACK tape should be addressed to

ARGONNE CODE CENTER
Bldg. 221, Room C-235
Argonne National Laboratory
Argonne, Illinois 60439, USA

In Europe requests should be sent to

ENEA Computer Programme Library
Euratom CCR
Espra (Varese), Italy.

High-quality Fortran programs for eigenvalue calculations are also available from IMSL (International Mathematical and Statistical Libraries), which contain far more than matrix programs. Their eigenvalue programs are straightforward translations of some of the Handbook codes into Fortran. Inquires should be addressed to

IMSL
Sixth Floor - GNB Bldg.
7500 Bellaire,
Houston, Texas 77036, USA

In Britain a collaborative, ongoing activity creating and maintaining high-quality mathematical programs, in Algol and Fortran, has its headquarters at

NAG
7 Banbury Rd.
Oxford, England 0X2 6NN

The NAG (Numerical Algorithms Group) eigenvalue codes are modelled on EISPACK.

2-9 REPRESENTATIVE TIMINGS

The following excerpts from timings given in the first EISPACK guide give an indication of the state of the art in 1974 for small matrices.

All times are in seconds.

n	*Machine*	*Task*		
		All λ's and z's	2 λ's and 2 z's	1 λ
10	1	0.08	0.04	0.02
	2	0.28	0.13	0.04
	3	—	—	—
	4	0.13	0.06	0.04
40	1	3.08	0.56	0.49
	2	14.28	2.52	2.22
	3	0.74	0.14	0.12
	4	4.62	0.87	0.75
80	1	22.78	3.68	3.46
	2	112.00	17.28	16.22
	3	5.25	0.94	0.88
	4	32.64	5.24	4.92

1. IBM 360/75, Fortran H, OPT = 2.
2. IBM 360/67, Fortran G.
3. IBM 370/165, Fortran H, OPT = 2.
4. CDC 6400, FTN, OPT = 2.

2-10 ALTERNATIVE COMPUTER ARCHITECTURE

2-10-1 VECTOR COMPUTERS

Scientific computation seems to be dominated by operations on vectors, in particular the formation of linear combinations $\alpha\mathbf{x} + \mathbf{y}$ and the evaluation of inner products $\mathbf{y}^*\mathbf{x}$. As computations grow in ambition the vectors grow in length (i.e., dimension). Vector computers have special hardware that greatly facilitates these operations, for example, one element of the output vector can be produced at each so-called minor cycle of the computer (at 10^8 per second). The price for this desirable feature is that there is a severe start-up overhead cost for each vector operation. This start-up penalty cancels the rapidity of the subsequent execution for small vectors ($n < 100$). Nevertheless, for large sparse problems ($n > 1000$) the gains in speed are impressive and methods should be examined for the ease with which they can be "vectorized."

Three examples of vector computers are Control Data's Star-100 and the Texas Instrument ASC (Advanced Scientific Computer), and most recently the Cray-1.

2-10-2 PARALLEL COMPUTERS

There are one or two computers which have a number of arithmetic units (32, 64, 128) which can all obey the same instruction simultaneously.† The linking of the units is a complicated business, as is the programming of them. Nevertheless, for certain types of problems they offer tempting performance advantages.

It seems to be the case that most of the sophisticated methods which have been developed for the traditional serial machines (one multiplication at a time) perform badly on parallel machines. A fresh approach is called for but that call is not answered in this book. It seems likely that parallel machines will find their niche in handling very special problems.

†If there are n^2 units then, in principle, the product of two matrices can be formed in the time required for one inner product, but the data accessing problems are severe.

3

Counting Eigenvalues

3-1 TRIANGULAR FACTORIZATION

It is easy to solve triangular systems of equations because the unknowns can be calculated one by one provided only that the equations are taken in the proper order. For this reason it is an important fact that most square matrices can be written as a product of triangular matrices as shown in Fig. 3-1-1.

$$\underset{\mathsf{B}}{\begin{bmatrix} 4 & 12 & 16 \\ 8 & 26 & 28 \\ -4 & -6 & -27 \end{bmatrix}} = \underset{\mathsf{L}}{\begin{bmatrix} 1 & & \\ 2 & 1 & \\ -1 & 3 & 1 \end{bmatrix}} \underset{\mathsf{D}}{\begin{bmatrix} 4 & & \\ & 2 & \\ & & 1 \end{bmatrix}} \underset{\mathsf{U}}{\begin{bmatrix} 1 & 3 & 4 \\ & 1 & -2 \\ & & 1 \end{bmatrix}}$$

Zero elements have been suppressed for clarity.

The leading principal submatrices of B are

$$\mathsf{B}_1 = [4], \quad \mathsf{B}_2 = \begin{bmatrix} 4 & 12 \\ 8 & 26 \end{bmatrix}, \quad \mathsf{B}_3 = \mathsf{B}.$$

Figure 3-1-1 Triangular factorization

For theoretical work it is convenient to separate out D and U from the product (DU).

The standard Gauss elimination technique for solving systems of linear equations is best thought of as a process for computing the triangular factorization of the coefficient matrix. The multipliers which are needed during the elimination are the elements of the L-factor and the result is DU, (Ex. 3-1-1). By means of this factorization the direct solution of a full system $\mathbf{Bx} = \mathbf{b}$ is reduced to the solution of two triangular systems.

Algorithm for the solution of $\mathbf{Bx} = \mathbf{b}$.

1. $\mathbf{B} = \mathbf{LDU}$ (Triangular factorization),
2. Solve $\mathbf{Lc} = \mathbf{b}$ for $\mathbf{c}$ (Forward substitution),
3. Solve $(\mathbf{DU})\mathbf{x} = \mathbf{c}$ for $\mathbf{x}$ (Back substitution).

That is all there is to it for small matrices.

There do exist matrices B for which Step 1 fails. Such matrices are not necessarily strange or pathological. The simplest of them is

$$\mathbf{A} = \begin{bmatrix} 0 & 1 \\ 1 & 0 \end{bmatrix}.$$

This example suggests that after rearranging A's rows triangular factorization might be possible. Indeed whenever A is nonsingular such a rearrangement can be found but it will destroy symmetry (Exs. 3-1-3 and 3-1-4).

In practice the best way to determine whether triangular factorization is possible is to try it and see whether or not the process breaks down. For theoretical work it is useful to know conditions under which factorization is guaranteed and it is also convenient to normalize L and U by putting 1's on their diagonals.

The theorem behind triangular factorization employs the leading principal submatrices of B. These are denoted by $\mathbf{B}_j$ and are exhibited in Fig. 3-1-1. Since D is not a symmetric letter we will replace it by Δ from now on. Unfortunately U is a symmetric letter, but it will soon be replaced by $\mathbf{L}^*$.

LDU THEOREM If $\det \mathbf{B}_j \neq 0$ for $j = 1, 2, \ldots, n-1$, then unique normalized triangular factors $\mathbf{L}, \Delta, \mathbf{U}$ exist such that $\mathbf{B} = \mathbf{L}\,\Delta\mathbf{U}$. Conversely, if $\det \mathbf{B}_j = 0$ for $j \leqslant n-1$ then the factorization may not exist, but even if it does, the factors are not unique.

3-1-1 REMARKS

1. det B = 0 is permitted. Singular matrices of rank $n - 1$ may or may not permit triangular factorization.
2. If $A = L\Delta U$ then $U = L^*$, since $A^* = A$.
3. If A is positive definite then $A = L\Delta L^*$ and Δ is also positive definite. The factorization may be written

$$A = (L\Delta^{\frac{1}{2}})(L\Delta^{\frac{1}{2}})^* \equiv C^*C$$

where $C \equiv \Delta^{\frac{1}{2}}L^*$ is called the Choleski factor of A. Regrettably C is sometimes called the square root of A, but that term should be reserved solely for the unique positive definite symmetric solution of $X^2 = A$.

4. If A is banded and $A = L\Delta L^*$ then L inherits A's band structure; that is, if $a_{ij} = 0$ when $|i - j| > m$ then $l_{ij} = 0$ when $i - j > m$. Example:

$$A = \begin{bmatrix} 4 & 4 & 0 \\ 4 & 6 & 2 \\ 0 & 2 & 1 \end{bmatrix}, \qquad L = \begin{bmatrix} 1 & & \\ 1 & 1 & \\ 0 & 1 & 1 \end{bmatrix}$$

The number m is the half-bandwidth of A.

5. The block form of triangular factorization is often useful. There is much economy in the apt use of partitioned matrices; i.e., matrices whose elements are matrices. For example if B_{11} is invertible then

$$\begin{bmatrix} B_{11} & B_{12} \\ B_{21} & B_{22} \end{bmatrix} = \begin{bmatrix} I & O \\ B_{21}B_{11}^{-1} & I \end{bmatrix}\begin{bmatrix} B_{11} & O \\ O & \hat{B}_{22} \end{bmatrix}\begin{bmatrix} I & B_{11}^{-1}B_{12} \\ O & I \end{bmatrix},$$

where

$$\hat{B}_{22} = B_{22} - B_{21}B_{11}^{-1}B_{12}.$$

Sometimes $\hat{B}_{22}$ is called the Gauss transform of B_{22}, sometimes the Schur complement.

3-1-2 COST

Triangular factorization is such a useful tool that is warrants further consideration. Is it expensive? Is it reliable? The answers come in this and the next section.

Table 3-1-1 Arithmetic operations

Type	*Divs*	*Mults and adds*	$n \longrightarrow \infty$
General, n by n, dense	$\frac{1}{2}n(n-1)$	$\frac{1}{6}n(n-1)(2n-1)$	$\frac{1}{3}n^3$
Symmetric, n by n, half-bandwidth m	$m(n-m)+\frac{1}{2}m(m-1)$	$\frac{1}{2}m(m+1)(n-m)+\frac{1}{6}m(m+1)(m+2)$	$\frac{1}{2}m(m+1)n$

Storage. For full general matrices an n by n array can be used to hold L, Δ, and U. In the absence of interchanges the bandwidth of a matrix is preserved. Within the band, fill-in may occur. For A's with halfband-width m a conventional n by $(m + 1)$ array is often used for L and Δ. For large sparse matrices with a peculiar pattern of nonzero elements a special data structure may be needed.

Table 3-1-1 exhibits the number of arithmetic operations required in two important cases. The reader who has never done such a count is urged to verify the numbers in the table (by writing out the algorithm and inspecting the inner loop). The dramatic change from an $O(n^3)$ process (the general case) to an $O(n)$ process (symmetric matrices with narrow band) shows that the sparsity structure plays a vital role in the efficiency of algorithms. When A is tridiagonal the computation of $A^{-1}u$ costs little more than the computation of Au!

Storage costs show the same sort of advantage for matrices of narrow bandwith. The problem of "fill-in" has received a great deal of attention and has become a special field known as **sparse matrix technology**. The "fill" in any matrix transformation is the set of those elements initially zero, which become nonzero in the course of the transformation.

Exercises on Sec. 3-1

3-1-1. Carry out the standard Gauss elimination process on the matrix in Fig. 3-1-1 and verify that L does hold the multipliers and the resulting triangular matrix is ΔU.

3-1-2. By equating the (i, j) elements on each side of $B = L\Delta U$ by columns obtain the Crout algorithm which generates the elements of L, Δ, and U directly without computing any intermediate reduced matrices $B^{(j)}$, $j = 1, \ldots, n - 1$.

3-1-3. Show that there are nonsingular matrices A such that no symmetric permutation of A, i.e., PAP*, permits triangular factorization.

3-1-4. Exhibit one nondiagonal 3 by 3 matrix of rank 2 which permits triangular factorization and another which does not. Hint: work backward.

3-1-5. Show that if $A = L\Delta U$ then $U = L^*$. Recall that $A^* = A$.

3-1-6. Show that $A = L\Delta L^*$ for all positive definite A. One approach is to use a 2 by 2 block matrix and use induction. What about the semi-definite case when A has rank $n - 1$?

3-1-7. Compute the Choleski factorization of

$$\begin{bmatrix} 1 & 1/2 & 1/3 \\ 1/2 & 1/3 & 1/4 \\ 1/3 & 1/4 & 1/5 \end{bmatrix}.$$

Use a calculator if possible.

3-1-8. Prove that if $a_{ij} = 0$ when $|i - j| > m$ then $l_{ij} = 0$ when $i - j > m$.

3-1-9. Let $\mathsf{B} = \begin{bmatrix} \mathsf{B}_{11} & \mathsf{B}_{12} \\ \mathsf{B}_{21} & \mathsf{B}_{22} \end{bmatrix}$ where B_{11} and B_{22} are n by n. Suppose that B_{11} is nonsingular. Find a formula for det B as a product of two determinants of order n. Find another formula for the case when B_{11} is singular but B_{22} is not.

3-1-10. Verify the multiplication count for triangular factorization of a full matrix. What is the count for a symmetric matrix assuming that full advantage is taken of symmetry?

3-1-11. Verify the operation count for a symmetric matrix of half bandwidth m.

3-1-12. Show the "fill-in" during triangular factorization of a positive definite matrix with the indicated nonzero elements.

$$\left[\begin{array}{cccc|cccc} x & x & & & x & & & \\ x & x & x & & & x & & \\ & x & x & x & & & x & \\ & & x & x & & & & x \\ \hline x & & & & x & x & & \\ & x & & & x & x & x & \\ & & x & & & x & x & x \\ & & & x & & & x & x \end{array}\right]$$

3-1-13. Use induction and Remark 5 to prove the first assertion of the LDU theorem. Exhibit 2 by 2 matrices which exemplify the converse.

3-2 ERROR ANALYSIS OF TRIANGULAR FACTORIZATION

The triangular factorization of some well-conditioned matrices can be ruined by the roundoff errors in two or three places in the calculation.

The decomposition is so useful that it is worthwhile to understand how it can fail. Fortunately a complete error analysis is not needed to reveal when and how disaster can strike. It suffices to look carefully at the first step in the reduction process. Let A be written in partitioned form

$$\mathsf{A} = \begin{bmatrix} \alpha & \mathsf{c}^* \\ \mathsf{c} & \mathsf{M} \end{bmatrix}, \qquad \mathsf{M} \text{ is } (n-1) \text{ by } (n-1). \tag{3-2-1}$$

Using the 2 by 2 block factorization shown in Remark 5 of Sec. 3-1, we get

$$\mathsf{A} = \begin{bmatrix} 1 & \mathsf{o}^* \\ \mathsf{b} & \mathsf{I} \end{bmatrix} \begin{bmatrix} \alpha & \mathsf{o}^* \\ \mathsf{o} & \mathsf{A}^{(1)} \end{bmatrix} \begin{bmatrix} 1 & \mathsf{b}^* \\ \mathsf{o} & \mathsf{I} \end{bmatrix}, \qquad \mathsf{b} = \mathsf{c}/\alpha, \tag{3-2-2}$$

where

$$\mathsf{A}^{(1)} = \mathsf{M} - \mathsf{b}\alpha\mathsf{b}^* = \mathsf{M} - \mathsf{c}\mathsf{c}^*/\alpha, \tag{3-2-3}$$

or, descending to the element level

$$a_{ij}^{(1)} = a_{i+1,j+1} - b_i \alpha b_j, \qquad i, j = 1, \ldots, n-1. \tag{3-2-4}$$

$\mathsf{A}^{(1)}$ is called the **first reduced matrix** in the process.

At the next step, the same reduction is performed on $\mathsf{A}^{(1)}$. To see the role of roundoff, let b and $\mathsf{A}^{(1)}$ now denote the quantities actually stored in the computer and suppose that, by some stroke of luck, each $a_{ij}^{(1)}$ is computed with a minimal relative error (see the model of arithmetic in Chap. 2 for more details). Thus,

$$\begin{aligned} a_{ij}^{(1)} &= fl(a_{i+1,j+1} - b_i \alpha b_j), \\ &= (a_{i+1,j+1} - b_i \alpha b_j)(1 - \rho_{ij}), \qquad |\rho_{ij}| < \epsilon. \end{aligned} \tag{3-2-5}$$

Here ϵ is the unit roundoff. It is convenient to rewrite $1 - \rho_{ij}$ as $1/(1 + \eta_{ij})$. So $\eta_{ij} = \rho_{ij}/(1 - \rho_{ij}) \approx \rho_{ij}$. The key observation is that (3-2-5) can now be rewritten as

$$a_{ij}^{(1)} + b_i \alpha b_j = a_{i+1,j+1} - \eta_{ij} a_{ij}^{(1)}, \qquad i, j = 1, \ldots, n-1, \quad i \geqslant j. \tag{3-2-6}$$

In matrix notation (3-2-6) says

$$\begin{bmatrix} 1 & \mathsf{o}^* \\ \mathsf{b} & \mathsf{I} \end{bmatrix} \begin{bmatrix} \alpha & \mathsf{o}^* \\ \mathsf{o} & \mathsf{A}^{(1)} \end{bmatrix} \begin{bmatrix} 1 & \mathsf{b}^* \\ \mathsf{o} & \mathsf{I} \end{bmatrix} \equiv \mathsf{A} - \mathsf{H}^{(1)}, \qquad h_{ij}^{(1)} = \eta_{ij} a_{ij}^{(1)}. \tag{3-2-7}$$

Roundoff error made the algorithm factor $\mathsf{A} - \mathsf{H}^{(1)}$ instead of A. That is not satisfactory if $\mathsf{A}^{(1)}$ is huge compared to A, even under the favorable assumption that each arithmetic step is done with minimal error. It is not hard to prove the converse without making any favorable assumptions, namely if $\|\mathsf{A}^{(1)}\| \approx \|\mathsf{A}\|$ then $\|\mathsf{H}^{(1)}\|$ is small, like roundoff, compared to $\|\mathsf{A}\|$. When all rounding errors are included, the expression for $h_{ij}^{(1)}$ includes a multiple of $b_i b_j a_{11}$, but this simply makes $\|\mathsf{H}^{(1)}\|$ tiny compared to $\max(\|\mathsf{A}\|, \|\mathsf{A}^{(1)}\|)$. See Ex. 3-2-2.

At the next step, the computed quantities $\mathsf{b}^{(2)}$ and $\mathsf{A}^{(2)}$ turn out to be part of the decomposition not of $\mathsf{A}^{(1)}$ but of $\mathsf{A}^{(1)} - \hat{\mathsf{H}}^{(2)}$. And so on. At the end the factors L and Δ are seen to satisfy

$$\mathsf{L}\Delta\mathsf{L}^* = \mathsf{A} - \mathsf{H}^{(1)} - \mathsf{H}^{(2)} - \cdots - \mathsf{H}^{(n-1)} \equiv \mathsf{A} - \mathsf{H}. \tag{3-2-8}$$

where $\mathsf{H}^{(k)} \equiv \mathsf{O}_{k-1} \oplus \hat{\mathsf{H}}^{(k)}$ is tiny, like roundoff, compared with the larger of $\mathsf{A}^{(k)}$ and $\mathsf{A}^{(k-1)}$. If any of the $\|\mathsf{A}^{(k)}\|$ are huge compared with $\|\mathsf{A}\|$, then the algorithm has simply factored the wrong matrix.†

Note that small pivots, $\delta_k = a_{kk}^{(k)}$, may or may not provoke a large $\mathsf{A}^{(k+1)}$ and, likewise, large multipliers in $\mathsf{b}^{(k)}$ may or may not provoke large

---

†Refer to the list of notations on the inside covers for a description of the symbol $\oplus$.

$A^{(k+1)}$. These traditional scapegoats, whenever factorization fails, are simply not the relevant quantities. It is their outer products $(b^{(k)}b^{(k)*})a_{kk}^{(k)}$ which matter.

Triangular factorization, without any pivoting, can indeed fail and a failure is always indicated by element growth, i.e., $\|A^{(k)}\| \gg \|A\|$ for some $k < n$. It is easy to monitor this growth and so there is no need for failure to go undetected. The next section pursues this topic.

For positive definite matrices $\|A^{(k)}\| < \|A^{(k-1)}\|$ and so triangular factorization is very stable (Ex. 3-2-5).

The use of pivoting to achieve a stable factorization is not central to our purposes and the reader is referred to [Forsythe and Moler, 1967] for this information. One-sided pivoting spoils symmetry and symmetric pivoting (using matching row and column interchanges) usually spoils the band structure of sparse matrices.

Exercises on Sec. 3-2

3-2-1. Follow the error analysis numerically on the given matrix using 4 decimal floating point arithmetic (chopped). Exhibit L, Δ, and H.

$$A = \begin{bmatrix} 10^{-3} & 10 \\ 10 & 14 \end{bmatrix}.$$

3-2-2. If a calculator is available factorize the given matrices using 4 decimal arithmetic. Then form the product $L \Delta L^*$ exactly and compare with the original matrix.

$$H = \begin{bmatrix} 1.000 & 0.5000 & 0.3333 & 0.2500 \\ 0.5000 & 0.3333 & 0.2500 & 0.2000 \\ 0.3333 & 0.2500 & 0.2000 & 0.1666 \\ 0.2500 & 0.2000 & 0.1666 & 0.1428 \end{bmatrix}$$

$$W = \begin{bmatrix} 5 & 1 & & & & \\ 1 & 3 & 1 & & & \\ & 1 & 1 & 1 & & \\ & & 1 & -1 & 1 & \\ & & & 1 & -3 & 1 \\ & & & & 1 & -5 \end{bmatrix}$$

Next compute L and Δ in full precision and compare them with the earlier versions.

3-2-3. Prove that if B is nonsingular there always is a permutation matrix P, possibly many, such that PB permits triangular factorization. Hint: First show that a row interchange can always be made at each step in the factorization to ensure that it does not break down. Second (this is harder) show why it is permissible to pretend that all the interchanges have been done in advance before factorization begins.

3-2-4. By Sylvester's inertia theorem show that $A^{(k)}$ is positive definite if $A^{(k-1)}$ is positive definite.

3-2-5. Show that the k-th pivot $\delta_k = a_{kk}^{(k)} < a_{kk}^{(k-1)}$ for all k. Use the result of Ex. 3-2-4 to deduce that $\|A^{(k)}\| \leqslant \|A^{(k-1)}\|$. Here A is positive definite.

3-3 SLICING THE SPECTRUM

There is an elegant way to compute the number of A's eigenvalues that are less than any given real number σ. Since the result is an integer the technique appears to offer deliverance from the machinations of roundoff error and the extent to which this hope is justified will be examined below. The technique has been used by theoretical physicists for many years, certainly since the early 1950s.

The method is a corollary of Sylvester's inertia theorem (Fact 1-6) which states the invariance of $\nu(W)$, the number of negative eigenvalues of W, under congruence transformations. Of great use is the fact that the technique is directly applicable to the general eigenvalue problem $A - \lambda M$.

THEOREM Suppose that $A - \sigma M$ permits triangular factorization $A - \sigma M = L_\sigma \Delta_\sigma L_\sigma^*$ where Δ_σ is diagonal and M is positive definite. Then

$$\nu(\Lambda - \sigma I) = \nu(A - \sigma M) = \nu(\Delta_\sigma)$$

where $\Lambda = \text{diag}(\lambda_1, \lambda_2, \ldots, \lambda_n)$ and λ_i is an eigenvalue of the pair (A, M).

Proof. Since L_σ is unit lower triangular it is invertible and so $A - \sigma M$ is congruent to Δ_σ. By (15-3-3), the simultaneous reduction of two quadratic forms, there is an invertible matrix F such that

$$F^*(A - \sigma M)F = \Lambda - \sigma I,$$

whence $A - \sigma M$ is congruent to $\Lambda - \sigma I$. The result follows from Sylvester's inertia theorem applied to the congruent diagonal matrices $\Lambda - \sigma I$ and Δ_σ. □

On one hand $\nu(\Delta_\sigma)$ is simply the number of negative elements on Δ_σ's diagonal. On the other hand $\nu(\Lambda - \sigma I)$ is the number of eigenvalues of the pencil (A, M) which are less than σ.

The computation of Δ_σ and $\nu(\Delta_\sigma)$ reveals how σ slices the spectrum into two pieces. We call $\nu(\Delta_\sigma)$, or $\pi(\Delta_\sigma)$ if you prefer to dwell on the positive, **the spectrum slicer** and show it at work in Example 3-1-1.

EXAMPLE 3-3-1 Slicing the Spectrum

$$\mathsf{A} = \begin{bmatrix} 1 & 2 & 1 & 3 \\ 2 & 3 & -2 & 0 \\ 1 & -2 & -1 & -7 \\ 3 & 0 & -7 & 0 \end{bmatrix}, \qquad \mathsf{M} = \mathsf{I}$$

$$(\mathsf{A} - 0) = \begin{bmatrix} 1 & & & \\ 2 & 1 & & \\ 1 & 4 & 1 & \\ 3 & 6 & 1 & 1 \end{bmatrix} \cdot \begin{bmatrix} 1 & & & \\ & -1 & & \\ & & 14 & \\ & & & 13 \end{bmatrix} \cdot \begin{bmatrix} 1 & 2 & 1 & 3 \\ & 1 & 4 & 6 \\ & & 1 & 1 \\ & & & 1 \end{bmatrix}$$

$$(\mathsf{A} - 2) = \begin{bmatrix} 1 & & & \\ -2 & 1 & & \\ -1 & 0 & 1 & \\ -3 & 1.2 & 2 & 1 \end{bmatrix} \cdot \operatorname{diag} \begin{bmatrix} -1 \\ 5 \\ -2 \\ 7.8 \end{bmatrix} \cdot \begin{bmatrix} 1 & -2 & -1 & -3 \\ & 1 & 0 & 1.2 \\ & & 1 & 2 \\ & & & 1 \end{bmatrix}$$

Count: $\nu(\Delta_0) = 1, \quad \nu(\Delta_2) = 2.$

Conclusion: $\lambda_1[\mathsf{A}] < 0 < \lambda_2[\mathsf{A}] < 2 < \lambda_3[\mathsf{A}] \leqslant \lambda_4[\mathsf{A}].$

3-3-1 THE TRIDIAGONAL CASE

In this important application it turns out that the elements of L and Δ need not be stored. Only one cell of work space is needed. (See Ex. 3-3-5.)

For future reference we give the procedure in detail. Let α hold the diagonal elements, let γ hold the squares of the off diagonal elements, let δ be the extra cell, and let σ be the sample point (or origin shift). The purpose is to compute $\nu = \nu[A - \sigma]$.

Initialize: $\delta \leftarrow \alpha_1 - \sigma$

$$\nu \leftarrow \begin{cases} 1, & \text{if } \delta < 0 \\ 0, & \text{otherwise.} \end{cases}$$

Loop: for $k = 2, \ldots, n$ repeat

$$\begin{cases} \delta \leftarrow (\alpha_k - \gamma_{k-1}/\delta) - \sigma, \\ \text{if } \delta = 0 \text{ then } \delta \leftarrow \epsilon(|\alpha_k| + |\sigma| + \epsilon), \\ \text{if } \delta < 0 \text{ then } \nu \leftarrow \nu + 1. \end{cases} \tag{3-3-1}$$

Actually I prefer: If $\delta = 0$ then change σ slightly, and start again.

3-3-2 ACCURACY OF THE SLICE

For any symmetric A consider $\mathsf{A} - \sigma$ as σ varies over the real numbers. The LDU theorem shows that the factorization fails to exist if, and only if, one or more of the leading principal submatrices, $\mathsf{A}_k - \sigma$, is singular. By Cauchy's Interlace Theorem (Sec. 10-1) the eigenvalues $\lambda_i^{(k)}$ of A_k interlace those of A_{k+1}. Note that $\mathsf{A}_n = \mathsf{A}$. Consequently there are $\sum_{k=1}^{n-1} k = n(n-1)/2$ values of σ, not necessarily distinct, for which $\mathsf{A} - \sigma$ does not permit an $\mathsf{L\Delta L^*}$ factorization. For σ in tiny intervals around each of these values the factorization is unreliable in the face of roundoff. To see why this is so let A be partitioned as follows,

$$\mathsf{A} = \begin{bmatrix} \mathsf{V} & \mathsf{C}^* \\ \mathsf{C} & \mathsf{M} \end{bmatrix}, \tag{3-3-2}$$

and suppose that σ agrees with some $\lambda_j[\mathsf{V}]$ to p significant decimals. Then one of the reduced matrices encountered in the factorization process is, see Ex. 3-3-2,

$$\hat{\mathsf{M}} = \mathsf{M} - \sigma - \mathsf{C}(\mathsf{V} - \sigma)^{-1}\mathsf{C}^* \tag{3-3-3}$$

and, because σ is so close to an eigenvalue of V (Ex. 3-3-3),

$$\begin{aligned} \|(\mathsf{V} - \sigma)^{-1}\| &> 10^p/|\lambda_j[\mathsf{V}]| \\ &> 10^p/\|\mathsf{A}\|. \end{aligned} \tag{3-3-4}$$

It is certainly possible to have $\|\mathsf{C}\| \approx \|\mathsf{A}\|$ and, without favorable cancellation,

$$\|\mathsf{C}(\mathsf{V} - \sigma)^{-1}\mathsf{C}^*\| \approx 10^p\|\mathsf{A}\|. \tag{3-3-5}$$

The memorable fact is that there are only $n(n-1)/2$ danger spots for σ (namely the eigenvalues of the principal submatrices A_j, $j = 1, \ldots, n-1$). The probability of failure in the triangular factorization of $\mathsf{A} - \sigma$ is low. This suggests the following simple strategy. If intolerable element growth occurs in factoring $\mathsf{A} - \sigma$ then change σ by .01% and try again.

The computed factors L_σ and Δ_σ satisfy

$$\mathsf{L}_\sigma \Delta_\sigma \mathsf{L}_\sigma^* = (\mathsf{A} - \sigma) - \mathsf{H}_\sigma \qquad \text{(from Sec. 3-2)} \tag{3-3-6}$$

and so

> Spectrum slicing is impervious to the roundoff errors in triangular factorization provided that
>
> $$\nu(\mathsf{A} - \mathsf{H}_\sigma - \sigma) = \nu(\mathsf{A} - \sigma). \tag{3-3-7}$$

When no element growth occurs then H_σ is tiny relative to A and the slice is certainly as good as the precision of the calculation warrants. However, the two ν values may still be equal even when H_σ is not tiny. Thus inaccurate triangular factorization does not necessarily ruin the accuracy of the slice.

Conversely, let us note the implications of a wrong count. By the Weyl monotonicity theorem (Sec. 10-3 or Fact 1-11.)

$$|\lambda_j[\mathsf{A} - \mathsf{H}_\sigma] - \lambda_j[\mathsf{A}]| < \|\mathsf{H}_\sigma\|, \qquad j = 1, \ldots, n. \tag{3-3-8}$$

So if $\nu(\mathsf{A} - \mathsf{H}_\sigma - \sigma) \neq \nu(\mathsf{A} - \sigma)$ then

$$\|\mathsf{H}_\sigma\| > \min_j |\lambda_j[\mathsf{A}] - \sigma|. \tag{3-3-9}$$

The precision with which eigenvalues can be located by slicing is least satisfactory when, for some i and j, $\lambda_i[\mathsf{A}]$ and $\lambda_j[\mathsf{V}]$ agree too closely. V is shown in (3-3-2).

3-3-3 FAILURE OF SLICING PROCEDURE

There are two subtly different ways to compute Δ_σ. The most natural is to compute $\mathsf{U} = \mathsf{L}^{-1}\mathsf{A}$ by elimination and use its diagonal elements. However, when A is stored in compact form there may be no room for U and instead the elements of A are modified by the formula

$$a'_{ij} \leftarrow a_{ij} - \sum_{k=1}^{\min(i,j)-1} l_{ik}\,\delta_k\,l_{jk}.$$

The difference can be striking as Example 3-3-2 shows. The calculations in Example 3-3-2 were done in simulated 24-bit floating point arithmetic on the CDC 6400.

EXAMPLE 3-3-2 Failure of Slicing

A

$$\begin{bmatrix} 1.0E+0 & 2.8E-2 & 1.0E+1 & 1.5E+1 & 1.0E+0 & 1.0E+1 \\ 2.8E-2 & 2.7E+1 & 2.7E+1 & 1.3E+1 & 2.7E+1 & 2.0E+0 \\ 1.0E+1 & 2.7E+1 & 1.3E+2 & 1.9E+1 & 2.8E+1 & 1.2E+1 \\ 1.5E+1 & 1.3E+1 & 1.9E+1 & -3.0E+2 & -8.7E+1 & 2.8E+1 \\ 1.0E+0 & 2.7E+1 & 2.8E+1 & -8.7E+1 & 2.7E+1 & 1.2E+1 \\ 1.0E+1 & 2.0E+0 & 1.2E+1 & 2.8E+1 & 1.2E+1 & 1.0E+1 \end{bmatrix}$$

Please compare the Δ_σ and **L** produced by the factorizations shown by the **LU** technique, and for the **L(LΔ)*** technique on the facing page.

LU Technique

Δ_σ

$$\left(1.0E+0,\quad 2.6E+1,\quad 8.5E-7,\quad \underline{2.4E+11},\quad 1.5E+1,\quad 1.4E+2\right)$$

$$\nu(\Delta_\sigma) = 1$$

A − LU

$$\begin{bmatrix} 0 & 0 & 0 & 0 & 0 & 0 \\ 0 & 0 & 0 & 0 & 0 & 0 \\ 0 & 0 & 0 & 0 & 0 & 0 \\ 0 & 0 & 0 & -4.9E+2 & 6.4E+1 & -1.1E+2 \\ 0 & 0 & 0 & -8.4E+1 & -2.9E+0 & 1.2E+1 \\ 0 & 0 & 0 & -4.3E+2 & 3.2E+0 & -4.8E+2 \end{bmatrix}$$

L

$$\begin{bmatrix} 1.0E+0 & & & & & \\ 2.8E-2 & 1.0E+0 & & & & \\ 1.0E+1 & 1.0E+0 & 1.0E+0 & & & \\ 1.5E+1 & 4.8E-1 & -1.7E+8 & 1.0E+0 & & \\ 1.0E+0 & 1.0E+1 & -1.1E+7 & 6.2E-2 & 1.0E+0 & \\ 1.0E+1 & 6.6E-2 & -1.1E+8 & 6.2E-1 & \boxed{6.0E+1} & 1.0E+0 \end{bmatrix}$$

EXAMPLE 3-3-2 (Cont.)

Eigenvalues of A

$$(-3.3E+2,\quad -5.9E+0,\quad 1.1E+1,\quad 1.7E+1,\quad 5.0E+1,\quad 1.5E+2)$$

$$\sigma = -8.5E-6$$

Please compare the Δ_σ and L produced by the following technique with those produced by the LU technique on the facing page.

L(LΔ)* Technique

Δ_σ

$$\left(1.0E+0,\quad 2.6E+1,\quad 8.5E-7,\quad \underline{-2.4E+10},\quad 1.8E+1,\quad -2.3E+2\right)$$

$$\nu(\Delta_\sigma) = 2$$

A − L(LΔ)*

$$\begin{bmatrix} 0 & 0 & 0 & 0 & 0 & 0 \\ 0 & 0 & 0 & 0 & 0 & 0 \\ 0 & 0 & 0 & 0 & 0 & 0 \\ 0 & 0 & 0 & -4.9E+2 & 6.4E+1 & -1.1E+2 \\ 0 & 0 & 0 & 3.9E+1 & 9.0E-1 & 7.3E+0 \\ 0 & 0 & 0 & -4.3E+2 & 3.1E+0 & -4.8E+2 \end{bmatrix}$$

L

$$\begin{bmatrix} 1.0E+0 & & & & & \\ 2.8E-2 & 1.0E+0 & & & & \\ 1.0E+1 & 1.0E+0 & 1.0E+0 & & & \\ 1.5E+1 & 4.8E-2 & -1.7E+8 & 1.0E+0 & & \\ 1.0E+0 & 1.0E+0 & -1.1E+7 & 6.2E-2 & 1.0E+0 & \\ 1.0E+1 & 6.6E-2 & -1.1E+3 & 6.2E-1 & \boxed{4.8E+0} & 1.0E+0 \end{bmatrix}$$

Exercises on Sec. 3-3

3-3-1. Strictly speaking $\nu(A)$ denotes the number of negative elements in any diagonal matrix congruent to A. By using the spectral factorization of A show that $\nu(A)$ is the number of negative eigenvalues of A.

3-3-2. Establish (3-3-3) by invoking the uniqueness of triangular factorization.

3-3-3. Establish (3-3-4).

3-3-4. In Example 3-3-2 change σ by 1%, recompute Δ_σ and $\nu(\Delta_\sigma)$ in both cases and compare element growth in all three cases.

3-3-5. Let T be tridiagonal with k-th row $(\cdots 0, \beta_{k-1}, \alpha_k, \beta_k, 0 \cdots)$. Write out the algorithm for factoring $T - \sigma = L\Delta L^*$ and then show how the elements of L can be eliminated to yield the algorithm given in this section.

3-3-6. Suppose that A and W are both tridiagonal except that $a_{1n} = a_{n1} \neq 0$, $w_{1n} = w_{n1} \neq 0$. Let DA and DW be n-vectors holding the diagonal elements; let EA and EW be n-vectors holding the other nonzero elements. Write an algorithm which computes $\nu(A - \sigma W)$ without using any more arrays, just 6 ot 7 extra simple variables.

3-4 RELATION TO STURM SEQUENCES

Let $\chi_j(\tau)$ denote the characteristic polynomial of the leading principal j by j submatrix of A. Thus, $\chi_1(\tau) = \tau - a_{11}$. Let $\chi_0(\tau) = 1$. By the Cauchy Interlace Theorem (Sec. 10-3) the sequence $\{\chi_0, \chi_1, \ldots, \chi_n\}$ is a **Sturm sequence** of polynomials; that is, the zeros of consecutive χ's interlace each other. A rather careful argument† then shows that the number of sign **agreements** in consecutive terms of the numerical sequence $\{\chi_j(\sigma), j = 0, 1, \ldots, n\}$ equals the number of zeros of χ_n which are less than σ.‡

For general A there is no cheap way to compute $\chi_j(\sigma)$ from the $\chi_i(\sigma)$, $i < j$, but when A is tridiagonal there is a well-known three-term recurrence

$$\chi_{j+1}(\sigma) = (\sigma - a_{j+1,j+1})\chi_j(\sigma) - a_{j+1,j}^2\chi_{j-1}(\sigma) \qquad (3\text{-}4\text{-}1)$$

which is also discussed in Chap. 7. This recurrence was used to count the number of eigenvalues less than σ in the trend-setting report [Givens, 1954].

---

†This argument is given in [Wilkinson, 1965, p. 300].

‡A lot of research went into the vexed question of the right sign to be attributed to one, or even two, zero value of $\chi_j(\sigma)$. This led to such esoterica as **Grundelfinger's rule** which can be found in [Browne, 1930].

If $A - \sigma = L\Delta L^*$ (triangular factorization) then (Ex. 3-4-1)

$$(-1)^j\chi_j(\sigma) = \delta_1 \cdots \delta_j = \det \Delta_j, \quad j = 1, \ldots, n, \qquad (3\text{-}4\text{-}2)$$
$$\delta_j = -\chi_j(\sigma)/\chi_{j-1}(\sigma).$$

Thus, sign agreements in the sequence $\{\chi_j(\sigma)\}$ correspond to negative values in the sequence $\{\delta_j\}$. However, the rational functions δ_j are far more sedate than the polynomials χ_j and their use is to be preferred in finite precision computations. Problems with overflow and underflow recede when triangular factorization is used in place of (3-4-1).

The continual emphasis on Sturm sequences (i.e., 3-4-1) delayed the application of spectrum slicing to banded matrices and the general eigenvalue problem by numerical analysis. Indeed some authors call spectrum slicing a Sturm sequence technique although it never computes the χ_j.

Exercises on 3-4

3-4-1. Derive (3-4-2) and then obtain from (3-4-1) a three-term recurrence for the δ_j which requires 1 division and no multiplications at each step.

3-4-2. For the example in the preceding section compute the Sturm sequence $\{\chi_j(\sigma)\}$ and contrast with $\{\delta_j\}$.

3-5 BISECTION AND SECANT METHODS

3-5-1 BISECTION

If a half open interval $[\alpha, \beta)$ is known to contain at least one eigenvalue then the slicing techniques in Sec. 3-3 can be used to find whether $[\alpha, (\alpha + \beta)/2)$ also contains one. The process can be repeated again and again to locate an eigenvalue to an accuracy limited only by the errors in the triangular factorization.

This idea has been implemented with great care in the program **bisect** (p. 249 of the Handbook), but for tridiagonal matrices only.

Cost. One triangular factorization per slice. The operation counts are given in Sec. 3-1.

Usage. Bisection is efficient when low accuracy is required. It is also good for finding a few eigenvalues of matrices with narrow bandwidth, particularly when all the eigenvalues in a given interval are required. Bisection can often deliver results correct to full working precision when asked to do so.

The technique is not recommended when more than $n/4$ eigenvalues are wanted, or when the matrix is rather full.

3-5-2 *THE SECANT METHOD*

When an interval $[\alpha, \beta)$ has been found to contain a single eigenvalue λ then bisection is a rather primitive and slow way to pin down λ to high accuracy. The classical problem of finding the zero of a polynomial known to lie in a given interval has enjoyed a lot of attention for many years and some first-class algorithms have been developed.

For a mere $(n - 1)$ extra multiplications the factorization program can evaluate the polynomial itself because

$$\chi(\sigma) = \delta_1 \cdots \delta_n, \qquad \text{(watch out for over/under flow)},$$

and so any variation of the Secant method can be implemented. The order of convergence is 1.618, as against 1 for bisection; so convergence should be accelerated. For tridiagonal matrices $\chi(\sigma)$ can be calculated directly from (3-4-1) but for fatter matrices triangular factorization is the natural approach, and the extra cost of forming $\delta_1 \cdots \delta_{\hat{n}}$ is small. An important alternative to χ is discussed in Sec. 3-6.

The formula for one step of the Secant Iteration is

$$\xi_{j+1} = \xi_j - \omega\chi(\xi_j)/\Delta\chi(\xi_j, \xi_{j-1}) \tag{3-5-1}$$

where

$$\Delta\chi(\alpha, \beta) \equiv (\chi(\alpha) - \chi(\beta))/\,(\alpha - \beta) \tag{3-5-2}$$

is the first order divided difference of χ. The value $\omega = 1$ gives the standard formula, $\omega = 2$ gives the more practical Double Secant Iteration. See Ex. 3-5-4 for more details.

In good programs the Secant formula is not used blindly; it provides one among other candidates for the next approximation. By making the program more complicated and checking for various difficult cases the performance of zero finders has become exemplary. Clearly, if the Secant method's approximation lies outside the smallest interval known to contain λ then it should not be used. What should take its place? That is an interesting question. It is tempting to return to bisection but there are better solutions than that. The interested reader is referred to [Brent, 1973] and to [Bus and Dekker, 1978] for a full discussion of this topic.

Warning. Polynomial zero finders do not have at their disposal anything comparable to the spectrum slicer $\nu(\Delta_\sigma)$ and so they have to be very careful not to jump over two zeros (leaving no sign change). Eigenvalue codes should not be so cautious. See [Bathé and Wilson, 1976, Chap. 11] for the application of these ideas in computations with large matrices.

With tridiagonal matrices it is possible to obtain recurrences for evaluating $\chi'(\sigma)$ and even higher derivatives. Thus the whole gamut of zero finders could be considered for finding λ. However, the recurrences do not generalize nicely to matrices with greater bandwidth and this avenue will not be explored.

Exercises on Sec. 3-5

3-5-1. Assume that the Secant iteration converges to λ. Find an expression which relates the error in one step to the product of the previous two errors. Then deduce that the order of convergence is $(1 + \sqrt{5})/2$.

3-5-2. Write a simple program which combines bisection with Secant acceleration and test it on the examples given in this chapter. Compare your results with those from simple bisection and include the relative costs of one step in the two techniques.

3-5-3. Project. Take the algorithm **bisect** in the Handbook (p. 249) and replace the Sturm sequence technique by triangular factorization of a banded matrix.

3-5-4. Let $\lambda_1 \leqslant \lambda_2 \leqslant \cdots \leqslant \lambda_n$ be the zeros of χ. Let $\mu_1 \leqslant \mu_2 \leqslant \cdots \leqslant \mu_{n-1}$ be zeros of χ', the derivative of χ. If $\xi_{j-1} < \xi_j < \lambda_1$ show that for Equation (3-5-1)
(a) if $\omega = 1$ then $\xi_{j+1} < \lambda_1$,
(b) if $\omega = 2$ then $\xi_{j+1} < \mu_1$ (unpublished result of Kahan).
In neither case can ξ_{j+1} jump over **two** zeros of χ.

3-6 HIDDEN EIGENVALUES

By (3-4-2) the final pivot $\delta_n(\sigma)$ in the factorization of $\mathsf{A} - \sigma$ is equal to $-\chi_n(\sigma)/\chi_{n-1}(\sigma)$ and so has the same zeros as χ_n. Consequently the Secant method, or any of its variations, can be applied to the rational function $\delta_n(\sigma)$ thereby avoiding the $(n - 1)$ multiplications required to evaluate $\chi_n(\sigma)$ from the known pivots $\delta_i(\sigma)$, $i = 1, \ldots, n$.

An interesting phenomenon can easily mar a straightforward implementation of this replacement of χ by δ. It can happen that λ is a well-isolated simple zero of χ_n and is also extremely close to a zero of χ_{n-1}, χ_{n-2}, χ_{n-3}, etc. In other words $|\chi_n'(\lambda)|$ is large while $|\chi_{n-1}(\lambda)|$ is very small. The effect is that for almost all values of σ the pole of δ_n very near λ cancels out, or conceals, the zero at λ and persuades a good number of root finders that there is no treasure to be found near λ. Fig. 3-6-1 gives a picture of a hidden zero. Wilkinson constructed a beautiful example: the tridiagonal matrix W_{21}^- defined by

$$w_{ii} = 11 - i, \qquad \text{for } i = 1, \ldots, 21$$

$$w_{i,i+1} = w_{i+1,i} = 1, \qquad \text{for } i = 1, \ldots, 20.$$

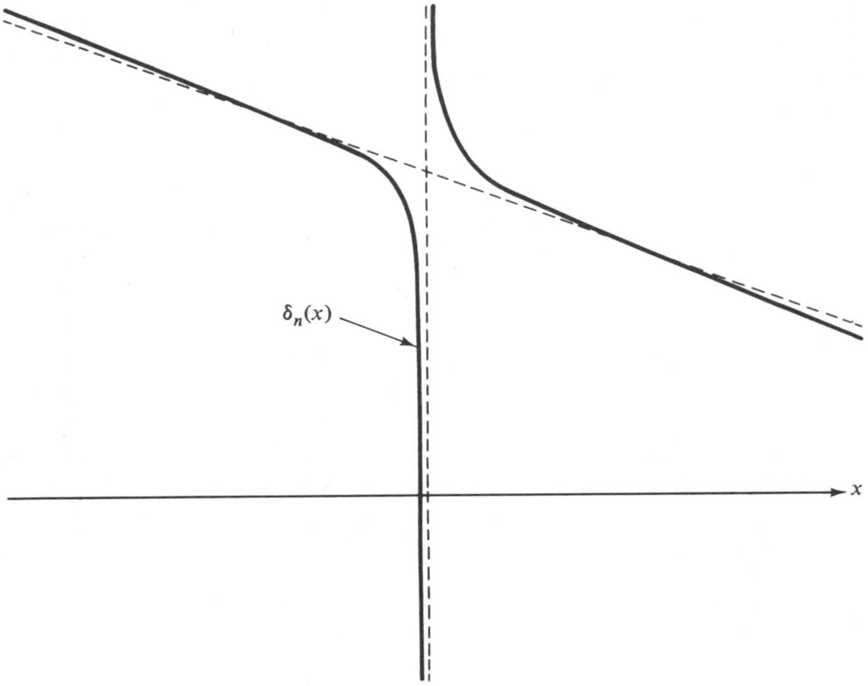

Figure 3-6-1 A hidden zero

The largest eigenvalue λ_{21} (= 10.746 . . .) agrees with the largest eigenvalue of the leading submatrix of order 20 in its first *fifteen* decimal digits. The graph of $\delta_{21}(\sigma)$ is close to -20 on all of the interval [10, 11] except on a subinterval of width less than 10^{-13} at λ_{21}. On many computer systems the critical subinterval is not detectable.

It must be emphasized that W_{21}^- is not a pathological matrix. If the algorithm in Sec. 3-3 is run in reverse (for $k = n, \ldots, 2$) then λ_{21} is not hidden from the new $\delta(\sigma)$ and is easily found. The point is that some rational functions, in practice, conceal their zeros from prying eyes.

Exercises on Sec. 3-6

3-6-1. Find the smallest value of n such that either $\lambda_i[\mathsf{W}_n^-]$ or $\lambda_n[\mathsf{W}_n^-]$ cannot be detected by evaluating δ_n on your computer system. Print out the zeros of $\chi_n, \chi_{n-1}, \chi_{n-2}$, etc.

3-7 THE CHARACTERISTIC POLYNOMIAL

If χ_A is given, or can be obtained easily, then the matrix problem reduces to the classical task of computing some or all zeros of a polynomial and the reader is referred to [Brent, 1973], [Traub, 1964], or [Wilkinson, 1964] which treat that problem. However if the matrix **A** is given then there is no incentive to compute the coefficients of χ_A because the zeros of a polynomial are extraordinarily sensitive functions of these coefficients. If $n = 300$ (medium order matrix) it is not clear how many hundred decimal digits in the coefficients are needed to determine all the zeros to two or three decimal places. We will not give further consideration to the coefficients of χ_A.

On the other hand $\chi_A(\sigma)$ can be evaluated satisfactorily by triangular factorization as described in Sec. 3-5.

A fundamental constraint on all algorithms for computing eigenvalues may be mentioned here. Over a hundred years ago Galois proved that there cannot be a finite procedure, utilizing only basic arithmetic operations and the extraction of roots, that will yield a zero of an arbitrary fifth- (or higher) degree polynomial. However the coefficients of χ_A can be computed from the elements of **A** by various finite schemes some of which are described in [Faddeev and Faddeeva, 1963]. It follows that, in the context of exact arithmetic, there can be no finite computer algorithm which will produce the eigenvalues of any given matrix if its order n is permitted to exceed four. Consequently all eigenvalue programs contain somewhere or other an iterative component. A method is sometimes called *direct* if it employs a finite number of similarity transformations (perhaps none at all), but this distinction does not seem to be useful.

4

Simple Vector Iterations

The power method is no longer a serious technique for computing eigenvectors. Nevertheless it warrants study because it is intimately related to current algorithms and so a good grasp of it is helpful in understanding more complicated methods. In particular inverse iteration is a useful technique although it is not used now in quite the way it was originally envisaged. This chapter first looks at the methods theoretically, in the context of exact arithmetic, and then practically, in the context of limited precision.

The most important variation on inverse iteration is the Rayleigh quotient iteration. We shall see that it converges for almost all starting approximations, however bad, and that ultimately—usually after two or three steps—the number of correct digits triples with each iteration.

4-1 EIGENVECTORS OF RANK ONE MATRICES

There is a class of full symmetric matrices whose eigenvalues are easy to find, namely those of rank one.

EXAMPLE 4-1-1 A Full Matrix of Rank One

$$\mathsf{A} = \begin{bmatrix} 8.41 & -6.09 & 3.19 \\ -6.09 & 4.41 & -2.31 \\ 3.19 & -2.31 & 1.21 \end{bmatrix} = \begin{bmatrix} 2.9 \\ -2.1 \\ 1.1 \end{bmatrix} \begin{bmatrix} 2.9 & -2.1 & 1.1 \end{bmatrix}$$

Suppose that, unknown to the user, $A = vv^*$. Then the first step is to take any $x \neq o$ and compute

$$y \equiv Ax[= v(v^*x)]. \tag{4-1-1}$$

If $y = o$ then x is an eigenvector belonging to the eigenvalue 0. Otherwise y is an eigenvector belonging to eigenvalue (v^*v) because, using (4-1-1),

$$Ay = Av(v^*x) = v(v^*v)(v^*x) = y(v^*v). \tag{4-1-2}$$

The computer, human or not, does not know this yet. The second step is

1. compute $z = Ay$,
2. compute $\min(z_i/y_i)$ and $\max(z_i/y_i)$ over all i for which y_i is not 0 (nor tiny). All the ratios will be v^*v.

Of course, if A is known to be of rank one then any nonzero column of A will be a dominant eigenvector. Without prior knowledge only *two* matrix-vector products are needed to discover it. Any vector orthogonal to v is an eigenvector with 0 eigenvalue (i.e., a null vector).

One might hope that one more matrix-vector product would yield the dominant eigenvector of rank two matrices. Unfortunately that is not the case as Ex. 4-1-2 shows.

EXAMPLE 4-1-2

$$A = \begin{bmatrix} -4 & 10 & 8 \\ 10 & -7 & -2 \\ 8 & -2 & 2 \end{bmatrix}, \qquad \text{Rank}(A) = 2.$$

Successive iterates, normalized to have the biggest element equal to one, are shown in Table 4-1-1.

Table 4-1-1

Step	0	1	2	3	4
	1.0	1.0	0.1429	1.0	?
Iterate	1.0	0.0714	0.9286	0.5	?
	1.0	0.5714	1.0	0.1	?

4-2 DIRECT AND INVERSE ITERATION

The special case of rank one matrices is relevant to the general case because, for large k, a normalized multiple of A^k is close to a rank one matrix (Ex. 4-2-1). Thus it is only necessary to find $y = A^k x$ and then check the accuracy by comparing $\mathbf{y}$ and $\mathbf{Ay}$. Fortunately it is not necessary to compute $\mathbf{A}^k$ explicitly since

$$\mathbf{A}^5\mathbf{x} = \mathbf{A}(\mathbf{A}(\mathbf{A}(\mathbf{A}(\mathbf{Ax})))).$$

> THE POWER METHOD (PM). Pick a unit vector $\mathbf{x}_0$ then for $k = 1, 2, 3, \ldots$
>
> 1. form $\mathbf{y}_k = \mathbf{Ax}_{k-1}$,
> 2. normalize, $\mathbf{x}_k = \mathbf{y}_k/\|\mathbf{y}_k\|$,
> 3. test for convergence of $\mathbf{x}_k$.

Table 4-1-1 shows a few steps of the process.

4-2-1 CONVERGENCE

Recall our standard notation:

$$\mathbf{Az}_i = \mathbf{z}_i\lambda_i, \ \|\mathbf{z}_i\| = 1, \quad \text{and}$$
$$\lambda_1 \leqslant \lambda_2 \leqslant \cdots \leqslant \lambda_n. \tag{4-2-1}$$

Note that $\|\mathbf{A}\|$ is either $-\lambda_1$ or λ_n. We shall assume that it is the latter, just to be definite, and we shall ignore roundoff effects.

The analysis is essentially two-dimensional. At step k there is a unique plane containing $\mathbf{x}_k$ and the wanted eigenvector $\mathbf{z}_n$. Let $\mathbf{u}_k$ be the unit vector in that plane which is orthogonal to $\mathbf{z}_n$ as shown in Fig. 4-2-1. As the algorithm proceeds the plane flaps on its one fixed axis $\mathbf{z}_n$ like an unlatched window. With the geometry established we can write

$$\mathbf{x}_k = \mathbf{z}_n \cos\theta_k + \mathbf{u}_k \sin\theta_k \tag{4-2-2}$$

and $\theta_k \equiv \angle(\mathbf{x}_k, \mathbf{z}_n)$ is the error angle.

In some ways θ_k is a more natural measure of error than the usual $\|\mathbf{x}_k - \mathbf{z}_n\| (= 2\sin(\theta_k/2))$ which so often brings on unnecessary normalizations. We hope that the power sequence $\{\mathbf{x}_0, \mathbf{x}_1, \mathbf{x}_2, \ldots\}$ converges to $\mathbf{z}_n$.

> THEOREM If λ_n is the unique dominant eigenvalue of $\mathbf{A}$ and if $\mathbf{z}_n^*\mathbf{x}_0 \neq 0$ then, as $k \longrightarrow \infty$, $\mathbf{x}_k \longrightarrow \mathbf{z}_n$ linearly with convergence factor $\max\{\lambda_{n-1}/\lambda_n, |\lambda_1|/\lambda_n\}$. (4-2-3)

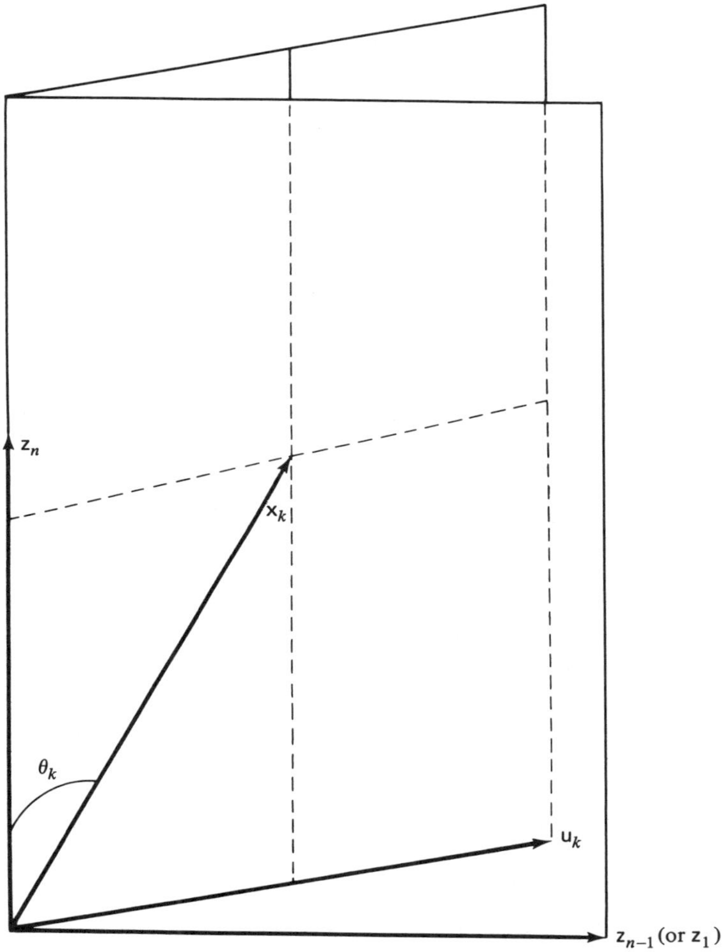

Figure 4-2-1 Representation of x_k

When convergence is linear, interest focuses on the **convergence factor** defined, in our problem, as $\lim_{k\to\infty} \theta_{k+1}/\theta_k$.

Proof. Consider the $(k + 1)$-th step wherein Ax_k is formed. Pre-multiply (4-2-2) by A and use (4-2-1) to find

$$Ax_k = z_n\lambda_n \cos\theta_k + \left(\frac{Au_k}{\|Au_k\|}\right)\|Au_k\| \sin\theta_k. \tag{4-2-4}$$

The key observation (Ex. 4-2-4) is that Au_k, like u_k, is orthogonal to z_n. Since x_{k+1} is a multiple of Ax_k, (4-2-4) provides an orthogonal decomposition of x_{k+1}. Compare (4-2-4) with (4-2-2) to obtain

$$u_{k+1} = Au_k/\|Au_k\|, \tag{4-2-5}$$

and

$$\tan\theta_{k+1} = \|Au_k\| \sin\theta_k/\lambda_n \cos\theta_k. \tag{4-2-6}$$

For all k, u_k is confined to the invariant subspace $z_n^\perp$. Hence we can invoke $A^\perp$, the restriction of A to $z_n^\perp$ (see Sec. 1-4-1) to obtain a bound

$$\|Au_k\| = \|A^\perp u_k\| \leqslant \|A^\perp\| = \max\{-\lambda_1, \lambda_{n-1}\}. \tag{4-2-7}$$

From (4-2-6) and (4-2-7) comes the decisive inequality,

$$\frac{\tan\theta_{k+1}}{\tan\theta_k} = \frac{\|Au_k\|}{\lambda_n} \leqslant \frac{\|A^\perp\|}{\|A\|} \equiv \rho < 1. \tag{4-2-8}$$

By hypothesis $|\tan\theta_0| < \infty$ and so, as $k \longrightarrow \infty$,

$$|\theta_k| \leqslant |\tan\theta_k| \leqslant \rho^k \tan\theta_0 \longrightarrow 0. \tag{4-2-9}$$

Rapid convergence can come in two ways, small ρ and/or small θ_0.

It is left as Ex. 4-2-6 to show that almost always $\|Au_k\| \longrightarrow \|A^\perp\|$ as $k \longrightarrow \infty$. Thus the convergence factor usually achieves its upper bound ρ. □

4-2-2 *INVERSE ITERATION (INVIT)*

This is the power method applied to A^{-1}. There is no need to invert A; instead replace $y_k = Ax_{k-1}$ in PM by

$$1'\colon \quad \text{solve } Ay_k = x_{k-1} \text{ for } y_k.$$

The convergence properties follow from theorem (4-2-3).

> COROLLARY If λ_1 is the eigenvalue closest to 0, if $z_1^* x_0 \neq 0$ then, as $k \longrightarrow \infty$ in INVIT, $x_k \longrightarrow z_1$ linearly with convergence factor at worst $\nu = |\lambda_1/\lambda_2|$.

EXAMPLE 4-2-1 (INVIT)

$$A \equiv \begin{bmatrix} -4 & 10 & 8 \\ 10 & -7 & -2 \\ 8 & -2 & 3 \end{bmatrix} \qquad \begin{aligned} \lambda_1 &\approx -17.895 \\ \lambda_2 &\approx 0.425 \\ \lambda_3 &\approx 9.470 \end{aligned}$$

Table 4-2-1

k	0	1	2	3	$\cdots$	∞
$\mathbf{x}_k$	1.0 1.0 1.0	0.700 1.00 -0.720	0.559 1.00 -0.931	0.554 1.00 -0.943	$\cdots$ $\cdots$ $\cdots$	0.554 1.00 -0.944
$\dfrac{\sin\theta_{k+1}}{\sin\theta_k}$	0.1758	0.0439	0.0433			$0.0449 \approx \lambda_2/\lambda_3 \equiv \nu_0$

4-2-3 SHIFTS OF ORIGIN

The power method and inverse iteration can easily be used with $\mathsf{A} - \sigma$ in place of A. The new convergence factors are as follows (Ex. 4-2-3).

POWER METHOD: $\rho_\sigma = \max_{j \neq s} |\lambda_j - \sigma| / |\lambda_s - \sigma|$

where $|\lambda_s - \sigma| = \max_m |\lambda_m - \sigma|$.

INVERSE ITERATION: $\nu_\sigma = |\lambda_t - \sigma| / \min_{j \neq t} |\lambda_j - \sigma|$

where $|\lambda_t - \sigma| = \min_m |\lambda_m - \sigma|$.

Figure 4-2-2 reveals something special about the indices implicitly defined above. As σ varies over all real values s can take on only the value

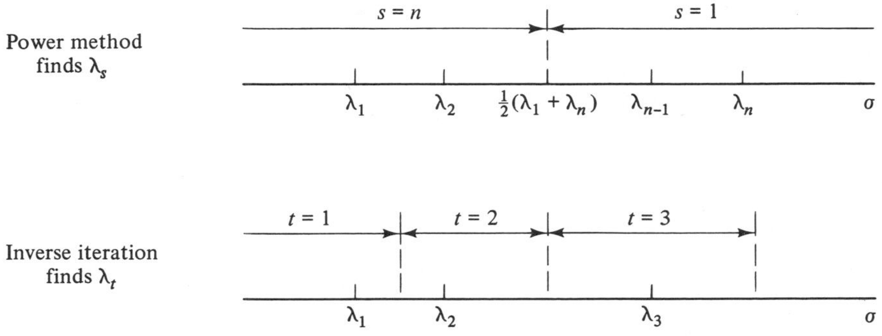

Figure 4-2-2 The effect of σ on the limit

1 or the value n. Thus the power method can only converge to z_1 or z_n and the choice of σ which minimizes ρ_σ is $(\lambda_2 + \lambda_n)/2$ in the former case and $(\lambda_1 + \lambda_{n-1})/2$ in the latter; neither gives a significant improvement over ρ_0. In contrast $\nu_\sigma \longrightarrow 0$ as $\sigma \longrightarrow \lambda_j$ and so INVIT converges rapidly when σ is chosen well.

4-2-4 COST

For a small full A each step of PM requires n^2 ops to form Ax, n ops for $\|\mathsf{Ax}\|^2$ and n divisions for y. Surprisingly each step of INVIT costs the same (approximately) provided that a triangular factorization $\mathsf{A} - \sigma = \mathsf{L\Delta L^*}$ is available. See Sec. 3.1 for more information on $\mathsf{L\Delta L^*}$. The new iterate y in 1′ is computed in two stages, $\mathsf{Lw} = \mathsf{x}$ and $\mathsf{\Delta L^* y} = \mathsf{w}$, each of which requires $n^2/2$ ops. The heavy initial investment in the triangular factorization, namely $n^3/3$ ops, can be amortized over all the steps taken. The catch in this argument is that INVIT is frequently run for only one step! In such cases σ is a computed eigenvalue correct to nearly full precision.

The case of banded matrices is left as an exercise (Ex. 4-2-5).

For large matrices it is hard to make a general statement about cost. For large banded matrices factorization is often feasible despite the fill-in within the band. For matrices which cannot be factored the equation $\mathsf{Ay}_k = \mathsf{x}_{k-1}$ is sometimes itself solved iteratively and so the cost cannot be bounded a priori.

Exercises on Sec. 4-2

4-2-1. Consider $\mathsf{A}^k/\|\mathsf{A}^k\|$ and give a bound for the norm of the difference between it and a close rank one matrix when $0 \leqslant \lambda_1 \leqslant \cdots \leqslant \lambda_{n-1} < \lambda_n$.

4-2-2. Show that $\mathsf{x}_{k+1} = \mathsf{A}^k\mathsf{x}_1/\|\mathsf{A}^k\mathsf{x}_1\|$.

4-2-3. Using Theorem (4-2-3) and its corollary establish the given equations for the convergence factors of the power method and inverse iteration.

4-2-4. With reference to (4-2-1), (4-2-2), and (4-2-4) show that if u_k is orthogonal to z_n then so is Au_k.

4-2-5. Compare the costs of one step of PM and INVIT when A has a half-band width m.

4-2-6. Show that the sequence $\{\mathsf{u}_k\}$ in the proof of Theorem (4-2-3) is produced by the power method using $\mathsf{A}^\perp$. What condition ensures that $\|\mathsf{Au}_k\| \longrightarrow \|\mathsf{A}^\perp\|$ as $k \longrightarrow \infty$?

4-2-7. Suppose that $\lambda_1 > 0$. What value must λ_n/λ_1 take in order that $\min_\sigma \rho_\sigma = \frac{1}{2}\rho_0$ when ρ_0 is close to 1?

4-3 ADVANTAGES OF AN ILL-CONDITIONED SYSTEM

Inverse iteration is often, but not always, used with a shift σ which is very close to some eigenvalue λ_j. The EISPACK programs INVIT and TINVIT use as σ a computed eigenvalue and it will often be correct to working accuracy. In these circumstances $\mathsf{A} - \sigma$ may have a condition number for inversion as large as 10^{14}. Now one of the basic results in matrix computations is that roundoff errors can give rise to completely erroneous "solutions" to very ill-conditioned systems of equations. So it seems that what is gained in the theoretical convergence rate by a good shift is lost in practice through a few roundoff errors. Indeed some textbooks have cautioned users not to take σ too close to any eigenvalue.

Fortunately these fears are groundless and furnish a nice example of confusing ends with means. In this section we explain why the EISPACK techniques work so well.

Let x denote the unit starting vector and y the computed result of one step of inverse iteration with shift σ. Because of roundoff, described in Sec. 3-2 or [Forsythe and Moler 1967], y satisfies

$$(\mathsf{A} - \sigma - \mathsf{H})\mathsf{y} = \mathsf{x} + \mathsf{f} \tag{4-3-1}$$

where, with a good linear equations solver, $\|\mathsf{H}\|$ is tiny compared to $\|\mathsf{A} - \sigma\|$ and $\|\mathsf{f}\|$ is tiny compared to $1(= \|\mathsf{x}\|)$. We can assume that $\|\mathsf{y}\|$ is huge (e.g., 10^{10}) compared to 1.

For the sake of analysis we make use of the vector $\mathsf{g} \equiv \mathsf{f} + \mathsf{H}\mathsf{y}$ so that we can rewrite (4-3-1) as

$$(\mathsf{A} - \sigma)\mathsf{y} = \mathsf{x} + \mathsf{g}. \tag{4-3-2}$$

Thus the "true" solution is $\mathsf{t} \equiv (\mathsf{A} - \sigma)^{-1}\mathsf{x}$ and the error $\mathsf{e} = (\mathsf{A} - \sigma)^{-1}\mathsf{g}$. Of course g involves y and so this is an unnatural approach for *proving* something about y. However a very weak assumption about g suffices to show why roundoff is not to be feared.

Figure 4-3-1 illustrates the situation.

Let σ be very close to λ_j and let $\psi = \angle(\mathsf{g}, \mathsf{z}_j)$. Then g can be decomposed orthogonally as

$$\mathsf{g} = (\mathsf{z}_j \cos\psi + \mathsf{u}\sin\psi)\|\mathsf{g}\| \tag{4-3-3}$$

where $\mathsf{u}^*\mathsf{z}_j = 0$. It turns out that $\|\mathsf{g}\|$ is irrelevant; all we need to assume about g's direction is some modest bound. In terms of Fig. 4-3-1, suppose that

$$|\tan\psi| \leqslant 100. \tag{4-3-4}$$

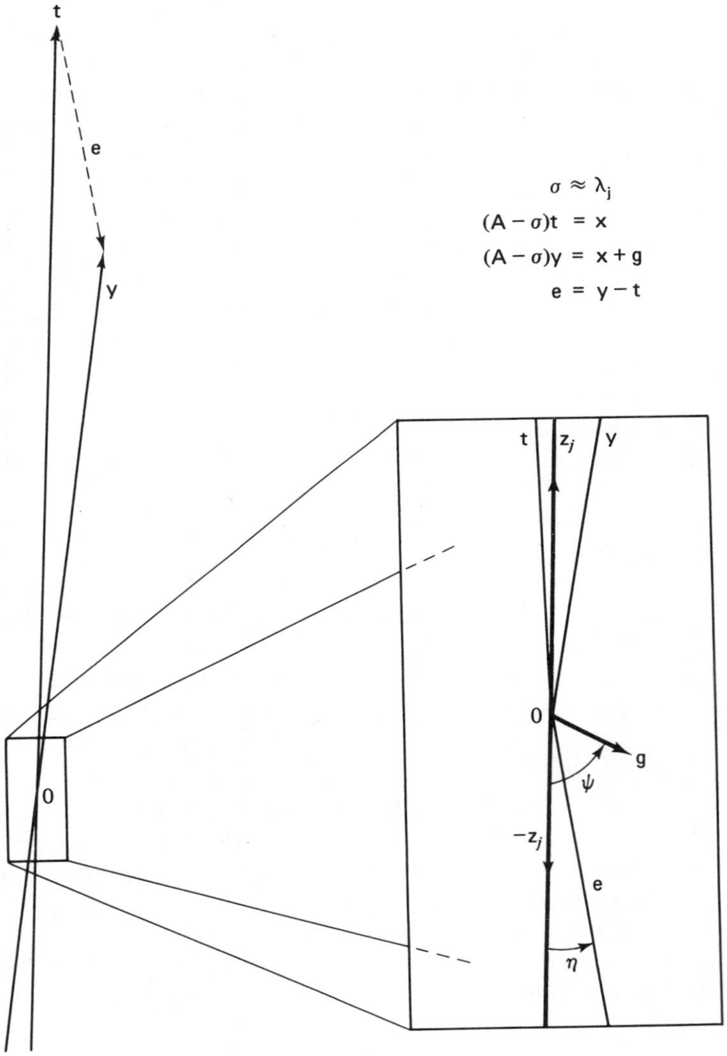

Figure 4-3-1 The error in inverse iteration

From (4-3-3) we obtain an orthogonal decomposition of the dreaded error,

$$\mathsf{e} = (\mathsf{A} - \sigma)^{-1}\mathsf{g} = \left[\mathsf{z}_j \cos\psi / (\lambda_j - \sigma) + \hat{\mathsf{u}}\|(\mathsf{A} - \sigma)^{-1}\mathsf{u}\| \sin\psi\right]\|\mathsf{g}\|, \tag{4-3-5}$$

where $\hat{\mathsf{u}}$ is the normalized version of $(\mathsf{A} - \sigma)^{-1}\mathsf{u}$ and is, like u, orthogonal to z_j (see Ex. 4-3-1). If η denotes the all important error angle between e and z_j then (4-3-5) says

$$\begin{aligned} \tan\eta &= \|(\mathsf{A} - \sigma)^{-1}\mathsf{u}\|(\lambda_j - \sigma)\tan\psi, \\ &\leqslant \frac{|\lambda_j - \sigma|}{\min\limits_{k \neq j}|\lambda_k - \sigma|}\tan\psi, \qquad \text{since } \mathsf{u} \perp \mathsf{z}_j. \end{aligned} \tag{4-3-6}$$

Our interest is precisely in the case when the factor multiplying $\tan\psi$ in (4-3-6) is very small (e.g., 10^{-10}). What (4-3-4) and (4-3-6) show is that

> The error e, which may be almost as large as the exact solution $(\mathsf{A} - \sigma)^{-1}\mathsf{x}$, is almost entirely in the direction of z_j.

This result is alarming if we had hoped for an accurate solution of $(\mathsf{A} - \sigma)\mathsf{y} = \mathsf{x}$ (the means) but is a delight in the search for z_j (the end).

When σ is correct to working precision then one step of inverse iteration is normally sufficient for convergence as Example 4-3-1 shows.

EXAMPLE 4-3-1 Inverse Iteration with a Good Shift

Same A and x as in Example 4-2-1.
$\sigma = 9.463$. The condition of $(\mathsf{A} - \sigma) = (\lambda_1 - \sigma)/(\lambda_3 - \sigma) = 10^5$.

Table 4-3-1

y	t	− e
3301.3	3598.1	296.8
1567.6	1708.6	141.0
3598.4	3921.8	323.4

Table 4-3-2

$\frac{y}{\lVert y \rVert}$	$\frac{t}{\lVert t \rVert}$	$\frac{-e}{\lVert e \rVert}$	*Eigenvector*
0.91744	0.91746	0.91775	0.91745
0.43564	0.43567	0.43500	0.43562
1.0000	1.0000	1.0000	1.0000

Exercises on Sec. 4-3

4-3-1. Show that if z is an eigenvector of A and if u is $z^{\perp}$ then so is $(A - \sigma)^{-1}u$.

4-3-2. (Calculator). Take the triangular matrix

$$B = \begin{bmatrix} 123.4 & 0.2273 & 0.1428 \\ 0 & 31.41 & -0.8571 \\ 0 & 0 & 2.718 \end{bmatrix}, \qquad \sigma = 123.3.$$

Take one step of inverse iteration with $x = (1, \frac{1}{6}, \frac{1}{36})^*$. First do it in 4-digit arithmetic (retain only the leading 4 digits of *each* intermediate quantity). Next do it to full precision. Take the difference of the two y's as an estimate of e before normalizing them.
A diagonal matrix is too special and a symmetric matrix requires more work than does the nonsymmetric B.

4-4 CONVERGENCE AND ORTHOGONALITY

The k-th step of inverse iteration computes y_k satisfying, in theory,

$$(A - \sigma)y_k = x_{k-1}, \qquad \lVert x_{k-1} \rVert = 1. \tag{4-4-1}$$

The **residual vector** of the approximate eigenpair (σ, y_k) is defined by

$$r_k \equiv (A - \sigma)y_k/\lVert y_k \rVert = x_{k-1}/\lVert y_k \rVert, \tag{4-4-2}$$

and, by (4-4-1), $\lVert r_k \rVert = 1/\lVert y_k \rVert$. Theorem (4-5-1) shows that σ differs from an exact eigenvalue of A by no more than $\lVert r_k \rVert$. Consequently the computed number $\lVert y_k \rVert$ is used as a measure of convergence.

The main, perhaps the only, weakness of inverse iteration is that the computed eigenvectors for two close eigenvalues may be acceptable (because their residuals are small) and yet not be mutually orthogonal. *This sounds like a contradiction* because the exact eigenvectors must be orthogonal. However a small residual vector r_k only guarantees accuracy for isolated eigenvalues as the gap theorems in Chap. 11 reveal.

The proper procedure in such a case is to take all the computed eigenvectors belonging to a cluster of eigenvalues (who shall decide on the clusters?) and compute the Rayleigh-Ritz approximations from their span as described in Chap. 11. After this facility has been incorporated into a program, however, the beautiful simplicity of inverse iteration has vanished and rival techniques, such as the QR algorithm of Chap. 8, become increasingly attractive, at least for small matrices.

A simpler remedy for clusters is to orthogonalize each approximate eigenvector, as soon as it is computed, against any eigenvectors already in the cluster. This technique and other devices are used in the subroutine TINVIT in EISPACK. Details are given under the name Tristurm in the last contribution to the Handbook (II/18).

4-5 SIMPLE ERROR BOUNDS

Let x be an approximate eigenvector and let $y = Ax$. The simplest approximation to an eigenvalue which can be derived from x and y is $\sigma = y_i/x_i$, where x_i is a maximal element of x. A better but more expensive approximation will be given below. First we ask how good are σ and x?

THEOREM For any scalar σ and any nonzero vector x there is an eigenvalue λ of A satisfying

$$|\lambda - \sigma| \leqslant \|Ax - x\sigma\|/\|x\|. \qquad (4\text{-}5\text{-}1)$$

Proof. If $\sigma = \lambda$ the result is immediate. If $\sigma \neq \lambda$ then $A - \sigma$ is invertible. So $x = (A - \sigma)^{-1}(A - \sigma)x$ and

$$\begin{aligned} 0 \neq \|x\| &\leqslant \|(A-\sigma)^{-1}\| \cdot \|(A-\sigma)x\| \\ &= \left(1/\min_i|\lambda_i[A] - \sigma|\right)\|(A-\sigma)x\|. \qquad \square \end{aligned}$$

In most applications the vector $y(= Ax)$ and the number $\|x\|$ will be on hand in which case the error bound can be computed for the modest price of $2n$ ops. For another n ops the Rayleigh quotient $\rho = \rho(x) \equiv x^*(Ax)/\|x\|^2$ can be computed. Fact 1-9 in Chap. 1 shows that ρ is the scalar which minimizes, over all σ, the bound in Theorem (4-5-1). Consequently it is quite reasonable to define the residual of x by $r = r(x) \equiv Ax - x\rho$. Then, as shown in Sec. 1-4, $r^*x = 0$ and this yields an alternative and instructive derivation of the error bound in (4-5-1). The first step is

THEOREM Let x be any nonzero vector with Rayleigh quotient ρ and residual vector r. Then (ρ, x) is an eigenpair for a matrix $A - M$ and $\|M\| \equiv \mu \equiv \|r\|/\|x\| \geqslant 0$. (4-5-2)

Proof. The magic formula for M is $(\mathsf{xr}^* + \mathsf{rx}^*)/\|\mathsf{x}\|^2$. Verification that the nonzero eigenvalues of M are $\pm\mu$ and that M fulfills the claim of the theorem is left as an exercise. □

This result accords with the spirit of backward error analysis (Chap. 2). If μ is less than the uncertainty in A then the question of error becomes moot; the pair (ρ, x) is as good an eigenpair as A warrants. In these circumstances a new question arises: How sensitive are the eigenvalues and eigenvectors to changes to, or uncertainties in A?

Fact 1-11 in Chap. 1 states that the eigenvalues of symmetric matrices are robust, that they change by no more than the change in the elements of A. Applying this to Theorem (4-5-2) above yields $|\lambda_i[\mathsf{A}] - \rho| \leq \|\mathsf{A} - (\mathsf{A} - \mathsf{M})\| = \|\mathsf{M}\| = \mu$, for some value of i. This is the same result as Theorem (4-5-1). More refined bounds are given in Chap. 11.

There is no analogous error bound for x, essentially because eigenvectors are not uniquely defined for multiple eigenvalues as Example 4-5-1 shows.

EXAMPLE 4-5-1

$$\mathsf{A} = \begin{bmatrix} 1 & \epsilon \\ \epsilon & 1 \end{bmatrix} = \frac{1}{2}\begin{bmatrix} 1 & 1 \\ 1 & -1 \end{bmatrix}\begin{bmatrix} 1+\epsilon & 0 \\ 0 & 1-\epsilon \end{bmatrix}\begin{bmatrix} 1 & 1 \\ 1 & -1 \end{bmatrix}.$$

With $\mathsf{x} = \mathsf{e}_1$ then $\rho(\mathsf{x}) = 1$, and $\mathsf{r}(\mathsf{x}) = \epsilon\mathsf{e}_2$. So $\|\mathsf{r}\| = \mu = \epsilon$ and yet the error angle for x is $\pi/4$ or 45°! Of course the close matrix A − M of Theorem (4-5-2) is I which certainly has x as an eigenvector.

With more information available bounds can be placed on the error in computed eigenvectors. Some of these results are given in Chap. 11.

Exercise on Sec 4-5

4-5-1. Using the x and r in Theorem (4-5-2) and $\mu = \|\mathsf{r}\|/\|\mathsf{x}\|$ exhibit the eigenvectors of M which belong to eigenvalues $\pm\mu$. Verify that (ρ, x) is an eigenpair for A − M.

4-6 THE RAYLEIGH QUOTIENT ITERATION

A natural extension of inverse iteration is to vary the shift at each step. The error bounds of the previous section suggest that the best shift which can be derived from the current eigenvector approximation x is the **Rayleigh quotient** of x, namely $\rho(\mathsf{x}) = \mathsf{x}^*\mathsf{A}\mathsf{x}/\mathsf{x}^*\mathsf{x}$.

The Rayleigh Quotient Iteration. Pick a unit vector x_0; then, for $k = 0, 1, 2, \ldots,$ repeat the following:

RQI
1. Compute $\rho_k = \rho(x_k)$.
2. If $A - \rho_k$ is singular then solve $(A - \rho_k)x_{k+1} = o$ for unit vector x_{k+1} and stop. Otherwise, solve the equation $(A - \rho_k)y_{k+1} = x_k$ for y_{k+1}.
3. Normalize, i.e., $x_{k+1} = y_{k+1}/\|y_{k+1}\|$.
4. If $\|y_{k+1}\|$ is big enough then stop.

Definition. The **Rayleigh sequence** generated by x_0 is $\{x_k, k = 0, 1, \ldots\}$. It is not essential to use the Euclidean norm in (3) and (4).

In the course of computing the fundamental mode of a vibrating system Lord Rayleigh, in the 1870s, improved an approximate eigenvector x_1 by solving $[A - \rho(x_1)]y_1 = e_1$ for y_1. This technique is less powerful than RQI and, as far as we know, Lord Rayleigh never studied RQI. The name simply points to the shifts that are used to accelerate inverse iteration.

Cost. A new system of equations must be solved at each step. The increase in cost over inverse iteration with a fixed shift is large when A is full but modest when A is tridiagonal. See Ex. 4-6-2.

The rewards for the extra cost are considerable because the Rayleigh sequence converges very rapidly as Example 4-6-1 shows.

EXAMPLE 4-6-1

Same A and x as in Example 4-1-2.

k	0	1	2	∞
x_k	1.000 1.000 1.000	0.8686 0.3644 1.000	0.9176 0.4358 1.000	0.9175 0.4356 1.0000
ρ_k	8.000	9.444	9.468	9.468

Invariance Properties. Analysis of the behavior of the sequence $\{\rho_k, x_k\}$ is more difficult than for inverse iteration because RQI is a *nonstationary* iteration, i.e., the function which maps x_k to x_{k+1} changes at every step. Consequently the behavior of RQI cannot be described easily in terms of $A - \rho_0$. For example, $\{\rho_k\}$ need not converge to the eigenvalue closest to ρ_0 as $k \longrightarrow \infty$.

The following invariance properties of RQI are worth noting. Let A and x_0 produce the Rayleigh sequence $\{\rho_k, x_k: k = 0, 1, 2, \ldots\}$.

Scaling. The matrix αA, $\alpha \neq 0$, and x_0 produce $\{\alpha\rho_k, x_k\}$.

Translation. The matrix $A - \alpha$ and x_0 produce $\{\rho_k - \alpha, x_k\}$.

Orthogonal Similarity. The matrix QAQ^*, where $Q^* = Q^{-1}$, and x_0 produce $\{\rho_k, Qx_k\}$.

The last property says that RQI is coordinate free, i.e., its properties are independent of the basis chosen to represent A and x_k.

Exercises on Sec. 4-6

4-6-1. Take one step of RQI using $A = \text{diag}(4, 2, 1)$ and $x_0^* = (0.1, 1.0, 0.1)$.

4-6-2. Do an operation count for one step of RQI in two separate cases: (a) A full, n by n; (b) A with half-bandwidth m.

4-6-3. Verify the invariance properties of RQI.

4-7 LOCAL CONVERGENCE

When the Rayleigh sequence $\{x_k\}$ does converge to an eigenvector z the behavior is best described in terms of $\phi_k = \angle(x_k, z)$, the error angle. It turns out that as $k \longrightarrow \infty$, $\phi_k \longrightarrow 0$ to third order which ensures that the number of correct digits in x_k triples at each step, for k large enough—and this often means $k > 2$ (see Example 4-6-1).

The current iterate x_k can be written in terms of ϕ_k as

$$x_k = z \cos \phi_k + u_k \sin \phi_k \tag{4-7-1}$$

where $u_k^* z = 0$ and $\|u_k\| = 1 = \|z\|$.

In part the phenomenal convergence rate can be attributed to the stationarity of the Rayleigh quotient ρ at eigenvectors, in particular,

$$\lambda - \rho(x_k) = [\lambda - \rho(u_k)]\sin^2 \phi_k \tag{4-7-2}$$

where $Az = z\lambda$. Verification of (4-7-2) is left as Ex. 4-7-2.

We can now state the result formally. The analysis is similar to that for the power method and, if necessary, the reader should refer to Sec. 4-2.

> THEOREM Assume that the Rayleigh sequence $\{x_k\}$ converges to an eigenvector. As $k \longrightarrow \infty$ the error angles ϕ_k satisfy $\lim |\phi_{k+1}/\phi_k^3| \leq 1$. Equality holds almost always. (4-7-3)

Proof. We ignore the pleasant but unlikely possibility that the iteration terminates at finite k with ρ_k an eigenvalue and x_{k+1} its eigenvector.

First apply $(\mathsf{A} - \rho_k)^{-1}$ to (4-7-1) and obtain

$$\mathsf{y}_{k+1} = \mathsf{z} \cos \phi_k / (\lambda - \rho_k) + \mathsf{u}_{k+1} \sin \phi_k \|(\mathsf{A} - \rho_k)^{-1}\mathsf{u}_k\|, \tag{4-7-4}$$

where $\mathsf{u}_{k+1} = (\mathsf{A} - \rho_k)^{-1}\mathsf{u}_k / \|(\mathsf{A} - \rho_k)^{-1}\mathsf{u}_k\|$ and is orthogonal to z. Since x_{k+1} is a multiple of y_{k+1} (see Step (3) of RQI), (4-7-4) says

$$\begin{aligned} \tan \phi_{k+1} &= \sin \phi_k \|(\mathsf{A} - \rho_k)^{-1}\mathsf{u}_k\| / \cos \phi_k (\lambda - \rho_k)^{-1}, \\ &= (\lambda - \rho_k)\|(\mathsf{A} - \rho_k)^{-1}\mathsf{u}_k\| \tan \phi_k, \\ &= [\lambda - \rho(\mathsf{u}_k)]\|(\mathsf{A} - \rho_k)^{-1}\mathsf{u}_k\| \tan \phi_k \sin^2 \phi_k. \end{aligned} \tag{4-7-5}$$

The line above made use of (4-7-2). The cleanest way to bound the last norm on the right of (4-7-5) is to invoke the restriction of $(\mathsf{A} - \rho_k)$ to the invariant subspace $\mathsf{z}^\perp$. Thus

$$\begin{aligned} \|(\mathsf{A} - \rho_k)^{-1}\mathsf{u}_k\| &= \|[(\mathsf{A} - \rho_k)^{-1}]^\perp \mathsf{u}_k\|, \text{ since } \mathsf{u}_k \in \mathsf{z}^\perp, \\ &\leqslant \|[(\mathsf{A} - \rho_k)^{-1}]^\perp\|, \text{ since } \|\mathsf{u}_k\| = 1, \\ &= 1 / \min_{\lambda_i \neq \lambda} |\lambda_i - \rho_k|, \text{ using Sec. 1-4-1.} \end{aligned} \tag{4-7-6}$$

The multiplicity of λ itself is irrelevant.

The hypothesis that $\{\mathsf{x}_k\} \longrightarrow \mathsf{z}$ is used to bound the right-hand side in (4-7-6). Let the **gap** γ be defined by $\gamma \equiv \min |\lambda_i - \lambda|$ over all $\lambda_i \neq \lambda$. Since $\phi_k \longrightarrow 0$, (4-7-2) shows that $\rho_k \equiv \rho(\mathsf{x}_k) \longrightarrow \lambda$ as $k \longrightarrow \infty$. Thus

$$|\lambda_i - \rho_k| \geqslant \gamma/2 \text{ for large enough } k. \tag{4-7-7}$$

Cubic convergence of ϕ_k to 0 follows from (4-7-2), (4-7-5), (4-7-6), and (4-7-7) but further consideration of u_k reveals, surprisingly, that the gap γ does not affect the asymptotic convergence rate itself; a small value of γ merely delays the onset of the asymptotic regime. Further consideration of (4-7-4) reveals that the sequence $\{\mathsf{u}_k\}$ is produced by inverse iteration with a variable shift ρ_k which is converging to λ. For large enough k the transformation from u_k to u_{k+1} is arbitrarily close to a step of inverse iteration in the subspace $\mathsf{z}^\perp$ with fixed shift λ.

Case 1. $\{\mathsf{u}_k\}$ converges. The limit vector $\hat{\mathsf{z}}$ must be an eigenvector of A in $\mathsf{z}^\perp$. Its eigenvalue $\hat{\lambda}$ will almost always be the one closest to λ but that is not germane. As $k \longrightarrow \infty$ the key terms in

(4-7-5) satisfy

$$[\lambda - \rho(u_k)]\|(A - \rho_k)^{-1}u_k\| \longrightarrow$$

$$\pm \|(\lambda - \hat{\lambda})\hat{z}/(\hat{\lambda} - \lambda)\| = \pm 1. \qquad (4\text{-}7\text{-}8)$$

On substituting (4-7-8) into (4-7-5) the limit is seen to be 1.

Case 2. $\{u_k\}$ does not converge. It is left to Exs. 4-7-3 and 4-7-4 to show that there are two eigenvalues of A equidistant from λ and the accumulation points of $\{u_k\}$ are two vectors in the plane spanned by the matching eigenvectors, say $\alpha z_p \pm \beta z_q$ where $\alpha^2 + \beta^2 = 1$, $\alpha \neq 0$, $\beta \neq 0$. It follows that, as $k \longrightarrow \infty$, $\rho(u_k)$ converges and $|\phi_{k+1}/\phi_k^3| \longrightarrow |\alpha^2 - \beta^2| < 1$. □

EXAMPLE 4-7-1 Cubic Convergence

Same data as in Example 4-6-1.

k	0	1	2
ρ_k	8.000	9.444	9.468
ϕ_k	0.3073	0.4954×10^{-1}	0.1204×10^{-3}
$\dfrac{\phi_{k+1}}{\phi_k^3}$	1.708	0.9903	—

Exercises on Sec. 4-7

4-7-1. Prove that $\|(A - \sigma)^{-1}u\| \leqslant 1/\gamma$ if u is a unit vector and γ is the gap between σ and the eigenvalues of A.

4-7-2. Verify (4-7-2).

4-7-3. Let $\{u_k\}$ be generated by $(A - \sigma)u_{k+1} = u_k\tau_k$ where τ_k ensures that $\|u_{k+1}\| = 1$. Assume that $\{u_k\}$ does not converge as $k \longrightarrow \infty$ and show, by an eigenvector expansion or otherwise, that the accumulation points of $\{u_k\}$ are of the form $\alpha z_p \pm \beta z_q$ where $\lambda_p = \sigma + \delta$, $\lambda_q = \sigma - \delta$. Deduce that $\rho(u_k) \longrightarrow \sigma + (\alpha^2 - \beta^2)\delta$ and $\|(A - \sigma)^{-1}u_k\| \longrightarrow 1/\delta$.

4-7-4. Let $\{u_k\}$ be generated by $(A - \sigma_k)u_{k+1} = u_k\tau_k$ where τ_k ensures that $\|u_{k+1}\| = 1$. Assume that (a) $\sigma_k \longrightarrow \sigma$ and (b) $\{u_k\}$ does not converge. Show that the regime described in Ex. 4-7-3 must hold in this case too.

4-8 MONOTONIC RESIDUALS

The best computable measure of the accuracy of (ρ_k, x_k) as an eigenpair for A is their residual vector $\mathsf{r}_k \equiv (\mathsf{A} - \rho_k)\mathsf{x}_k$. The key fact, not appreciated until 1965, is that however poor the starting vector x_0 the residuals always decrease.

THEOREM For the RQI $\|\mathsf{r}_{k+1}\| \leqslant \|\mathsf{r}_k\|$, for all k. Equality holds if, and only if, $\rho_{k+1} = \rho_k$ and x_k is an eigenvector of $(\mathsf{A} - \rho_k)^2$. (4-8-1)

In general an eigenvector of M^2 need not be an eigenvector of M. Thus x_k in theorem (4-8-1) need not be, indeed cannot be, an eigenvector of A.

Proof.

$$\begin{aligned}
\|\mathsf{r}_{k+1}\| &\equiv \|(\mathsf{A} - \rho_{k+1})\mathsf{x}_{k+1}\|, \text{ by definition,}\\
&\leqslant \|(\mathsf{A} - \rho_k)\mathsf{x}_{k+1}\|, \text{ by Fact 1-9,}\\
&= |\mathsf{x}_k^*(\mathsf{A} - \rho_k)\mathsf{x}_{k+1}|, \text{ since } \mathsf{x}_k \text{ is a multiple of } (\mathsf{A} - \rho_k)\mathsf{x}_{k+1},\\
&\leqslant \|(\mathsf{A} - \rho_k)^*\mathsf{x}_k\| \cdot \|\mathsf{x}_{k+1}\|, \text{ by Cauchy-Schwarz inequality,}\\
&= \|\mathsf{r}_k\|, \text{ since } \mathsf{A} - \rho_k \text{ is symmetric and } \|\mathsf{x}_j\| = 1 \text{ for all } j.
\end{aligned}$$

Equality holds in the first $\leqslant$ only if $\rho_{k+1} = \rho_k$ and in the second instance only if r_k is a multiple of x_{k+1}, i.e., only if $(\mathsf{A} - \rho_k)\mathsf{x}_k = (\mathsf{A} - \rho_k)^{-1}\mathsf{x}_k\nu_k$ for some ν_k. □

EXAMPLE 4-8-1

$$\mathsf{A} = \operatorname{diag}(3, 2, 1).$$

Table 4-8-1 Normal case

k	0	1	2	3	4
x_k	0.3333 0.6667 0.6667	0.1374 0.5494 −0.8242	0.0322 0.3242 0.9455	0.0019 0.0411 −0.9992	0.2×10^{-5} 0.7×10^{-4} 1.0
ρ_k	1.6667	1.340	1.107	1.002	1.0
$\|\mathsf{r}_k\|$	0.6667	0.5119	0.3104	0.0413	0.7×10^{-4}

Table 4-8-2 Stagnant case

k	0	1	2
x_k	0 0.707 0.707	0 0.707 −0.707	0 0.707 0.707
ρ_k	1.5	1.5	1.5
$\|r_k\|$	1	1	1

*4-9 GLOBAL CONVERGENCE

Inverse iteration was introduced as a way of **improving** an approximate eigenvector. With the impact of automatic computation by digital computers it was natural to ask whether the method could **find** eigenvectors starting from scratch. Our analysis of inverse iteration with a fixed shift shows that convergence will occur from all starting vectors which are not orthogonal to **the** target eigenvector. However some qualification is necessary because if $(A - \sigma)$ has a $\pm$ pair of smallest eigenvalues then $\{x_k\}$ does not know to which of the rival eigenvectors it should converge, although it tends to the subspace spanned by them without any hesitation.

When variable shifts are used convergence can be accelerated but there is the nasty possibility that, when initial approximations are poor, the shifts might lead to endless cycling. Consequently it is important to know that for the Rayleigh quotient iteration (defined in Sec. 4-6) the probability of such a cycle is zero. The proof here is a minor variation on Kahan's original argument.

THEOREM Let $\{x_k\}$ be the Rayleigh sequence generated by any unit vector x_0. As $k \longrightarrow \infty$,

1. $\{\rho_k\}$ converges, and either
2. $(\rho_k, x_k) \longrightarrow (\lambda, z)$, cubically, where $Az = z\lambda$, or
3. $x_{2k} \longrightarrow x_+$, $x_{2k+1} \longrightarrow x_-$, linearly, where x_+ and x_- are the bisectors of a pair of eigenvectors whose eigenvalues have mean $\rho = \lim_k \rho_k$.

The situation in (3) is unstable under perturbations of x_k.

Proof. By the monotonicity of the residual norms, which was proved in the previous section,

$$\|\mathbf{r}_k\| = \|(\mathbf{A} - \rho_k)\mathbf{x}_k\| \longrightarrow \tau \geqslant 0 \text{ as } k \longrightarrow \infty. \tag{4-9-1}$$

Since the sequence $\{\mathbf{x}_k\}$ is confined to the unit sphere, a compact subset of $\mathcal{E}^n$, $\{\mathbf{x}_k\}$ must have one or more accumulation points (vectors which are grazed infinitely often by the sequence). It remains to characterize these points and thereby count them. Note that ρ_k is also confined to a compact subset of $\mathcal{R}$, namely $[-\|\mathbf{A}\|, \|\mathbf{A}\|]$.

Case 1: $\tau = 0$ (the usual case). Any accumulation point $(\bar{\rho}, \mathbf{z})$ of $\{\rho_k, \mathbf{x}_k\}$ is, by definition, the limit of a subsequence $\{(\rho_j, \mathbf{z}_j) : j \text{ in } \mathcal{J}\}$ for some index set $\mathcal{J}$. Let $j \longrightarrow \infty$ in $\mathcal{J}$ to see that, because $\rho(\cdot)$ is a continuous function on the unit sphere,

$$\rho(\mathbf{z}) = \lim_{\mathcal{J}} \rho(\mathbf{x}_j) = \lim_{\mathcal{J}} \rho_j = \bar{\rho}, \tag{4-9-2}$$

and

$$\|(\mathbf{A} - \bar{\rho})\mathbf{z}\| = \lim_{\mathcal{J}} \|\mathbf{r}_j\| = \tau = 0. \tag{4-9-3}$$

Thus $(\bar{\rho}, \mathbf{z})$ must be an eigenpair of $\mathbf{A}$. By the local convergence theorem in Sec. 4-7, as soon as both $|\rho_j - \bar{\rho}|$ and $\|\mathbf{x}_j - \mathbf{z}\|$ are small enough then, as $k \longrightarrow \infty$ through all subsequent integer values, whether in $\mathcal{J}$ or not, $|\rho_k - \bar{\rho}| \longrightarrow 0$ and $\|\mathbf{x}_k - \mathbf{z}\| \longrightarrow 0$ very rapidly. Consequently $\lim_{k\to\infty} \mathbf{x}_k$ exists and is the eigenvector $\mathbf{z}$.

There appears to be no simple description of how $\mathbf{z}$ depends on $\mathbf{x}_0$. See Ex. 4-9-1.

In order to analyze the harder case $\tau > 0$ we write the defining equation for RQI in the form

$$(\mathbf{A} - \rho_k)\mathbf{x}_{k+1} = \mathbf{x}_k\tau_k, \quad \tau_k = 1/\|(\mathbf{A} - \rho_k)^{-1}\mathbf{x}_k\|. \tag{4-9-4}$$

Let $\theta_k = \angle(\mathbf{r}_k, \mathbf{x}_{k+1})$. Pre-multiply (4-9-4) by $\mathbf{x}_k^*$ to see that

$$0 < \tau_k = \mathbf{r}_k^*\mathbf{x}_{k+1} = \|\mathbf{r}_k\| \cos\theta_k, \tag{4-9-5}$$

and so θ_k is acute.

Case 2: $\tau > 0$. The characteristic feature of this case is that $\|\mathbf{r}_{k+1}\|/\|\mathbf{r}_k\| \longrightarrow \tau/\tau = 1$ as $k \longrightarrow \infty$ and so the two equality conditions of the monotonic residual theorem must hold in the limit, namely as $k \longrightarrow \infty$,

$$|\rho_{k+1} - \rho_k| \longrightarrow 0 \tag{4-9-6}$$

(which does not of itself imply that $\{\rho_k\}$ converges) and, since θ_k is acute,

$$\|\mathbf{r}_k - (\mathbf{x}_{k+1}\|\mathbf{r}_k\|)\| \longrightarrow 0, \tag{4-9-7}$$

i.e., $\theta_k \longrightarrow 0$. Now (4-9-4), (4-9-5), and (4-9-7) yield an important result; as $k \longrightarrow \infty$,

$$\left\|\left[(\mathsf{A} - \rho_k)^2 - \|\mathsf{r}_k\|^2 \cos\theta_k\right]\mathsf{x}_k\right\| = \|(\mathsf{A} - \rho_k)(\mathsf{r}_k - \mathsf{x}_{k+1}\|\mathsf{r}_k\|)\|$$
$$\leqslant \|\mathsf{A} - \rho_k\| \cdot \|\mathsf{r}_k - \mathsf{x}_{k+1}\|\mathsf{r}_k\|\,\| \longrightarrow 0. \tag{4-9-8}$$

So any accumulation point $\bar{\rho}$ of the bounded sequence $\{\rho_k\}$ must satisfy

$$\det\left[(\mathsf{A} - \bar{\rho})^2 - \tau^2\right] = 0, \tag{4-9-9}$$

whether or not $\{\mathsf{x}_k\}$ converges. In other words, $\bar{\rho} = \lambda_i \pm \tau$ for one or more eigenvalues λ_i of A. This gives only a finite number of possible accumulation points and with this limitation (4-9-6) does imply that $\{\rho_k\}$ actually converges,† i.e., $\rho_k \longrightarrow \bar{\rho}$ as $k \longrightarrow \infty$.

Turning to $\{\mathsf{x}_k\}$ we see that any accumulation point x must be an eigenvector of $(\mathsf{A} - \bar{\rho})^2$ without being an eigenvector of A (Case 1). This can only happen (Ex. 4-9-2) when orthogonal eigenvectors belonging to distinct eigenvalues $\bar{\rho} + \tau$ and $\bar{\rho} - \tau$ become a basis for a whole eigenspace of $(\mathsf{A} - \bar{\rho})^2$. [When $\rho + \tau$ and $\rho - \tau$ are simple eigenvalues of A then their eigenvectors span an eigenplane of $(\mathsf{A} - \rho)^2$. For example, if $\mathsf{A} = \begin{pmatrix} 0 & 1 \\ 1 & 0 \end{pmatrix}$ then $\mathsf{A}^2\mathsf{v} = \mathsf{v}$ for all v.]

The sequence $\{\mathsf{x}_k\}$ cannot converge since, by definition of ρ_k, $\mathsf{x}_k \perp \mathsf{r}_k$ and, by (4-9-7), $\mathsf{r}_k/\|\mathsf{r}_k\| \longrightarrow \mathsf{x}_{k+1}$, as $k \longrightarrow \infty$. This suggests that we consider $\{\mathsf{x}_{2k}\}$ and $\{\mathsf{x}_{2k-1}\}$ separately. Let x_+ be any accumulation point of $\{\mathsf{x}_{2k}\}$. Then, as $2j \longrightarrow \infty$ through the appropriate subsequence, (4-9-4), (4-9-5), and (4-9-7) show that

$$\mathsf{x}_{2j-1} = (\mathsf{A} - \rho_{2j-1})\mathsf{x}_{2j}/\tau_{2j-1} \longrightarrow (\mathsf{A} - \bar{\rho})\mathsf{x}_+/\tau \equiv \mathsf{x}_-. \tag{4-9-10}$$

Multiply (4-9-10) by $(\mathsf{A} - \bar{\rho})$ and use (4-9-8) to get

$$(\mathsf{A} - \bar{\rho})\mathsf{x}_- = (\mathsf{A} - \bar{\rho})^2\mathsf{x}_+/\tau = \tau\mathsf{x}_+. \tag{4-9-11}$$

Finally (4-9-10) and (4-9-11) combine to yield

$$(\mathsf{A} - \bar{\rho})(\mathsf{x}_+ \pm \mathsf{x}_-) = \pm\tau(\mathsf{x}_+ \pm \mathsf{x}_-). \tag{4-9-12}$$

The last equation shows that x_+ and x_- are indeed bisectors of eigenvectors of A belonging to $\bar{\rho} + \tau$ and $\bar{\rho} - \tau$. If these eigenvalues are simple then x_+ and x_- are unique (to within $\pm$) and so must actually be the limits of their respective sequences $\{\mathsf{x}_{2k}\}$ and $\{\mathsf{x}_{2k-1}\}$. Even if $\bar{\rho} + \tau$ and $\bar{\rho} - \tau$ are multiple eigenvalues there is still a unique normalized eigenvector associated with each eigenvalue and together the eigenvectors serve to

---

†Ultimately the sequence $\{\rho_k\}$ cannot jump a gap between two accumulation points because of (4-9-6).

fix x_+ and x_- unambiguously. These eigenvectors are (Ex. 4-9-3) the projections of x_0 onto the eigenspaces of $\bar{\rho} + \tau$ and $\bar{\rho} - \tau$.

It remains to show the linearity of the convergence of $\{x_{2k}\}$ to x_+ and its instability under perturbations.

In the limit $\{x_{2k}\}$ behaves like inverse iteration with the matrix $(A - \bar{\rho})^2$. If τ^2 is its smallest eigenvalue then convergence will be linear and the reduction factor will depend on the next smallest eigenvalue of $(A - \bar{\rho})^2$. If there are eigenvalues smaller than τ^2 then any components of x_{2k} in the corresponding directions will have to vanish in the limit to permit convergence to x_+. The details are left to Ex. 4-9-4.

To verify the instability it is simplest to check that if $v = x_+ + \epsilon x_-$ then (Ex. 4-9-5)

$$\|(A - \rho(v))v\|^2 = \tau^2 - 3\epsilon^2\tau^2 + O(\epsilon^3), \qquad \|v\|^2 = 1 + \epsilon^2,$$
$$< \tau^2 \text{ for small enough } \epsilon.$$

Thus x_+ is a saddle point of the residual norm and perturbations of x_k which upset the balance between x_+ and x_- will drive the norm below τ and into Case 1. Perturbations orthogonal to both x_+ and x_- do not upset the regime of Case 2 (Ex. 4-9-6) and keep the residual norm above τ. □

Rayleigh quotient shifts can be used with inverse iteration right from the start without fear of preventing convergence, but we cannot say to which eigenvector $\{x_k\}$ will converge.

Exercises on Sec. 4-9

4-9-1. Show, by a 3 by 3 example, that RQI can converge to an eigenvalue which is not the closest to ρ_0 and to an eigenvector which is not closest to x_0.

4-9-2. Show that any eigenvector of A is an eigenvector of A^2. Using an eigenvector expansion, or otherwise, show that the converse fails only when distinct eigenvalues of A are mapped into the same eigenvalue of A^2.

4-9-3. By exercising the choice in the selection of eigenvectors for multiple eigenvalues show that there is no loss of generality in assuming that all eigenvalues are simple for the analysis of simple vector iterations.

4-9-4. See what conditions must obtain so that $x_{2k} \longrightarrow x_+$, $\rho_k \longrightarrow \bar{\rho}$, $\|r_k\| \longrightarrow \tau$ even when $(A - \bar{\rho})^2$ has eigenvalues smaller than τ^2. Is it essential that x_0 have no components in the corresponding eigenvectors?

4-9-5. Recall that $(A - \bar{\rho})x_\pm = \pm\, \tau x_\mp$. Compute $\rho(x_+ + \epsilon x_-)$ and then

$$\|[A - \rho(v)]v\|^2$$

with $v = x_+ + x_-$ retaining all terms through ϵ^4. Evaluate the gradient and the Hessian of $\|(A - \rho)v\|^2$ at $v = x_+$.

4-9-6. Show that perturbations ϵu of x_+ increase the residual norm if $u \perp x_+$ and $u \perp x_-$.

Notes and References

[Householder, 1964] and [Wilkinson, 1965] give valuable references to the development of the power method and inverse iteration during the 1950's and even earlier. Wilkinson was instrumental in exposing the myth that the proximity of $A - \sigma$ to a singular matrix, when σ is a computed eigenvalue, will spoil the powerful convergence of inverse iteration.

The simple error bounds in Sec. 4-4 have been derived by many people but are still not as well known as they should be.

Ostrowski [1958, 1959] devoted two difficult papers to the Rayleigh quotient iteration for symmetric matrices and gave a rigorous proof of asymptotic cubic convergence. See also [Crandall, 1951] and [Temple, 1952]. The key observation that the norms of the residual vectors are monotone decreasing is due to Kahan, as is the related global convergence theorem [Parlett and Kahan, 1969]. The proof of the monotone decrease in the residuals comes from [Parlett, 1974].

5

Deflation

When eigenvectors, or eigenvalues, are computed one by one it is necessary to prevent the algorithm from computing over again the quantities which it has already produced. In other words it is essential to get rid of each eigenvector immediately it has been found. The established word for this banishment is **deflation**. Various ways of doing it are described below.

In each case the vector to be banished is a unit vector $\hat{\mathbf{z}}$ which makes a small angle η with some unit eigenvector $\mathbf{z}$, so it can be written

$$\hat{\mathbf{z}} = \mathbf{z}\cos\eta + \mathbf{w}\sin\eta, \ \mathbf{w}^*\mathbf{z} = 0, \ \|\mathbf{w}\| = 1. \tag{5-1}$$

5-1 DEFLATION BY SUBTRACTION

The spectral theorem (Fact 1-4 in Chap. 1) expresses $\mathbf{A}$ as $\sum_{i=1}^{n} \lambda_i(\mathbf{z}_i\mathbf{z}_i^*)$. If λ_n and $\mathbf{z}_n$ were known then it would be tempting to work with the new n by n matrix $\bar{\mathbf{A}}$ defined by

$$\bar{\mathbf{A}} = \mathbf{A} - \lambda_n\mathbf{z}_n\mathbf{z}_n^* = \sum_{i=1}^{n-1} \lambda_i(\mathbf{z}_i\mathbf{z}_i^*), \tag{5-1-1}$$

which has traded $\mathbf{z}_n$'s old eigenvalue λ_n for a new one $\bar{\lambda}_n = 0$. If $|\lambda_n| > |\lambda_{n-1}| > |\lambda_{n-2}| > \cdots$ then, for example, the power method applied to $\bar{\mathbf{A}}$

will converge to $\mathbf{z}_{n-1}$ and the deflation process may be repeated again to yield $\bar{\bar{\mathsf{A}}}$ whose largest eigenvalue is λ_{n-2}, and so on. That is the formal theory but what happens in practice? Consider deflation of A by the vector $\hat{\mathbf{z}}$ given in (5-1) above. First observe that

$$\hat{\mathbf{z}}\hat{\mathbf{z}}^* = \mathbf{z}\mathbf{z}^* \cos^2 \eta + \frac{1}{2} \sin 2\eta(\mathbf{z}\mathbf{w}^* + \mathbf{w}\mathbf{z}^*) + \mathbf{w}\mathbf{w}^* \sin^2 \eta$$

$$= \mathbf{z}\mathbf{z}^* + \mathsf{W}, \text{ defining } \mathsf{W}, \tag{5-1-2}$$

and $\|\mathsf{W}\| = \sin \eta$ (Ex. 5-1-1). Given only $\hat{\mathbf{z}}$ the best approximation to $\mathbf{z}$'s eigenvalue λ is the Rayleigh quotient (Sec. 4-6),

$$\mu \equiv \rho(\hat{\mathbf{z}}) = \lambda - [\lambda - \rho(\mathbf{w})] \sin^2 \eta. \tag{5-1-3}$$

Even if the subtraction of $\mu\hat{\mathbf{z}}\hat{\mathbf{z}}^*$ were performed exactly the result would be $\hat{\mathsf{A}} \equiv \mathsf{A} - \mu\hat{\mathbf{z}}\hat{\mathbf{z}}^*$ instead of $\bar{\mathsf{A}}$. It is left as Ex. 5-1-2 to show that as $\eta \longrightarrow 0$,

$$\|\bar{\mathsf{A}} - \hat{\mathsf{A}}\| = \mu\eta + 0(\eta^2). \tag{5-1-4}$$

By Fact 1-11 some eigenvalues may be damaged by as much as $|\mu\eta|$. If μ approximates the smallest eigenvalue and η is tiny then (5-1-4) is very satisfactory.

If μ approximates the dominant eigenvalue λ_n and/or η is not tiny (say $\eta = 10^{-4}$ rather than $\eta = 10^{-13}$) then (5-1-4) shows that $\hat{\mathsf{A}}$ is not very close to $\bar{\mathsf{A}}$ and we may fear that by using $\hat{\mathsf{A}}$ we have already lost accuracy in the small eigenvalues. This fear is unnecessary as the following analysis suggests.

Let $(\mu, \hat{\mathbf{z}})$ approximate $(\lambda_n, \mathbf{z}_n)$ and let us see how close $(\lambda_1, \mathbf{z}_1)$ is to an eigenpair of $\hat{\mathsf{A}} = \mathsf{A} - \mu\hat{\mathbf{z}}\hat{\mathbf{z}}^* + \mathsf{H}$ where H accounts for the roundoff error incurred in the subtractions $a_{ij} - \mu\xi_i\xi_j$. Assume that $\|\mathsf{H}\| \leqslant 2\epsilon\|\mathsf{A}\|$. The residual vector is

$$\hat{\mathsf{A}}\mathbf{z}_1 - \mathbf{z}_1\lambda_1 = (\mathsf{A}\mathbf{z}_1 - \mathbf{z}_1\lambda_1) - \mu\hat{\mathbf{z}}(\hat{\mathbf{z}}^*\mathbf{z}_1) + \mathsf{H}\mathbf{z}_1 \tag{5-1-5}$$

and

$$\hat{\mathbf{z}}^*\mathbf{z}_1 = (\mathbf{z}_n \cos \eta + \mathbf{w}_n \sin \eta)^*\mathbf{z}_1 = (\mathbf{w}_n^*\mathbf{z}_1) \sin \eta. \tag{5-1-6}$$

By theorem (4-5-1) there is an eigenvalue $\hat{\lambda}_1$ of $\hat{\mathsf{A}}$ satisfying

$$|\hat{\lambda}_1 - \lambda_1| \leqslant \|\hat{\mathsf{A}}\mathbf{z}_1 - \mathbf{z}_1\lambda_1\| \leqslant \mu|\mathbf{w}_n^*\mathbf{z}_1| \sin \eta + 2\epsilon\|\mathsf{A}\|. \tag{5-1-7}$$

The new feature, missing in (5-1-4), is $\mathbf{w}_n^*\mathbf{z}_1$. Section 4-2 shows that $\mathbf{w}_n \approx \mathbf{z}_{n-1}$ and so $|\mathbf{w}_n^*\mathbf{z}_1| \approx \epsilon$. Thus it is the second term in (5-1-7) which dominates the error.

> The change in distant eigenvalues and eigenvectors caused by deflation of an approximate eigenpair is the same as the change induced by perturbing the elements of A in the last place held.

Example 5-1-1 bears this out.

EXAMPLE 5-1-1 Deflation in 6-Decimal Arithmetic

A = 6 by 6 Hilbert Matrix, $a_{ij} = 1/(i+j-1)$	$\hat{A} = A - \lambda_6 z_6 z_6^*$
$\{\lambda_i[A]\}$	$\{\lambda_i[\hat{A}]\}$
0.11193×10^{-6}	0.11832×10^{-6}
0.12568×10^{-4}	0.12569×10^{-4}
0.61576×10^{-3}	0.61576×10^{-3}
0.16322×10^{-1}	0.16322×10^{-1}
0.24236×10^{0}	0.24236×10^{0}
0.16189×10^{1}	0.24544×10^{-7}

Table 5-1-1

Eigenvector	z	$\hat{z}$	$\angle(z, \hat{z})$
$\lambda, \hat{\lambda}$	0.11193×10^{-6}	0.11833×10^{-6}	
	-0.1246×10^{-2}	-0.8704×10^{-1}	
	-0.3553×10^{-1}	-0.8728×10^{-1}	
$(Az = \lambda z)$	0.2406×10^{0}	0.2010×10^{0}	6.6°
	-0.6254×10^{0}	-0.6510×10^{0}	
$(\hat{A}\hat{z} = \hat{\lambda}\hat{z})$	0.6899×10^{0}	0.6602×10^{0}	(0.12 radian)
	-0.2717×10^{0}	-0.2912×10^{0}	
$\lambda, \hat{\lambda}$	0.12568×10^{-4}	0.12569×10^{-4}	$\angle(z, \hat{z})$
	0.1114×10^{-1}	0.1238×10^{-1}	
	-0.1797×10^{0}	-0.1790×10^{0}	
$(Az = \lambda z)$	0.6042×10^{0}	0.6043×10^{0}	0.63°
	-0.4437×10^{0}	-0.4434×10^{0}	
$(\hat{A}\hat{z} = \hat{\lambda}\hat{z})$	-0.4414×10^{0}	-0.4409×10^{0}	(0.011 radian)
	0.4591×10^{0}	0.4593×10^{0}	

The differences in the other eigenvectors were not significant.

Exercises on Sec. 5-1

5-1-1. Show that the matrix W of (5-1-2) satisfies $\|W\| = \sin \eta$.

5-1-2. By using Ex. 5-1-1 and (5-1-3) show that (5-1-4) holds as $\eta \longrightarrow 0$.

5-1-3. (Computer project.) Perturb elements in A of Example 5-1-1, recompute λ_1 and z_1, and compare with Example 5-1-1.

5-2 DEFLATION BY RESTRICTION

If $Az = z\lambda$ and z is known then it is natural to continue working with $A^{\perp}$, the restriction of A to the invariant subspace $z^{\perp}$. See Sec. 1-4 for basic information on $z^{\perp}$. $A^{\perp}$ has all of A's eigenpairs except for (λ, z). To use the power method or inverse iteration with $A^{\perp}$ it is only necessary, in exact arithmetic, to choose a starting vector orthogonal to z. Of course the product $(A^{\perp})x$ is at least as expensive as Ax however many eigenvectors have been removed, an inevitable consequence of using the original A.

In practice we have $\hat{z}$ instead of z and roundoff will ensure that each computed product Ax will have a tiny but nonzero component of those eigenvectors already found. If ignored these components will grow until eventually they become dominant again. In fact it is a toss up whether the current sequence of vectors will converge to a new eigenvector before the dominant eigenvector pulls the sequence toward itself. This is the numerical analyst's version of the race between the hare and the tortoise.

In order to avoid these games it is wise to orthogonalize the current vector x against the computed eigenvectors from time to time. The current eigenvalue estimate can be used to determine the frequency with which the known vectors should be suppressed. See Ex. 5-2-2.

By these means the power method, or inverse iteration, can be made to converge to each eigenvector in turn. The accuracy is limited only by the accuracy with which the previous eigenvectors have been computed and the care taken to keep down their components in subsequent calculation. The price paid for this nice feature is a small increase in cost. To suppress the component of $\hat{z}$ in x it is necessary to compute $\hat{z}^*x$ and then replace x by $\hat{x} \equiv x - \hat{z}(\hat{z}^*x) = (I - \hat{z}\hat{z}^*)x$. This requires $2n$ ops. If all the eigenvectors were to be found by the power method and this mode of deflation then the cost of the starting vectors for the later eigenvectors would become quite heavy. However the point is that this technique is capable of giving accurate results and is attractive for large sparse matrices.

Exercises on Sec. 5-2

5-2-1. If the power method or inverse iteration produces the sequence $\{x_k\}$ then, for any eigenvector z, $x_1^*z = 0$ implies $x_k^*z = 0$ for $k > 1$. Show this and then examine the effect of explicitly orthogonalizing x_k against $\hat{z}$ as given in (5-1).

5-2-2. Let (λ_n, z_n) be a computed eigenpair. Let (μ, x) be the current estimates of a different eigenpair. Show that the increase in the component of z_n at the next step of the power method is approximately $|\lambda_n/\mu|$. If this component is to be kept less than $\sqrt{\epsilon}$ find a formula for the frequency with which z_n should be purged from the current approximation.

5-3 DEFLATION BY SIMILARITY TRANSFORMATIONS

The domain of the operator $\mathsf{A}^{\perp}$ mentioned in Sec. 5-2 has dimension $n - 1$ and so it is not unreasonable to consider finding an $(n - 1)$ by $(n - 1)$ matrix which represents $\mathsf{A}^{\perp}$ rather than continuing to work with A. Formally such a matrix is easy to find. Take any orthogonal matrix P whose first column is the eigenvector $\mathbf{z}$. Thus $\mathsf{P}^* = \mathsf{P}^{-1}$ and $\mathsf{P}\mathbf{e}_1 = \mathbf{z}$. The desired representation is the matrix $\mathsf{A}^{(1)}$ shown in (5-3-1). Consider the orthogonal similarity transformation of A induced by P,

$$\begin{aligned} \mathsf{P}^*\mathsf{A}\mathsf{P} &= \mathsf{P}^*(\mathbf{z}\lambda, \ldots) \\ &= \begin{bmatrix} \lambda & \mathbf{0}^* \\ \mathbf{0} & \mathsf{A}^{(1)} \end{bmatrix}, \text{ since } \mathsf{P}^*\mathbf{z} = \mathbf{e}_1. \end{aligned} \tag{5-3-1}$$

If $\mathsf{A}^{(1)}\mathbf{u} = \mathbf{u}\alpha$ then $\mathsf{P}\binom{0}{\mathbf{u}}$ is an eigenvector of A with eigenvalue α. Thus computation may proceed with $\mathsf{A}^{(1)}$ provided that P is preserved in some way. In Chaps. 6, 7, and 8 it will be shown how to pick a simple P, carry out the transformation, and preserve P. It must be admitted that at first sight this appears to be a complicated and costly technique. However if all eigenvalues are wanted then the advantage of reducing the order of the matrix at each deflation is a definite attraction.

A subtle and important feature of the method can be brought out without going into great detail. In practice only the vector $\hat{\mathbf{z}}$ given in Sec. 5-1 will be on hand, not the eigenvector $\mathbf{z}$. Thus we must consider the orthogonal matrix $\hat{\mathsf{P}}$ with $\hat{\mathsf{P}}\mathbf{e}_1 = \hat{\mathbf{z}}$ and the associated similarity transformation,

$$\hat{\mathsf{P}}^*\mathsf{A}\hat{\mathsf{P}} = \begin{bmatrix} \mu & \mathbf{c}^* \\ \mathbf{c} & \hat{\mathsf{A}}^{(1)} \end{bmatrix}. \tag{5-3-2}$$

It can be shown (Ex. 5-3-1) that if $\eta = \angle(\mathbf{z}, \hat{\mathbf{z}})$ is small enough then $\|\mathbf{c}\| \leqslant \|\mathsf{A} - \mu\|\eta + 0(\eta^2)$. The point is that explicit computation of $\hat{\mathsf{P}}^*\mathsf{A}\hat{\mathsf{P}}$ may sometimes reveal that $\mathbf{c}$ is not small enough to be neglected. That is valuable information because, instead of deflating automatically by simply ignoring $\mathbf{c}$, the algorithm can seek further orthogonal similarities which will reduce the (2, 1) and (1, 2) blocks as described in Sec. 7-4. When this has been done (5-3-2) is replaced by

$$\mathsf{Q}^*\hat{\mathsf{P}}^*\mathsf{A}\hat{\mathsf{P}}\mathsf{Q} = \begin{bmatrix} \hat{\mu} & \mathbf{d}^* \\ \mathbf{d} & \bar{\mathsf{A}}^{(1)} \end{bmatrix}, \tag{5-3-3}$$

where $\mathbf{d}$ is negligible and $\hat{\mu}$ is a probably undetectable improvement in μ. The advantage is that $\bar{\mathsf{A}}^{(1)}$ is known to be an adequate representation of $\mathsf{A}^{\perp}$ and the computation can proceed safely. This is exactly what occurs in the tridiagonal QL algorithm of Chap. 8. Similarity transformations are continued until the matrix **declares itself reduced** (to working accuracy) and then the deflation occurs naturally.

Exercise on Sec. 5-3

5-3-1. Show that $\hat{P}^*(A\hat{z} - \hat{z}\mu)\begin{pmatrix} 0 \\ c \end{pmatrix}$. Simplify $A\hat{z} - \hat{z}\mu$ using (5-1) and then conclude that $\|c\| = \|(A - \mu)w\| \sin \eta + O(\eta^2)$. Take $\mu = \rho(\hat{z}) = \hat{z}A\hat{z}$.

Notes and References

Deflation by subtraction is often attributed to Hotelling [Hotelling, 1943] although the idea is quite natural. Several early papers of Wilkinson analyzed deflation and the first and third methods are discussed thoroughly in [Wilkinson, 1965] and also in introductory texts such as [Fox, 1964]. Mention should be made of the influential paper [Householder, 1961].

6

Useful Orthogonal Matrices (Tools of the Trade)

6-1 ORTHOGONALS ARE IMPORTANT

Table 6-1-1 indicates the importance of orthogonal matrices.

Table 6-1-1

Property	*Transformations Which Preserve the Property*	*Symbol*
Eigenvalues	Similarities	$A \longrightarrow FAF^{-1}$
Symmetry	Congruences	$A \longrightarrow FAF^*$ F invertible

Each of the properties is so valuable that we restrict attention to those transformations which preserve them both. This entails

$$F^* = F^{-1}.$$

In the real case this forces F to be orthogonal and the transformations induced by such F are called **orthogonal congruences** or, using one more syllable, **orthogonal similarities**.

Definition. A real matrix F is **orthogonal** if

$$F^*F = FF^* = I.$$

An orthogonal F is **proper** if det F = + 1. The definition of orthogonality can be interpreted as saying that the columns of F are mutually orthonormal and so are the rows.

The next step is to find classes of easy to use orthogonal matrices which are rich enough so that any orthogonal matrix can be obtained by forming products of the simple ones.

6-2 PERMUTATIONS

A permutation of an ordered set of n objects is a rearrangement of those objects into another order; in other words a permutation is a one-to-one transformation of the set onto itself. One notation for a permutation π is

$$(\pi_1, \pi_2, \ldots, \pi_n) \tag{6-2-1}$$

where π_j is the new position of the object initially in position j.

EXAMPLE 6-2-1

$$n = 3, \qquad \pi = (3, 1, 2), \qquad O = (O_1, O_2, O_3),$$
$$\pi O = (O_2, O_3, O_1).$$

If an ordered set O of n objects is permutated first by $\pi^{(1)}$ and then by $\pi^{(2)}$ the result is another permutation of O called the composition or product of $\pi^{(1)}$ and $\pi^{(2)}$ and written like multiplication as

$$\pi^{(2)}\pi^{(1)}O. \tag{6-2-2}$$

The permutations form a group under composition; in particular each permutation has an inverse.

EXAMPLE 6-2-2

$$\pi = (3, 1, 2), \qquad \pi^{-1} = (2, 3, 1).$$

If the objects are the columns of a matrix B then a permutation of them can be effected by post-multiplying B by a special matrix P, called a **permutation matrix**. P is obtained from I by permuting its **columns** by π.

Thus, if $\pi = (3, 1, 2)$ then

$$\pi(\mathsf{b}_1, \mathsf{b}_2, \mathsf{b}_3) = (\mathsf{b}_2, \mathsf{b}_3, \mathsf{b}_1) = \mathsf{BP}, \qquad \mathsf{P} = \begin{bmatrix} 0 & 0 & 1 \\ 1 & 0 & 0 \\ 0 & 1 & 0 \end{bmatrix}. \quad (6\text{-}2\text{-}3)$$

Note that P is also obtained by rearranging the rows of I according to π^{-1}.

The n by n permutation matrices form a group under matrix multiplication. Each such matrix can be represented compactly by the corresponding permutation π.

Most permutation matrices which occur in eigenvalue computations are products of very simple matrices, called **interchanges**, or **swaps**, which swap a pair of rows or columns and leave all the others unchanged. It is a fact that all permutations can be written as a product of interchanges, usually in several ways.

EXAMPLE 6-2-3

$(3 \;\; 1 \;\; 2) = (1 \;\; 3 \;\; 2)(2 \;\; 1 \;\; 3)$ (multiplication is from right to left).

A more compact way of representing a permutation is as a product of disjoint cycles. However this representation does not seem to be used in matrix computations, presumably because the conversion to and from interchanges to cycles does not seem to be warranted.

Every interchange is its own inverse and so a sequence of swaps simultaneously represents both a permutation and its inverse. The inverse is obtained by performing the swaps in reverse order.

Although permutations involve no floating point arithmetic operations they are useful tools. In the solution of large sparse sets of linear equations the ordering of the equations has a decisive influence on the cost of Gauss elimination.

For more information on the implementation of permutations in a computer see [Knuth, 1969, Vol. I, Sec. 1-3-3].

Exercises on Sec. 6-2

6-2-1. If P is the permutation matrix corresponding to π prove that P* corresponds to π^{-1}.

6-2-2. Often in matrix applications a sequence of interchanges has the special form $(1, \nu_1), (2, \nu_2), \ldots, (n-1, \nu_{n-1})$ where $\nu_j \geqslant j$. This sequence can be represented compactly by the array $(\nu_1, \nu_2, \ldots, \nu_{n-1}) = \mathsf{n}$.

(a) Write an algorithm to effect the inverse of the permutation represented by n.

(b) Write an algorithm to convert from the representation n to a product of disjoint cycles.

6-3 REFLECTIONS (OR SYMMETRIES)

A basic result in real Euclidean geometry is that any rigid motion which leaves the origin fixed (i.e., any orthogonal transformation) can be represented as a **product** of **reflections**. The mirror for a reflection in $\mathcal{E}^n$ is an $(n - 1)$ space or hyperplane, which is most easily characterized by the direction orthogonal (or **normal**) to it.

Definition. The **hyperplane** normal to u is $\{\mathsf{x}: \mathsf{u}^*\mathsf{x} = 0\}$.

To each u there corresponds a unique reflection which reverses u and leaves invariant any vector orthogonal to u.

Definition. The matrix $\mathsf{H}(\mathsf{u})$ which effects the reflection $\mathsf{x} \longrightarrow \mathsf{H}(\mathsf{u})\mathsf{x}$ is the **reflector which reverses** u;

$$\mathsf{H}(\mathsf{u})\mathsf{v} = \begin{cases} -\mathsf{v}, & \text{if} \quad \mathsf{v} = \alpha\mathsf{u}, \\ \mathsf{v}, & \text{if} \quad \mathsf{u}^*\mathsf{v} = 0. \end{cases}$$

Note that $\mathsf{H}(\alpha\mathsf{u}) = \mathsf{H}(\mathsf{u})$ for any $\alpha \neq 0$. It is easily verified that the matrix representation of $\mathsf{H}(\mathsf{u})$ is

$$\mathsf{H}(\mathsf{u}) = \mathsf{I} - \gamma\mathsf{u}\mathsf{u}^*, \qquad \gamma = \gamma(\mathsf{u}) = 2/\mathsf{u}^*\mathsf{u}.$$

The basic properties of $\mathsf{H}(\mathsf{u})$ are covered in Exs. 6-3-1 and 6-3-2. Note that $\mathsf{H}(\mathsf{u})$ has everything: It is elementary, symmetric, orthonormal, and its own inverse!

By a sequence of reflections we can build up any orthogonal transformation. No other tool is necessary (Ex. 6-3-3). There is however one small blemish; I is not a reflector (but $\mathsf{H}^2 = \mathsf{I}$). If a sequence of orthogonal matrices converges to I then the reflector factors of each matrix will not converge to I. In Ex. 6-3-6 we present a proper orthogonal matrix (determinant $+1$) which is analogous to $\mathsf{H}(\mathsf{u})$.

Standard Task: Given b find c such that $\mathsf{H}(\mathsf{c})\mathsf{b} = \mathsf{e}_1\mu$.

Solution: $\mu = \mp \|\mathsf{b}\|$, $\mathsf{c} = \mathsf{b} \pm \mathsf{e}_1\mu$ (see Ex. 6-3-4).

6-3-1 COMPUTATIONAL ASPECTS

Let $\mathsf{b} = (\beta_1, \ldots, \beta_n)^*$, $\mathsf{c} = (\gamma_1, \ldots, \gamma_n)^*$. The only arithmetic step in computing c is the calculation $\gamma_1 = \beta_1 \pm \mu$.

Case 1. $\mu = -\|\mathsf{b}\| \operatorname{sign}(\beta_1)$. The quantity $\gamma_1 = (|\beta_1| + \|\mathsf{b}\|) \operatorname{sign}(\beta_1)$ involves addition of positive numbers and will have a low relative error always.

Case 2. $\mu = \|\mathsf{b}\| \operatorname{sign}(\beta_1)$. The formula $\gamma_1 = (|\beta_1| - \|\mathsf{b}\|)\operatorname{sign}(\beta_1)$ involves genuine subtraction which will result in a high relative error whenever $|\beta_1| \doteqdot \|\mathsf{b}\|$. Example 6-3-1 shows the dire effects of using the obvious formula in this case.

It is sometimes said that Case 2 is unstable and should not be used. This is not right. It is the **formula** $(|\beta_1| - \|\mathsf{b}\|)$ which is dangerous. Below we give a formula which always gives low relative error because the subtraction is done analytically.

$$\begin{aligned} |\beta_1| - \|\mathsf{b}\| &= (\beta_1^2 - \|\mathsf{b}\|^2)/(|\beta_1| + \|\mathsf{b}\|) \\ &= -\sigma/(|\beta_1| + \|\mathsf{b}\|), \end{aligned}$$

where $\sigma = \beta_2^2 + \cdots + \beta_n^2$. The extra cost is one division and one addition.

EXAMPLE 6-3-1 Reflections in Case 2

Consider 4-decimal arithmetic ($1 + 10^{-4} \longrightarrow 1$).

$$\mathsf{b} = \begin{bmatrix} 1 \\ 10^{-2} \end{bmatrix}, \qquad \|\mathsf{b}\| = 1, \qquad \mathsf{c} = \begin{bmatrix} 0 \\ 10^{-2} \end{bmatrix},$$

$$\mathsf{H(c)} = \mathsf{I} - \gamma\mathsf{cc}^* = \begin{bmatrix} 1 & 0 \\ 0 & -1 \end{bmatrix}$$

$$\mathsf{H(c)b} = \begin{bmatrix} 1 \\ 10^{-2} \end{bmatrix} \neq \mathsf{e}_1!!$$

Use of the proper formula yields $\mathsf{c} = \begin{bmatrix} 10^{-4}/2 \\ 10^{-2} \end{bmatrix}$ which is correct to working precision.

In Case 1 b is reflected in the external bisector of b and e_1; in Case 2 it is the internal bisector. For real vectors Case 1 (which is the popular one) yields the vector $-\|\mathsf{b}\|\mathsf{e}_1$. This is a mild nuisance because the QR factorization (Sec. 6-7) is defined with a positive diagonal for R and Case 1 makes it negative.

Exercises on Sec. 6-3

6-3-1. Show that $\mathsf{H(u)}^* = \mathsf{H(u)}$, $\mathsf{H}^2\mathsf{(u)} = \mathsf{I}$, and $\mathsf{H(u)}$ is elementary.

6-3-2. Find the eigenvalues and eigenvectors of $\mathsf{H(u)}$. Show that $\det[\mathsf{H}] = -1$.

6-3-3. Express $\begin{bmatrix} \cos\theta & \sin\theta \\ -\sin\theta & \cos\theta \end{bmatrix}$ as a product of two reflectors. Draw a picture to show the mirrors.

6-3-4. Derive the formula $\mathsf{u} = \mathsf{v} - \mathsf{e}_1\mu$ which ensures $\mathsf{H(u)v} = \mathsf{e}_1\mu$.

6-3-5. How should u be chosen so that $\mathsf{H(u)v} = \mathsf{w}$ when $\mathsf{w^*w} = \mathsf{v^*v}$?

6-3-6. Let s be an $(n-1)$-vector and let $\gamma^2 + \|\mathsf{s}\|^2 = 1$. How must ν be chosen so that the partitioned matrix

$$\mathsf{R}(\mathsf{s}) = \begin{bmatrix} \gamma & -\mathsf{s}^* \\ \mathsf{s} & \mathsf{I} - \nu\mathsf{s}\mathsf{s}^* \end{bmatrix}$$

is orthogonal? What is det $\mathsf{R}(\mathsf{s})$? How must s be chosen so that $\mathsf{R}(\mathsf{s})\mathsf{b} = e_1\|\mathsf{b}\|$?

6-4 PLANE ROTATIONS

The matrix representation of the transformation which rotates $\mathcal{E}^2$ through an angle θ (counterclockwise) is

$$\mathsf{R}(\theta) \equiv \begin{bmatrix} \cos\theta & -\sin\theta \\ \sin\theta & \cos\theta \end{bmatrix}, \qquad \mathsf{R}^{-1}(\theta) = \mathsf{R}(-\theta)$$

There are two different tasks for R:

1. Find θ so that $\mathsf{R}(\theta)\mathsf{b} = \mathsf{e}_1\mu$ given $\mathsf{b} = (\beta_1, \beta_2)^*$. The solution is given by $\tan\theta = -\beta_2/\beta_1$ (Ex. 6-4-1).
2. Find θ so that $\mathsf{R}(\theta)\mathsf{A}\mathsf{R}(-\theta)$ is diagonal. The two solutions, $0 \leqslant \theta < \pi$, are found from $\cot 2\theta = (a_{22} - a_{11})/2a_{12}$. (Ex. 6-4-2).

In n-space we use a **plane rotation** $\mathsf{R}(i, j, \theta)$ which rotates the (i, j) plane, that is the plane spanned by e_i and e_j, through an angle θ and leaves the orthogonal complement of this plane invariant. So $\mathsf{R}(i, j, \theta)$ is the identity matrix except that $r_{ii} = r_{jj} = \cos\theta$, $-r_{ij} = r_{ji} = \sin\theta$ $(i < j)$. Here is a class of proper orthogonal matrices which includes I. Each $\mathsf{R}(i, j, \theta)$ is an elementary matrix (Ex. 6-4-3).

The analogue of Task 1 above is to transform a given b into a multiple of e_1 by a **sequence** of plane rotations. Either of the following choices will work:

$$\mathsf{R}(1, n, \theta_n) \cdots \mathsf{R}(1, 3, \theta_3)\mathsf{R}(1, 2, \theta_2)\mathsf{b} = \mathsf{e}_1\mu$$

$$\mathsf{R}(1, 2, \phi_2) \cdots \mathsf{R}(1, n-1, \phi_{n-1})\mathsf{R}(1, n, \phi_n)\mathsf{b} = \mathsf{e}_1\nu$$

where

$$|\mu| = |\nu| = \|\mathsf{b}\|.$$

Other sequences of planes are possible. In general each sequence produces a different orthogonal transformation of b into $\pm\mathsf{e}_1\|\mathsf{b}\|$. These solutions are rivals to the two reflections which accomplished the same task in Sec. 6-3.

Task 2 has two variants. See Fig. 6-4-1.

$$
\overset{\mathsf{R}(3,\,5,\,\theta)^*}{\begin{bmatrix}
1 & & & & & \\
 & 1 & & & & \\
 & & c & & s & \\
 & & & 1 & & \\
 & & -s & & c & \\
 & & & & & 1
\end{bmatrix}}
\quad
\overset{\mathsf{A}}{\begin{bmatrix}
* & * & a'_{13} & * & a'_{15} & * \\
* & * & a'_{23} & * & a'_{25} & * \\
a'_{31} & a'_{32} & a''_{33} & a'_{34} & a''_{35} & a'_{36} \\
* & * & a'_{43} & * & a'_{45} & * \\
a'_{51} & a'_{52} & a''_{53} & a'_{54} & a''_{55} & a'_{56} \\
* & * & a'_{63} & * & a'_{65} & *
\end{bmatrix}}
\quad
\overset{\mathsf{R}(3,\,5,\,\theta)}{\begin{bmatrix}
1 & & & & & \\
 & 1 & & & & \\
 & & c & & -s & \\
 & & & 1 & & \\
 & & s & & c & \\
 & & & & & 1
\end{bmatrix}}
$$

Key:
- $*$ Unchanged
- a' Changed once
- a'' Changed twice
- Blank is zero

Figure 6-4-1 Effect of plane rotation of A

6-4-1 JACOBI ROTATION

Find θ such that the (i, j) and (j, i) elements of $\mathsf{R}(i, j, \theta)\mathsf{A}\mathsf{R}(i, j, -\theta)$ are zero. These rotations are discussed more fully in Chap. 9.

6-4-2 GIVENS ROTATION

Find ϕ such that the (k, ℓ) and (ℓ, k) elements of $\mathsf{R}(i, j, \phi)\mathsf{A}\mathsf{R}(i, j, -\phi)$ are zero. Since rows i and j are the only ones to change we must have one of the pair i, j equal to one of the pair k, ℓ.

Jacobi actually used these rotations in 1846 but the reflections and Givens rotations were not introduced until the 1950s. In many ways Givens rotations are more useful than Jacobi rotations. The initial attraction of the Jacobi rotation arises from the fact that, of all choices for θ, with given i, j, it produces the biggest decrease in the sum of the squares of the off diagonal elements.

6-4-3 OPERATION COUNT

If $\mathsf{B}' = \mathsf{R}(i, j, \theta)\mathsf{B}$ then

$$
\begin{aligned}
b'_{ik} &= b_{ik}\cos\theta - b_{jk}\sin\theta = \cos\theta(b_{ik} - b_{jk}\tan\theta), \\
b'_{jk} &= b_{ik}\sin\theta + b_{jk}\cos\theta = \cos\theta(b_{ik}\tan\theta + b_{jk})
\end{aligned}
$$

for all k. This operation requires $4n$ multiplications. The count is identical for the formation of RAR^{-1} provided that advantage is taken of symmetry. A square root is required for the calculation of $\cos \theta$.

For scaled versions of $\mathsf{R}(i, j, \theta)$ see Sec. 6-8.

6-4-4 *COMPACT STORAGE*

When a long sequence of plane rotations must be recorded it is often advantageous to accept a little extra computation in order to encode each rotation with a single number rather than the pair $\cos \theta$, $\sin \theta$. Clearly θ itself is wasteful and $t = \tan \theta$ is the natural choice. Formally c and s are recovered from

$$c = 1/\sqrt{1 + t^2}, \qquad s = c \cdot t \qquad (-\pi/2 \leqslant \theta \leqslant \pi/2).$$

This scheme will fail when θ is $\pi/2$ or very close to it and a slightly more complicated process is called for. Stewart [1976] advocates saving the number ρ defined by

$$\rho = \begin{cases} 1, & \text{if } \sin\theta = 1 \\ \sin\theta, & \text{if } |\sin\theta| < \cos\theta \\ \sec\theta \cdot \operatorname{sign}(\sin\theta), & \text{if } |\sin\theta| \geqslant \cos\theta. \end{cases}$$

Exercises on Sec. 6-4

6-4-1. Show that $\mathsf{R}(\theta)\mathsf{v} = \pm \mathsf{e}_1\|\mathsf{v}\|$ if $\tan\theta = -\mathsf{v}_2/\mathsf{v}_1$.

6-4-2. Show that $\mathsf{e}_1^*\mathsf{R}(\theta)\mathsf{AR}(-\theta)\mathsf{e}_2 = 0$ if $\cot 2\theta = (\mathsf{a}_{22} - \mathsf{a}_{11})/2\mathsf{a}_{12}$. What is the relation between the two possible values of θ?

6-4-3. Show that $\mathsf{R}(i, j, \theta)$ is an elementary matrix. Hint: Think of $\theta/2$.

6-4-4. Find the Givens rotation in the (2, 3) plane which annihilates elements (1, 3) and (3, 1).

6-4-5. Write out the product $\mathsf{R}(2, 3, \theta)\mathsf{R}(2, 4, \theta)$ when $n = 5$.

6-4-6. Find a neat, stable algorithm to recover $\sin\theta$ and $\cos\theta$ from the encoding ρ given above.

6-5 ERROR PROPAGATION IN A SEQUENCE OF ORTHOGONAL CONGRUENCES

This section is of fundamental importance to the understanding of the stability of many popular eigenvalue computations based on transforming the given matrix to simpler form. The typical situation is that A_0 is given and A_k is computed, in principle, as the last term in the sequence

$$\mathsf{A}_j = \mathsf{B}_j^*\mathsf{A}_{j-1}\mathsf{B}_j, \qquad j = 1, \ldots, k,$$

where each B_j is orthogonal and depends on A_{j-1}.

This is not what happens in practice. Let us consider a typical step. From the current matrix A_{j-1} we compute B_j but it will not be exactly orthogonal. Let

$$B_j = G_j + \hat{F}_j \tag{6-5-1}$$

where G_j is an unknown orthogonal matrix very close to B_j. We will discuss the size of the error $\hat{F}_j$ later. Now we try to compute $B_j^* A_{j-1} B_j$ but fail. Instead the resulting matrix A_j will satisfy

$$A_j = B_j^* A_{j-1} B_j + \overline{W}_j \tag{6-5-2}$$

where $\overline{W}_j$ is the local error matrix which results from roundoff error in the similarity transformation. Sometimes it is useful to write $A_j = fl(B_j^* A_{j-1} B_j)$ to indicate the intention; *fl* is an acronym for "floating point result of." Because of the nice properties of orthogonal matrices we replace B_j by G_j to find

$$\begin{aligned} A_j &= (G_j + \hat{F}_j)^* A_{j-1}(G_j + \hat{F}_j) + \overline{W}_j, \\ &= G_j^* A_{j-1} G_j + W_j, \end{aligned} \tag{6-5-3}$$

where

$$W_j \equiv \overline{W}_j + G_j^* A_{j-1} \hat{F}_j + \hat{F}_j^* A_{j-1} G_j + \hat{F}_j^* A_{j-1} \hat{F}_j. \tag{6-5-4}$$

Here W_j is a name for a messy, but hopefully tiny, local error matrix. Now apply (6-5-3) for $j = k, k-1, \ldots, 2, 1$ in turn to find

$$\begin{aligned} A_k &= G_k^* A_{k-1} G_k + W_k \\ &= G_k^*(G_{k-1}^* A_{k-2} G_{k-1} + W_{k-1}) G_k + W_k, \\ &= \cdots \\ &= J_0^* A_0 J_0 + \sum_{j=1}^{k} J_j^* W_j J_j, \end{aligned} \tag{6-5-5}$$

where

$$J_j = G_{j+1} \cdots G_k, \qquad J_k = I. \tag{6-5-6}$$

The final step is to rewrite (6-5-5) in a simple form,

$$\boxed{A_k = J_0^*(A_0 + M_0) J_0} \tag{6-5-7}$$

where

$$M_0 \equiv J_0 \left(\sum_{j=1}^{k} J_j^* W_j J_j \right) J_0^*. \tag{6-5-8}$$

All we have done is to relate A_k to A_0 in terms of matrices which are orthogonal. Equation (6-5-7) says that the **computed** A_k is exactly orthogonally congruent to $A_0 + M_0$.

Definition. M_0 is called the **equivalent perturbation matrix** induced by the computation.

When A_k is finally written over A_{k-1} the best that can be done is to find the eigenvalues of $A_0 + M_0$; those of A_0 are beyond recall. Because the G_j, and therefore the J_j, are orthogonal we have

$$\|M_0\| \leqslant \sum_{j=1}^{k} \|W_j\| \tag{6-5-9}$$

Moreover, using (6-5-4) and the orthogonality of G, we get

$$\|W_j\| \leqslant \|\overline{W}_j\| + \|A_{j-1}\|(2\|\hat{F}_j\| + \|\hat{F}_j\|^2). \tag{6-5-10}$$

Only at this point do the details of the process need to be examined. The reflectors H(w) and plane rotations R_{ij} fail to be orthogonal only because, for the computed quantities γ, s, and c, it will turn out that $2 - \gamma(w^*w) \neq 0$ and $s^2 + c^2 - 1 \neq 0$ and when the arithmetic unit is good

$$\|\hat{F}_j\| \leqslant \epsilon = \text{the roundoff unit.} \tag{6-5-11}$$

We say that H and R_{ij} are orthogonal to working accuracy. Note that any error in w or θ ($s = \sin\theta$) is irrelevant to (6-5-11).

Some effort is required to bound $\overline{W}_j$ because it depends strongly on the form of B_j. Note that **if** A_j were simply the result of rounding the exact product $B_j^* A_{j-1} B_j$ we would have

$$\begin{aligned}\|\overline{W}_j\| &\leqslant \epsilon\|B_j^* A_{j-1} B_j\| \leqslant \epsilon\|B_j\|^2\|A_{j-1}\| \\ &\leqslant \epsilon(1+\epsilon)^2\|A_{j-1}\|, \qquad \text{using (6-5-11).}\end{aligned}$$

In practice the error is a little bigger. Suppose that for each j,

$$\|\overline{W}_j\| \leqslant n\epsilon\|A_{j-1}\|, \tag{6-5-12}$$

and

$$\|A_j\| \leqslant (1+\epsilon)^3\|A_{j-1}\|. \tag{6-5-13}$$

Putting (6-5-10) and (6-5-11) into (6-5-9) yields

$$\|M_0\| \leqslant (k+1)n\epsilon\|A_0\|.$$

Even if the errors did achieve these crude bounds the result is considered satisfactory because of the usual values of k, n, and ϵ. For example, if $k = 99$, $n = 100$, and $\epsilon = 10^{-14}$ then $\|M_0\|/\|A_0\| < 10^{-10}$. If computation with $\epsilon = 10^{-7}$ were compulsory then it would pay to show when the n in (6-5-12) can be replaced by a constant like 5.

With large matrices ($n > 400$) it is rare to employ a sequence of orthogonal similarity transformations because they tend to make the matrix full and then storage costs escalate.

6-6 BACKWARD ERROR ANALYSIS

Equation (6-5-7) expresses an exact relationship between the input A_0 and the output A_k; it is neither approximate nor asymptotic. Yet the reader might be forgiven for thinking that little has been said and the results are trivial. To a certain extent it is this elementary character which underlies their importance. Early attempts to analyze the effect of roundoff error made the task seem both difficult and boring. The new simplicity follows from a point of view developed mainly by Wilkinson in the 1950s and referred to as **backward error analysis**. This "backward" step was a great advance.

The following points are worth making:

1. No mention has been made of the error! No symbol has been designated for the true "A_k" which would have been produced with exact arithmetic.
2. Use of symbols such as $\oplus$ to represent the computer's addition operation have been abolished. (The author had to struggle with these pseudo-operations when he first studied numerical methods). Because $=$, $+$, etc. have their ordinary meaning all the nice properties such as associativity and commutativity can be invoked without special mention.
3. Equation (6-5-7) casts the roundoff error *back* as though it were a perturbation of the data. The effect of all these errors is the same *as if* an initial change M_0 were made in A_0 and then all calculations were done exactly with the orthogonal matrices G_j. If, by chance, M_0 is smaller, element for element, than the uncertainty in A_0 then roundoff error is inconsequential.
4. The (forward) error, $\|A_k - \text{"}A_k\text{"}\|$, could be very large even when M_0 is small relative to A_0. This may or may not matter, but it can only happen when the function taking A_0 into "A_k" is intrinsically sensitive to small changes in A_0. Intuitively we think of this function as having a large derivative and we say that the calculation is **ill conditioned**. In other words the backward approach shows clearly how the actual error depends on the algorithm, namely M_0, and how it depends on the task, namely its condition. It is easy to make the mistake of thinking that big error means bad method.
5. For eigenvalue calculations the computation of A_k may be an intermediate goal; for example, A_k might be a tridiagonal matrix. In this case M_0 is far *more* relevant than the actual error because the eigenvalues of A_k cannot differ from those of A_0 by more than $\|M_0\|$ however large the error in A_k may be.

6. Forward error analysis (bound the error) is not a defeated rival to backward error analysis. Both are necessary. In particular, users are primarily interested in estimating the accuracy of their output. The combination of a backward analysis (when it can be done) and perturbation theory often gives better bounds than a straightforward attempt to see how roundoff enlarges the intermediate error at each step.

6-7 THE QR FACTORIZATION AND GRAM-SCHMIDT

> Any nonnull rectangular m by n matrix B can be written as $\mathsf{B} = \mathsf{QR}$ with m by r Q satisfying $\mathsf{Q^*Q} = \mathsf{I}_r$, and r by n R upper triangular with nonnegative diagonal elements; $r = \text{rank}(\mathsf{B})$. Both Q and R are unique.

See Ex. 6-7-3 for the proof.

EXAMPLE 6-7-1

$$r = 2: \begin{bmatrix} 2 & 1 \\ -1 & 1 \\ 1 & 1 \end{bmatrix} = \begin{bmatrix} 2/\sqrt{6} & 1/\sqrt{21} \\ -1/\sqrt{6} & 4/\sqrt{21} \\ 1/\sqrt{6} & 2/\sqrt{21} \end{bmatrix} \begin{bmatrix} \sqrt{18} & \sqrt{2} \\ 0 & \sqrt{7} \end{bmatrix} \frac{1}{\sqrt{3}}$$

$$r = 1: \begin{bmatrix} 1 & -2 & 3 \\ -2 & 4 & -6 \end{bmatrix} = \begin{bmatrix} 1/\sqrt{5} \\ -2/\sqrt{5} \end{bmatrix} \begin{bmatrix} 1 & -2 & 3 \end{bmatrix} \sqrt{5}$$

The QR factorization is the matrix formulation of the Gram-Schmidt process for orthonormalizing the columns of B in the order $\mathbf{b}_1, \mathbf{b}_2, \ldots, \mathbf{b}_n$. In other words the set $\{\mathbf{q}_1, \mathbf{q}_2, \ldots, \mathbf{q}_j\}$ is one orthonormal basis of the subspace spanned by $\{\mathbf{b}_1, \ldots, \mathbf{b}_j\}$, for each $j = 1, 2, \ldots, n$; in symbols, $\text{span}(\mathsf{Q}_j) = \text{span}(\mathsf{B}_j)$. Here we assume that $r = n$.

When B has full rank, i.e., when $r = n$, then R is the (upper) Cholesky factor of $\mathsf{B^*B}$ since

$$\mathsf{R^*R} = \mathsf{R^*Q^*QR} = \mathsf{B^*B}.$$

In the full rank case another orthonormal matrix with the same column space as B is $\mathsf{B}(\mathsf{B^*B})^{-\frac{1}{2}}$. See Ex. 6-7-6.

In finite precision arithmetic it is unwise to form Q and R by a blind imitation of the formal Gram-Schmidt process, namely when $r = n$

$$\left.\begin{aligned} \tilde{q}_i &= b_i - \sum^{i-1} q_k(q_k^* b_i), \\ q_i &= \tilde{q}_i / \|\tilde{q}_i\|. \end{aligned}\right\} \quad i = 1, \ldots, n$$

Here are some alternatives.

6-7-1 *MODIFIED GRAM-SCHMIDT (MGS)*

As soon as q_j is formed deflate it from all remaining b's. Let $b_i^{(1)} = b_i$, $i = 1, \ldots, n$, and then, for $j = 1, \ldots, n$, form

$$\begin{aligned} q_j &= b_j^{(j)} / \|b_j^{(j)}\|, \\ b_i^{(j+1)} &= b_i^{(j)} - q_j(q_j^* b_i^{(j)}), \qquad i = j + 1, \ldots, n. \end{aligned}$$

This is a rearrangement of the standard Gram-Schmidt process. It is preferable when there is strong cancellation in the subtractions.

6-7-2 *HOUSEHOLDER'S METHOD*

Pre-multiply B by a sequence of reflectors H_i to reduce it to upper triangular form. At the first step form

$$H_1 B = \left[\begin{array}{c|c} r_{11} & r^* \\ \hline o & B^{(2)} \end{array}\right], \qquad H_1 = H(w_1), \qquad w_1 = b_1 \pm e_1\|b_1\|.$$

Then work on $\mathsf{B}^{(2)}$ and continue until

$$H_n \cdots H_1 B = \begin{bmatrix} R \\ O \end{bmatrix},$$

and then

$$Q = H_1 \cdots H_n E_n$$

where E_n denotes the first n columns of I_m ($n \leqslant m$).

Another technique, suitable for large m, is given in Sec. 6-8.

Exercises on Sec. 6-7

6-7-1. How are the coefficients which occur in *MGS* related to the elements of R?

6-7-2. Show that $\mathsf{B^*B}$ is positive definite when $r = n \leqslant m$.

6-7-3. Establish the existence of the *QR* factorization in the rank deficient case by modifying the Gram-Schmidt process to cope with zero vectors.

6-7-4. Find the QR factorization of the three matrices

$$(\mathsf{abc}), \qquad (\mathsf{bca}), \qquad (\mathsf{cab})$$

where

$$\mathsf{a} = \begin{bmatrix} 3 \\ 2 \\ 1 \end{bmatrix}, \qquad \mathsf{b} = \begin{bmatrix} 0 \\ 1 \\ 0 \end{bmatrix}, \qquad \mathsf{c} = \begin{bmatrix} 0 \\ -1 \\ 1 \end{bmatrix}.$$

6-7-5. Do an operation count for the formation of Q when the vectors w_i which determine the reflections H_i are given.

6-7-6. If X is positive definite then $\mathsf{X}^{\frac{1}{2}}$ is the unique symmetric positive definite matrix whose square is X. Verify that when the rank of m by n B is n then $\mathsf{B}(\mathsf{B}^*\mathsf{B})^{-\frac{1}{2}}$ exists and is orthonormal.

*6-8 FAST SCALED ROTATIONS

Consider multiplication of any B by a plane rotation $\mathsf{R}(k, j, \theta)$. Only rows k and j of B are changed so that we can simplify our discussion by writing $\xi_i = b_{ki}$, $\eta_i = b_{ji}$, $\gamma = \cos\theta$, $\sigma = \sin\theta$. The task is to compute, for $i = 1, 2, 3, \ldots, n$,

$$\begin{aligned} \xi_i' &= \gamma\xi_i - \sigma\eta_i, \\ \eta_i' &= \sigma\xi_i + \gamma\eta_i. \end{aligned} \tag{6-8-1}$$

A minor drawback of the conventional Givens transform is the square root needed to compute γ and σ. (If the square root is implemented in hardware it takes no more time than two or three multiplications.) The major drawback is the four products needed in (6-8-1) for each i. The goal is to get rid of half these multiplications and the trick is to hold vectors in factored form.

If we restrict attention to just one rotation then no improvement can be made. However in practice many plane rotations are done in sequence as when B is reduced to upper triangular form as discussed in Sec. 6-7.

To introduce the idea we rewrite (6-8-1) as a two-phase operation. Let $\tau = \tan\theta$, $\gamma = \cos\theta$. First compute

$$\hat{\xi}_i \equiv \xi_i - \tau\eta_i, \qquad \hat{\eta}_i \equiv \tau\xi_i + \eta_i, \qquad i = 1, \ldots, n, \tag{6-8-2}$$

and then note that

$$\xi_i' = \gamma\hat{\xi}_i, \qquad \eta_i' = \gamma\hat{\eta}_i. \tag{6-8-3}$$

The payoff comes from **postponing** the execution of (6-8-3). The price is that the new matrix B' is held in factored form $\Delta\hat{\mathsf{B}}$ where the diagonal matrix Δ must hold the multiplier γ in positions k and j. Our exposition is based on [Hammerling, 1974].

6-8-1 *DISCARDING MULTIPLICATIONS*

With the preceding remarks as motivation we begin again. Suppose that B is given in factored form as ΔC with Δ diagonal. Our task is to compute R(k, j, θ)B in the form Δ′C′ where Δ′ may be chosen at our convenience. Thus, looking at rows k and j, we see that

$$\begin{bmatrix} \mu' & 0 \\ 0 & \nu' \end{bmatrix} \begin{bmatrix} \xi_1' & \xi_2' & \cdots & \xi_n' \\ \eta_1' & \eta_2' & \cdots & \eta_n' \end{bmatrix} = \begin{bmatrix} \gamma & -\sigma \\ \sigma & \gamma \end{bmatrix} \begin{bmatrix} \mu & 0 \\ 0 & \nu \end{bmatrix} \times \begin{bmatrix} \xi_1 & \xi_2 & \cdots & \xi_n \\ \eta_1 & \eta_2 & \cdots & \eta_n \end{bmatrix} \tag{6-8-4}$$

and for each i

$$\begin{aligned} \mu'\xi_i' &= \gamma\mu\xi_i - \sigma\nu\eta_i, \\ \nu'\eta_i' &= \sigma\mu\xi_i + \gamma\nu\eta_i. \end{aligned} \tag{6-8-5}$$

There are several choices for μ' and ν' which permit ξ_i' and η_i' to be computed with only two multiplications:

$$\mu' = \gamma\mu, \qquad \nu' = \gamma\nu; \tag{6-8-6a}$$

$$\mu' = \sigma\nu, \qquad \nu' = \sigma\mu; \tag{6-8-6a'}$$

$$\mu' = \gamma\mu, \qquad \nu' \text{ forces } \gamma\nu/\nu' = \sigma\nu/\mu'; \tag{6-8-6b}$$

$$\mu' = \sigma\nu, \qquad \nu' \text{ forces } \gamma\mu/\mu' = \sigma\mu/\nu'. \tag{6-8-6c}$$

6-8-2 *AVOIDING SQUARE ROOTS*

In order to avoid taking a square root we must utilize the fact that a Givens rotation is specifically designed to annihilate a matrix element, typically η_1'. In this case (6-8-5) with $i = 1$ gives $\tau = \tan\theta$ as

$$\tau = \sigma/\gamma = -\nu\eta_1/\mu\xi_1, \tag{6-8-7}$$

and, from basic trigonometric identities

$$\gamma^2 = 1/(1 + \tau^2), \qquad \sigma^2 = \gamma^2\tau^2. \tag{6-8-8}$$

To be specific we choose μ' and ν' by (6-8-6a) above. Then, from (6-8-5)

$$\begin{aligned} \xi_i' &= \xi_i - \left(\frac{\sigma\nu}{\gamma\mu}\right)\eta_i, \\ \eta_i' &= \left(\frac{\sigma\mu}{\gamma\nu}\right)\xi_i + \eta_i. \end{aligned} \tag{6-8-9}$$

The key observation is that σ/γ is a multiple of ν/μ. By (6-8-7)

$$-\frac{\sigma\nu}{\gamma\mu} = \frac{\nu^2\eta_1}{\mu^2\xi_1}, \qquad -\frac{\sigma\mu}{\gamma\nu} = \frac{\eta_1}{\xi_1}. \tag{6-8-10}$$

Equation (6-8-10) shows that μ and ν are not needed and that μ^2 and ν^2 suffice! To prepare for the next Givens rotation we only need

$$(\mu')^2 = \gamma^2\mu^2, \qquad (\nu')^2 = \gamma^2\nu^2, \tag{6-8-11}$$

and there is no need to compute γ, σ, or τ. To summarize:

> The Fast Givens transform updates Δ^2 and C so that ΔC becomes the rotated matrix.

Using (6-8-6a) the actual algorithm gives:

$$\alpha \leftarrow \eta_1/\xi_1, \qquad \beta \leftarrow (\nu^2)\alpha/(\mu^2), \qquad \omega \leftarrow 1/(1+\alpha\beta),$$

$$(\mu^2) \leftarrow (\mu^2)\omega, \qquad (\nu^2) \leftarrow (\nu^2)\omega, \qquad \xi_1 \leftarrow \xi_1 + \beta\eta_1,$$

and, for $i = 2, \ldots, n$,

$$\lambda \leftarrow \xi_i, \qquad \xi_i \leftarrow \xi_i + \beta\eta_i, \qquad \eta_i \leftarrow \eta_i - \alpha\lambda. \tag{6-8-12}$$

EXAMPLE 6-8-1 FAST GIVENS

$$\Delta^2 \quad \begin{bmatrix} 1 & 0 \\ 0 & 4 \end{bmatrix} \qquad \mathrm{C} \quad \begin{bmatrix} 5 & 2 & 4 & 6 \\ 2 & 3 & 7 & 5 \end{bmatrix}$$

$$\alpha \leftarrow \eta_1/\xi_1 = 2/5 = 0.4, \qquad \beta \leftarrow (\nu^2)\alpha/(\mu^2) = 4(0.4)/1 = 1.6$$

$$\omega \leftarrow 1/(1+\alpha\beta) = 1/(1+(0.4)(1.6)) = 0.61$$

$$\mu'^2 \leftarrow (\mu^2)\omega = 1(0.61) = 0.61, \qquad \nu'^2 \leftarrow (\nu^2)\omega = 4(0.61) = 2.44$$

ξ_i' and ν_i' are calculated by using (6-8-12).

$$\Delta'^2 \quad \begin{bmatrix} 0.61 & 0 \\ 0 & 2.44 \end{bmatrix} \qquad \mathrm{C}' \quad \begin{bmatrix} 8.2 & 6.8 & 15.2 & 14. \\ 0 & 2.2 & 5.4 & 2.6 \end{bmatrix}$$

The elements of Δ can never increase but there is a possibility of underflow. To forestall such a calamity it is best not to settle on any one formula in (6-8-6) but rather to let the program choose, say (a) when $|\eta_1| \leqslant |\xi_1|$ and (a′) otherwise, to ensure that $(\mu')^2 \geqslant (\mu^2)/2$.

6-8-3 ERROR ANALYSIS

Algorithms such as (6-8-12) are robust in the face of roundoff. This is to be expected because (6-8-12) is just a scaled version of the standard Givens rotation and products are always computed with tiny relative error (barring underflow or overflow).

The situation is simple enough to warrant presentation. In order to suppress distracting details let ϵ denote a tiny number (e.g., $|\epsilon| = 10^{-14}$) which *may be different at every appearance*. Thus $(1 + \epsilon)^2$ is shorthand for $(1 + \epsilon_1)(1 + \epsilon_2)$. We denote computed quantities by overbars, $\bar{\alpha}$, $\bar{\beta}$, etc., and make the realistic assumption that in executing (6-8-12),

$$\bar{\alpha} = \alpha(1 + \epsilon), \qquad \bar{\beta} = \beta(1 + \epsilon),$$

$$\bar{\omega} = \omega(1 + \epsilon)^2 = (1 + \epsilon)^2 / (1 + \tau^2) = \left[(1 + \epsilon)\gamma\right]^2. \tag{6-8-13}$$

We worked hard to remove γ and σ from the computation in (6-8-12) but for analysis we want to bring them back. To do this we define $\bar{\mu}'$, which is never seen, as the exact square root of the new value of the variable μ^2 and assume realistically that

$$\bar{\mu}' = (1 + \epsilon)\mu\gamma, \qquad \bar{\nu}' = (1 + \epsilon)\nu\gamma. \tag{6-8-14}$$

Now we are ready to consider the main computation in (6-8-12), where

$$\bar{\xi}_i' = \left[\xi_i + \bar{\beta}\eta_i(1 + \epsilon)\right](1 + \epsilon),$$

because of the multiply and the add operations. Next multiply by $\bar{\mu}'$ and use (6-8-14), (6-8-13), and (6-8-10) to find

$$\bar{\mu}'\bar{\xi}' = \gamma\mu\xi_i(1 + \epsilon)^2 + \gamma\mu(\sigma\nu/\gamma\mu)\eta_i(1 + \epsilon)^3,$$

$$\bar{\nu}'\bar{\eta}_i' = -(\sigma/\gamma)\gamma\nu\xi_i(1 + \epsilon)^3 + \gamma\nu\eta_i(1 + \epsilon)^2. \tag{6-8-15}$$

Replace $(1 + \epsilon)^k$ by $1 + k\epsilon$ (remember ϵ's mercurial nature) and recall (6-8-5) to obtain the desired relation

$$\begin{bmatrix} \bar{\mu}'\bar{\xi}_i' \\ \bar{\nu}'\bar{\eta}_i' \end{bmatrix} = \begin{bmatrix} \mu'\xi_i' \\ \nu'\eta_i' \end{bmatrix} + \begin{bmatrix} 2\epsilon\gamma & 3\epsilon\sigma \\ -3\epsilon\sigma & 2\epsilon\gamma \end{bmatrix} \begin{bmatrix} \mu\xi_i \\ \nu\eta_i \end{bmatrix} \tag{6-8-16}$$

This equation is of exactly the same form as the equation for the errors incurred in a standard Givens rotation. The new formulas are as stable as the old ones—and underflow is avoided by allowing flexibility in the choice of formulas in (6-8-6).

6-8-4 ACCUMULATING THE PRODUCT

In some applications the product of all the plane rotations is needed. Normally this product is accumulated gradually; at each step the current product is pre-multiplied by the current plane rotation. In this situation the values of γ and σ are needed and it looks as though the square root must be taken after all! Of course it is still desirable to keep the two-multiplication formulas for updating Δ^2 and C.

Once again it turns out that the square root is not necessary. By computing a slightly unorthodox variant of the *QR* decomposition a scaled version of Q can be updated at each step without taking a square root. Let us illustrate how this is done with the (6-8-6a) choice of scaling. Denote by Q the current orthogonal matrix. The nontrivial part of the Givens rotation has been written as

$$\mathsf{R} = \gamma \mathsf{G} = \begin{bmatrix} \gamma & 0 \\ 0 & \gamma \end{bmatrix} \begin{bmatrix} 1 & -\tau \\ \tau & 1 \end{bmatrix}.$$

Suppose Q is scaled so that the two rows to be modified are

$$\mathsf{Q} = \Psi \mathsf{S} = \begin{bmatrix} \psi_1 & 0 \\ 0 & \psi_2 \end{bmatrix} \begin{bmatrix} \cdot & \cdot & \cdot & \cdots & \cdot \\ \cdot & \cdot & \cdot & \cdots & \cdot \end{bmatrix}$$

To preserve the two-multiplication update we must permute G and Ψ. Thus

$$\begin{aligned} \mathsf{RQ} &= \gamma \mathsf{G} \Psi \mathsf{S}, \\ &= \gamma(\Psi\Psi^{-1})\mathsf{G}\Psi\mathsf{S}, \\ &= (\gamma\Psi)(\bar{\mathsf{G}}\mathsf{S}), \end{aligned}$$

where

$$\bar{\mathsf{G}} \equiv \Psi^{-1}\mathsf{G}\Psi = \begin{bmatrix} 1 & -\tau\rho \\ \tau/\rho & 1 \end{bmatrix}, \qquad \rho = \psi_2/\psi_1,$$

has the same desirable form as G. For arbitrary positive ψ_1, ψ_2 this permutation of G and Ψ can be done and the savings in multiplications in the formation of $\bar{\mathsf{G}}\mathsf{S}$ can be achieved. However the quantities γ, τ, μ, and ν are not directly available from the algorithm in (6-8-12) and the formation of $\gamma\Psi$ seems to require the square root of ω. Fortunately the choice $\psi_1 = \mu$, $\psi_2 = \nu$ (i.e., $\Psi = \Delta$) removes this small blemish because, from (6-8-7), (6-8-11), and (6-8-12)

$$\begin{aligned} \tau\rho &= -(\nu\eta_1/\mu\xi_1)(\nu/\mu) = -\beta = \text{known}, \\ \tau/\rho &= -(\nu\eta_1/\mu\xi_1)(\mu/\nu) = -\alpha = \text{known}, \\ \gamma\psi_1 &= \gamma\mu = \mu', \qquad \gamma\psi_2 = \gamma\nu = \nu'. \end{aligned}$$

The result is that exactly the same scale factors are used for B as for the transforming matrix Q. In symbols

$$\mathsf{B} = \Delta \mathsf{C} \longrightarrow \mathsf{B}' = \mathsf{RB} = \Delta' \mathsf{C}',$$
$$\mathsf{Q} = \Delta \mathsf{S} \longrightarrow \mathsf{Q}' = \mathsf{RQ} = \Delta' \mathsf{S}',$$

where the work is done in forming

$$\mathsf{C}' = \bar{\mathsf{G}}\mathsf{C}, \qquad \mathsf{S}' = \bar{\mathsf{G}}\mathsf{S}, \qquad (\Delta')^2 = \omega\Delta^2,$$

and

$$\bar{\mathsf{G}} = \begin{bmatrix} 1 & \beta \\ -\alpha & 1 \end{bmatrix}.$$

* 6-9 ORTHOGONALIZATION IN THE FACE OF ROUNDOFF

The simplest orthogonalization calculation, to which all others can be reduced, is this: Given vectors $\mathsf{y} \neq \mathsf{o}$ and z, produce z's component p orthogonal to y. The formula is

$$\mathsf{p} = \mathsf{z} - \mathsf{y}(\mathsf{y}^*\mathsf{z}/\mathsf{y}^*\mathsf{y}) = [\mathsf{I} - \mathsf{y}(\mathsf{y}^*\mathsf{y})^{-1}\mathsf{y}^*]\mathsf{z}.$$

Roundoff complicates matters in two ways. First, we can only compute an approximation λ to $\mathsf{y}^*\mathsf{z}/\mathsf{y}^*\mathsf{y}$. Second, we can only compute an approximation x to $\mathsf{p} = \mathsf{z} - \lambda\mathsf{y}$. In particular, when z is almost parallel to

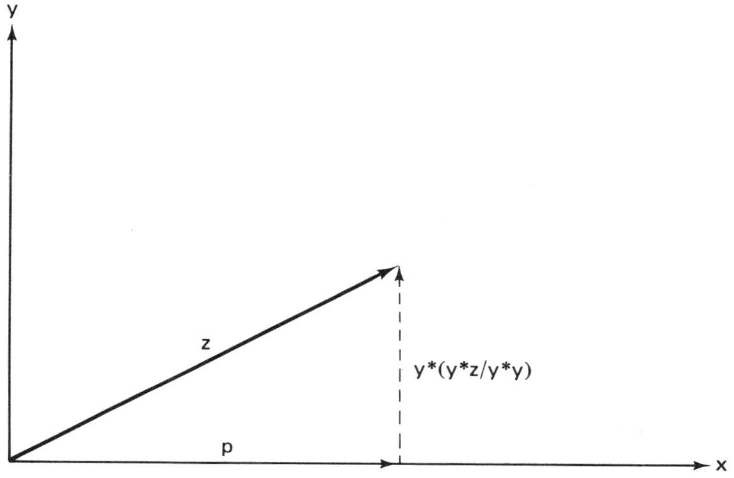

Figure 6-9-1 Orthogonality in the face of roundoff

y the latter approximation will largely cancel, leaving a handful of rounding errors in place of p as shown in Example 6-9-1.

EXAMPLE 6-9-1

The vertical rule separates the figures which remain from those which will be discarded.

Given $$\mathbf{y} = \begin{bmatrix} 0.3179 \\ 0.0253 \\ 0.0082 \end{bmatrix}, \qquad \mathbf{z} = \begin{bmatrix} 0.3170 \\ 0.0258 \\ 0.0085 \end{bmatrix},$$

Compute $$\lambda = \frac{\mathbf{y}^*\mathbf{z}}{\mathbf{y}^*\mathbf{y}} = -0.9973|3707$$

$$\lambda\mathbf{y} = \left[\begin{array}{c|c} 0.3170 & 5346 \\ 0.0252 & 3263 \\ 0.0081 & 7816 \end{array}\right]$$

$$\mathbf{p} \equiv \mathbf{z} - (\lambda\mathbf{y})_8 = \begin{bmatrix} -0.00005346 \\ 0.00056737 \\ 0.00032184 \end{bmatrix}, \qquad \frac{\mathbf{y}^*\mathbf{p}}{\|\mathbf{y}\|\|\mathbf{p}\|} = -6.634 \times 10^{-6}$$

$$\mathbf{x} \equiv \mathbf{z} - (\lambda\mathbf{y})_4 = \begin{bmatrix} 0.0000 \\ 0.0006 \\ 0.0004 \end{bmatrix}, \qquad \frac{\mathbf{y}^*\mathbf{x}}{\|\mathbf{y}\|\|\mathbf{x}\|} = 0.0802.$$

Most applications of orthogonalization require that p's approximation x have two properties:

I. x is orthogonal to y to working accuracy.
II. z is very nearly a linear combination of x and y.

Then x is very nearly the complement, orthogonal to y, of a vector very near z.

How close would we expect to come to satisfying these conditions if we naively compute

$$\lambda = \mathbf{y}^*\mathbf{z}/\|\mathbf{y}\|^2 \text{ with roundoff,} \qquad \mathbf{x} = \mathbf{z} - \lambda\mathbf{y} \text{ with roundoff?}$$

Then we shall have, as it turns out,

$$|\lambda - \mathbf{y}^*\mathbf{z}/\|\mathbf{y}\|^2| \leqslant \delta_1\|\mathbf{z}\|/\|\mathbf{y}\| \qquad \text{and} \qquad \|\mathbf{x} - (\mathbf{z} - \lambda\mathbf{y})\| \leqslant \delta_2\|\mathbf{z}\|$$

for some tiny positive δ_1 and δ_2 dependent on the precision carried and other computational details. The latter inequality satisfies property I above. But since

$$\mathbf{e} \equiv \mathbf{x} - \mathbf{p} = \mathbf{x} - (\mathbf{z} - \lambda\mathbf{y}) - (\lambda - \mathbf{y}^*\mathbf{z}/\|\mathbf{y}\|^2)\mathbf{y},$$

then

$$\|e\| \leqslant (\delta_1 + \delta_2)\|z\|,$$

and we find

$$|y^*x| = |y^*e| \leqslant \|y\| \cdot \|e\| \leqslant (\delta_1 + \delta_2)\|y\| \cdot \|x\| \cdot (\|z\|/\|x\|)$$

so property I may fail if $\|z\|/\|x\|$ is huge, as will happen whenever z and y are nearly parallel.

Nonetheless it is possible to compute an x which satisfies both I and II, despite roundoff, though at a price. The following ("twice is enough") algorithm and analysis are due to W. Kahan.

Suppose that a simple subprogram **orthog** is available which, given $y \neq o$ and z computes an approximation x' to $p \equiv z - y(y^*z/\|y\|^2)$. Let the error e' ($\equiv x' - p$) satisfy $\|e'\| \leqslant \varepsilon\|z\|$ for some tiny positive ε independent of y and z. Let κ be any fixed value in the range $[1/(0.83 - \varepsilon), 0.83/\varepsilon]$.

Algorithm. First call orthog(y, z, x') to get x'.

Case 1. If $\|x'\| \geqslant \|z\|/\kappa$ accept $x = x'$ and $e = e'$.

Otherwise call orthog(y, x', x'') to get x'' with error $e'' \equiv x'' - (x' - yy^*x'/\|y\|^2)$, satisfying $\|e''\| \leqslant \varepsilon\|x'\|$ and proceed to Case 2.

Case 2. If $\|x''\| \geqslant \|x'\|/\kappa$ accept $x = x''$, $e = x'' - p$.

Case 3. If $\|x''\| < \|x'\|/\kappa$ accept $x = o$, $e = -p$.

Note that when κ is small, like 1.25, the bounds are very good but Case 1 will be rarer and the algorithm will be more expensive than when $\kappa = 100$.

> LEMMA. The vector x computed by the algorithm ensures that $\|e\| \leqslant (1 + 1/\kappa)\varepsilon\|z\|$ and $|y^*x| \leqslant \kappa\varepsilon\|y\|\|x\|$.

Proof. We examine the three cases which can occur in the algorithm.

Case 1.

$$\|e\| = \|e'\| \leqslant \varepsilon\|z\| \leqslant \left(1 + \frac{1}{\kappa}\right)\varepsilon\|z\|,$$

$$|y^*x| = |y^*x'| = |y^*e'| \leqslant \|y\| \cdot \|e'\| \leqslant \varepsilon\|y\| \cdot \|z\| \leqslant \kappa\varepsilon\|y\| \cdot \|x\|.$$

Case 2.

$$|y^*x| = |y^*x''| = |y^*e''| \leqslant \|y\| \cdot \|e''\| \leqslant \varepsilon\|y\| \cdot \|x'\|$$
$$\leqslant \kappa\varepsilon\|y\| \cdot \|x\|,$$

$$\|e\| = \|x'' - p\| = \|p + (1 - yy^*/\|y\|^2)e' + e'' - p\|$$
$$\leqslant \|(1 - yy^*/\|y\|^2)e'\| + \|e''\|$$
$$\leqslant \|e'\| + \|e''\| \leqslant \varepsilon\|z\| + \varepsilon\|x'\| \leqslant \varepsilon\left(1 + \frac{1}{\kappa}\right)\|z\|.$$

Case 3. Since $x = o$, $|y^*x| = 0 \leqslant \kappa\varepsilon\|y\| \cdot \|x\|$. Next we must infer that $e = -p$ satisfies $\|p\| \leqslant \left(1 + \frac{1}{\kappa}\right)\varepsilon\|z\|$. Actually we shall infer more, namely $\|p\| < [1 - (\varepsilon + \kappa^{-1})^2]^{-1/2}\|e'\|$, from which the desired result will follow because $\|e'\| \leqslant \varepsilon\|z\|$ and $[1 - (\varepsilon + \kappa^{-1})^2]^{-1/2} \leqslant 1 + \kappa^{-1}$, the last inequality being the constraint that inspired and is implied by the lemma's simpler bounds on κ. See Ex. 6-9-2.

Write $e' = a + b$ where $a = (1 - yy^*/\|y\|^2)e'$ and $b = yy^*e'/\|y\|^2$. Then, with the aid of Ex. 6-9-1,

$$\kappa^{-1}\|x'\| > \|x''\| = \|p + a + e''\|$$
$$\geqslant \|p + a\| - \|e''\| \geqslant \|p + a\| - \varepsilon\|x'\|,$$

or

$$(\varepsilon + \kappa^{-1})\|x'\| = (\varepsilon + \kappa^{-1})\|p + a + b\| > \|p + a\|.$$

Since $b^*p = b^*a = 0$, squaring yields

$$(\varepsilon + \kappa^{-1})^2(\|p + a\|^2 + \|b\|^2) > \|p + a\|^2,$$

whence

$$\|p\| \leqslant \|p + a\| + \|a\| < \|a\| + \|b\|(\varepsilon + \kappa^{-1})/\sqrt{1 - (\varepsilon + \kappa^{-1})^2}$$
$$< \sqrt{1 + (\varepsilon + \kappa^{-1})^2/\left[1 - (\varepsilon + \kappa^{-1})^2\right]}\ \sqrt{\|a\|^2 + \|b\|^2}$$
$$= \|e'\|/\sqrt{1 - (\varepsilon + \kappa^{-1})^2}$$

as claimed. □

Example 6-9-2 shows the algorithm in action.

EXAMPLE 6-9-2

$\kappa = 10. \quad \varepsilon = 10^{-5} \quad$ (from Example 2-5-2)

$$\mathbf{y} = \begin{bmatrix} 0.16087 \\ -0.11852 \\ 0.98216 \times 10^{-1} \end{bmatrix} \quad \mathbf{z} = \begin{bmatrix} -0.50069 \times 10^{-1} \\ 0.36889 \times 10^{-1} \\ -0.30569 \times 10^{-1} \end{bmatrix} \quad \mathbf{x}' (= \mathbf{z} - \lambda \mathbf{y}) = \begin{bmatrix} -0.20000 \times 10^{-5} \\ 0.30000 \times 10^{-5} \\ -0.20000 \times 10^{-5} \end{bmatrix}$$

$\|\mathbf{y}\| = 0.22264 \qquad \|\mathbf{z}\| = 0.69297 \times 10^{-1} \qquad \|\mathbf{x}'\| = 0.41231 \times 10^{-5}$

$$\mathbf{x}'' (= \mathbf{x}' - \lambda' y) = \begin{bmatrix} 0.8353 \times 10^{-6} \\ 0.9110 \times 10^{-6} \\ -0.2689 \times 10^{-6} \end{bmatrix} \quad \mathbf{p} = \begin{bmatrix} 0.48705 \times 10^{-6} \\ 0.60812 \times 10^{-6} \\ -0.63987 \times 10^{-8} \end{bmatrix} \quad \mathbf{e} (= x'' - p) = \begin{bmatrix} 0.34825 \times 10^{-6} \\ 0.30288 \times 10^{-6} \\ -0.20431 \times 10^{-6} \end{bmatrix}$$

$$\|\mathbf{x}''\| = 0.12648 \times 10^{-5} \geqslant 0.41231 \times 10^{-6} = \|\mathbf{x}'\|/\kappa$$

$$\|\mathbf{e}\| = 0.50498 \times 10^{-6} \leqslant 0.76227 \times 10^{-6} = (1 + 1/\kappa)\varepsilon\|\mathbf{z}\|$$

$$|(\mathbf{x}'')^*\mathbf{y}| = 0.72914 \times 10^{-11} \leqslant 0.28160 \times 10^{-10} = \kappa\varepsilon\|\mathbf{x}''\|\|\mathbf{y}\|$$

Exercises on Sec. 6-9

6-9-1. Show that $\mathbf{x}'' = \mathbf{e}'' + \mathbf{a} + \mathbf{p}$.

6-9-2. Solve $1 = (1 + \mu)^2(1 - \mu^2)$ for $\mu > 0$. A calculator may help. For what value of μ is $[1 - (\varepsilon + \mu)^2](1 + \mu)^2 \geqslant 1$?

Notes and References

Plane rotations were used in [Jacobi, 1846] and became well-known tools after [Bargmann, Montgomery, and von Neumann, 1946]. It is odd that reflectors (or symmetries as they are called by group theorists), which are the only elementary matrices which can generate all orthogonal matrices, were not introduced as tools for computation until 1958 in [Householder, 1958]. Their rapid acceptance owes much to [Wilkinson, 1960]. Alternative formulas for reflections in external bisectors were given in [Parlett, 1971].

The possibility of avoiding half the multiplications in a sequence of plane rotations by adroit use of scaling was presented in [Gentleman, 1973]. Our presentation follows [Hammerling, 1974]. Very similar ideas are used in [Gill, Murray, and Saunders, 1975] where the Gram-Schmidt matrix **Q** in **B** = **QR** is itself held in scaled form. Mention should also be made of recent work on updating **Q** when **B** changes a little, in particular [Gill, et al., 1974].

Section 6-9 is based on unpublished notes by Kahan but the fact that "twice is enough" when orthogonalizing in the presence of roundoff is not as well known as it should be. The results in [Daniel et al., 1976] can be regarded as extensions of this analysis to the case of orthogonalizing against several nearly orthonormal vectors at the same time.

The error analysis in Secs. 6-5 and 6-6 is based on [Wilkinson, 1965, chap. 3].

7

Tridiagonal Form

7-1 INTRODUCTION

This chapter is concerned with the reduction of an arbitrary symmetric matrix A to a similar tridiagonal matrix T and also with the special properties which T enjoys.

Definition. T is **tridiagonal** if $t_{ij} = 0$ whenever $|i - j| > 1$. In order to simplify notation it is useful to write $t_{ii} = \alpha_i$, $t_{i,\,i+1} = \beta_i$ (sometimes it is more convenient to set $t_{i,\,i+1} = \beta_{i+1}$), and

$$\mathsf{T}_{\mu,\,\nu} = \begin{bmatrix} \alpha_\mu & \beta_\mu & & & \\ \beta_\mu & \alpha_{\mu+1} & \beta_{\mu+1} & & \\ & \beta_{\mu+1} & \cdot & \cdot & \\ & & \cdot & \alpha_{\nu-1} & \beta_{\nu-1} \\ & & & \beta_{\nu-1} & \alpha_\nu \end{bmatrix} \qquad (7\text{-}1\text{-}1)$$

Definition. T $(= \mathsf{T}_{1,\,n})$ is **unreduced** if $\beta_i \neq 0$, $i = 1, \ldots, n-1$.

If, for some k, $\beta_k = 0$ then T is a direct sum of two tridiagonal matrices, say T_1 of order k and T_2 of order $n - k$. The eigenvalues and eigenvectors of T can be recovered easily from those of T_1 and T_2 (Ex. 7-1-1) and in this way the computation may be reduced to finding the eigenvalues and eigenvectors of smaller tridiagonal submatrices which are unreduced. So there is no loss of generality in confining attention to the unreduced case.

The key facts are as follows:

1. The eigenvalues and eigenvectors of T can be found with significantly fewer arithmetic operations than are required for a full A.
2. Every A can be reduced to a similar T by a finite number of elementary orthogonal similarity transformations. In principle it requires an infinite number of transformations to diagonalize a matrix.
3. If T is unreduced its eigenvalues are distinct (though they may be very close) and its eigenvectors enjoy some useful special properties. See Secs. 7-7 and 7-9.

The tridiagonal form is not always the way to go; the smaller the bandwidth of a matrix the larger is the cost of reduction to tridiagonal form **relative** to the total calculation of eigenvalues and/or eigenvectors by other means.

Tridiagonal matrices sometimes occur as primary data. For example, they are associated with families of orthogonal polynomials and with special functions, like the Bessel functions, which satisfy three term recurrence relations. For example, the zeros of J_m are given by $2/\sqrt{\mu_k}$, $k = 1, 2, \ldots$ where the μ_k are the eigenvalues of $T_{1,\infty}$ and

$$\alpha_i = 2/\,(m + 2i - 1)(m + 2i + 1),$$

$$\beta_i^2 = 1/\,(m + 2i)(m + 2i + 1)^2(m + 2i + 2).$$

Exercise on Sec. 7-1

7-1-1. Let $T = \mathrm{diag}(T_1, T_2)$ and let

$$T_i = S_i \Theta_i S_i^*$$

be the spectral decomposition of T_i, $i = 1, 2$. What is the spectral decomposition of T? Suppose θ is an eigenvalue of T_1 and of T_2. Describe the eigenvectors of T belonging to θ.

7-2 UNIQUENESS OF REDUCTION

There are several ways to reduce an arbitrary A to tridiagonal form T but before discussing any of them in detail it is good to know to what extent the resulting T depends on the method. The answer is given in the theorem below but before presenting it some further normalization is necessary.

LEMMA Let T be unreduced and let T_+ be the matrix obtained by replacing each β_i by $|\beta_i|$, $i = 1, \ldots, n-1$. Then

$$\mathsf{T}_+ = \Delta\mathsf{T}\Delta = \Delta\mathsf{T}\Delta^{-1}$$

where $\Delta = \mathrm{diag}(\delta_1, \ldots, \delta_n)$ and $\delta_i = \pm 1$. (7-2-1)

The proof is left as Ex. 7-2-1. The lemma shows that there is no loss of generality in taking $\beta_i > 0$, $i = 1, \ldots, n-1$.

THEOREM Let $\mathsf{Q}^*\mathsf{A}\mathsf{Q} = \mathsf{T}_+$ with Q orthogonal. Then T_+ and $\mathsf{Q} \equiv (\mathsf{q}_1, \mathsf{q}_2, \ldots, \mathsf{q}_n)$ are uniquely determined by A and q_1 or by A and q_n. (7-2-2)

Proof. Since $\mathsf{Q}^* = \mathsf{Q}^{-1}$ the hypothesis can be rewritten as

$$\mathsf{Q}\mathsf{T}_+ = \mathsf{A}\mathsf{Q}. \tag{7-2-3}$$

Now equate the j-th column on each side of (7-2-3), use the fact that the j-th column of T_+ has only three nonzero elements, and rearrange terms to find an important relation,

$$\mathsf{q}_{j+1}\beta_j = \mathsf{A}\mathsf{q}_j - \mathsf{q}_j\alpha_j - \mathsf{q}_{j-1}\beta_{j-1} \equiv \mathsf{r}_j. \tag{7-2-4}$$

This holds for $j = 1, \ldots, n$ if we define $\beta_0 = \beta_n = 0$. Thus $\mathsf{r}_n = 0$ and q_0, q_{n+1} are undefined. Next use the orthogonality of Q in the form $\mathsf{q}_i^*\mathsf{q}_k = \delta_{ik}$ (Kronecker's δ symbol) to obtain

1. $0 = \mathsf{q}_j^*(\mathsf{q}_{j+1}\beta_j) = \mathsf{q}_j^*\mathsf{A}\mathsf{q}_j - 1 \cdot \alpha_j - 0 \cdot \beta_{j-1}$,
2. $\beta_j = \|\mathsf{q}_{j+1}\beta_j\| = \|\mathsf{r}_j\|$,
3. $\mathsf{q}_{j+1} = \mathsf{r}_j/\beta_j$, since $\beta_j > 0$ by hypothesis.

Thus β_{j-1}, q_{j-1}, and q_j determine, in turn, α_j, r_j, β_j, and q_{j+1} for $j = 1, 2, \ldots, n$. Since $\beta_0 = 0$, q_1 alone determines α_1, r_1, β_1, and q_2, and so, by finite induction, q_1 determines uniquely all the elements of T_+ and Q.

By rewriting (7-2-4) in the form

$$\mathsf{q}_{j-1}\beta_{j-1} = \mathsf{A}\mathsf{q}_j - \mathsf{q}_j\alpha_j - \mathsf{q}_{j+1}\beta_j \equiv \tilde{\mathsf{r}}_j$$

it can be shown that β_j, q_{j+1}, and q_j determine α_j, $\tilde{\mathsf{r}}_j$, β_{j-1} and q_{j-1}. Since $\beta_n = 0$, q_n also determines T_+ and Q uniquely. □

7-2-1 REMARKS

1. We have **not** shown that for a given A each unit vector q_1 determines a unique T_+ and Q. The uniqueness breaks down whenever some $r_j = 0$. Then $\beta_j = \|r_j\| = 0$ and T is reduced. This rare case of breakdown is not a curse but a blessing. Breakdown can occur only if q_1 is orthogonal to at least one eigenvector of A.

   If $\beta_j = 0$ then q_{j+1} can be taken as any unit vector orthogonal to the preceding q's and the process can then be continued. Only uniqueness has been lost.

2. In the proof β_{j-1} was defined by $\beta_{j-1} = \|r_{j-1}\|$. However if (7-2-4) is pre-multiplied by q_{j-1}^* it is clear that

$$\beta_{j-1} = q_{j-1}^* A q_j. \tag{7-2-5}$$

   These two expressions for β_{j-1} are equivalent in exact arithmetic but turn out to be very different when used by digital computers. It turns out that (7-2-5) is to be avoided, but this result is far from obvious and indeed (7-2-5) has often been recommended.

3. There is no suggestion that T and Q should necessarily be computed by the procedure used in the proof which happens to be the **Lanczos algorithm.** For large A the Lanczos process is very useful but has to be implemented carefully if all its benefits are to be reaped. See Chap. 13.

Exercises on Sec. 7-2

7-2-1. Prove Lemma (7-2-1). Given an eigenvector s of T_+ and Δ give an algorithm for overwriting s with the corresponding eigenvector of T.

7-2-2. Let

$$A = \begin{bmatrix} 3 & 0 & 1 \\ 0 & 2 & 0 \\ 1 & 0 & 1 \end{bmatrix}$$

Reduce A to tridiagonal form by following the proof of Theorem (7-2-2). Try two starting vectors e_1 and e_3.

7-3-3. Reduce diag(1, 10, 100) to tridiagonal form starting with $(1, 1, 1)^*/\sqrt{3}$.

7-3 MINIMIZING CHARACTERISTICS

Let $T = T_+ = Q^*AQ$. Since T and Q are determined by $q_1(= Qe_1)$ it is plausible that the formulas for at least some of the elements of T and Q are nice and simple. This section presents some such formulas and their derivation is little more than a systematic exploitation of the tridiagonal form. Consider, for instance, a 4 by 4 example:

$$\mathsf{T}\mathbf{e}_1 = \begin{bmatrix} \alpha_1 \\ \beta_1 \\ 0 \\ 0 \end{bmatrix}, \qquad \mathsf{T}^2\mathbf{e}_1 = \begin{bmatrix} \alpha_1^2 + \beta_1^2 \\ \beta_1(\alpha_1 + \alpha_2) \\ \beta_1\beta_2 \\ 0 \end{bmatrix},$$

$$\mathsf{T}^3\mathbf{e}_1 = \begin{bmatrix} \alpha_1^3 + (2\alpha_1 + \alpha_2)\beta_1^2 \\ \beta_1(\alpha_1^2 + \alpha_1\alpha_2 + \alpha_2^2 + \beta_1^2 + \beta_2^2) \\ \beta_1\beta_2(\alpha_1 + \alpha_2 + \alpha_3) \\ \beta_1\beta_2\beta_3 \end{bmatrix}.$$

It turns out that the character of the first columns of the powers of T gives a bizarre but useful characterization of the β's and $\mathbf{q}$'s in terms of certain polynomials.

Recall from Sec. 7-1 that the leading principal j by j submatrix of T may be written $\mathsf{T}_{1,j}$. Its characteristic polynomial is χ_j, so

$$\chi_j(\xi) \equiv \det[\xi - \mathsf{T}_{1,j}], \qquad j = 1, \ldots, n.$$

For convenience let $\chi_0(\xi) = 1$ and let $\mathcal{M}\mathcal{P}_k$ denote the set of monic polynomials of degree k.

> THEOREM Let tridiagonal T have positive subdiagonal elements as given in (7-1-1). Then, for $j = 1, \ldots, n-1$,
> (a) $\beta_1\beta_2 \cdots \beta_j = \|\chi_j(\mathsf{T})\mathbf{e}_1\| = \min \|\psi(\mathsf{T})\mathbf{e}_1\|$ over all $\psi \in \mathcal{M}\mathcal{P}_j$,
> (b) $\mathbf{e}_{j+1} = \chi_j(\mathsf{T})\mathbf{e}_1/(\beta_1 \cdots \beta_j)$. (7-3-1)

Proof. Observe that the last nonzero element in $\mathsf{T}^j\mathbf{e}_1$ is $(\beta_1 \cdots \beta_j)$ and is in row $j+1$ (Ex. 7-3-1). It follows that the same is true for $\psi(\mathsf{T})\mathbf{e}_1$ for any ψ in $\mathcal{M}\mathcal{P}_j$ because T^j is the only term which contributes a nonzero value in row $j+1$. Hence

$$\beta_1\beta_2 \cdots \beta_j \leqslant \|\psi(\mathsf{T})\mathbf{e}_1\|,$$

with equality only if $\psi(\mathsf{T})\mathbf{e}_1 = \mathbf{e}_{j+1}(\beta_1 \cdots \beta_j)$.

A closer look at the first j elements of $\psi(\mathsf{T})\mathbf{e}_1$, or rather of the $\mathsf{T}^k\mathbf{e}_1$, $k < j$, shows that they involve $\mathsf{T}_{1,j}$ only. More precisely, let

$\mathsf{E}_j = (\mathbf{e}_1, \ldots, \mathbf{e}_j)$ then (Ex. 7-3-2), for any $\psi \in \mathcal{MP}_j$,

$$\mathsf{E}_j^* \psi(\mathsf{T})\mathbf{e}_1 = \psi(\mathsf{T}_{1,j})\mathbf{e}_1. \qquad (7\text{-}3\text{-}2)$$

Note that $\mathbf{e}_1$ on the left is in $\mathcal{E}^n$ while $\mathbf{e}_1$ on the right is in $\mathcal{E}^j$. By the Cayley-Hamilton theorem $\chi_j(\mathsf{T}_{1,j}) = \mathsf{O}$ and so $\mathsf{E}_j^*\chi_j(\mathsf{T})\mathbf{e}_1 = \mathbf{o}$. By the second sentence in this proof $\chi_j(\mathsf{T})\mathbf{e}_1$ must be $\mathbf{e}_{j+1}(\beta_1 \cdots \beta_j)$ and thus both assertions are established. □

Of course, with T in front of us there is little incentive to characterize the β's or the $\mathbf{e}_i$ at all. However we can substitute $\mathsf{Q^*AQ}$ for T to get expressions involving A and $\mathbf{q}_1$. See Ex. 7-3-3.

COROLLARY Let $\mathsf{T} = \mathsf{T}_+ = \mathsf{Q^*AQ}$, with $\mathsf{Q} = (\mathbf{q}_1, \ldots, \mathbf{q}_n)$ orthogonal. Then, for $j = 1, \ldots, n - 1$,

$$\beta_1 \cdots \beta_j = \|\chi_j(\mathsf{A})\mathbf{q}_1\| = \min \|\psi(\mathsf{A})\mathbf{q}_1\| \text{ over all } \psi \in \mathcal{MP}_j, \qquad (7\text{-}3\text{-}3a)$$

$$\mathbf{q}_{j+1} = \chi_j(\mathsf{A})\mathbf{q}_1/(\beta_1 \cdots \beta_j). \qquad (7\text{-}3\text{-}3b)$$

There are situations in which A and $\mathbf{q}_1$ are known but T and Q are not. Neither is χ_j but it is interesting that the unknown β's must have the minimal property (7-3-3a). When A is fixed then χ_j is said to be the minimal polynomial for $\mathbf{q}_1$ in $\mathcal{MP}_j$. The χ_j's are sometimes called the Lanczos polynomials for $\mathbf{q}_1$.

Exercises on Sec. 7-3

7-3-1. By induction, or otherwise, show that $\mathbf{e}_k^*\mathsf{T}^j\mathbf{e}_1 = 0$ if $k > j + 1$, $= \beta_1 \cdots \beta_j$ if $k = j + 1$.

7-3-2. By induction, or otherwise, show that for $k \leqslant j$

$$\mathsf{E}_j^*\mathsf{T}^k\mathbf{e}_1 = \mathsf{T}_{1,j}^k\mathbf{e}_1.$$

Then conclude that $\mathsf{E}_j^*\psi(\mathsf{T})\mathbf{e}_1 = \psi(\mathsf{T}_{1,j})\mathbf{e}_1$ for all ψ in $\mathcal{MP}_j$.

7-3-3. Establish the corollary of Theorem (7-3-1).

7-3-4. Find an expression for α_j in terms of $\mathbf{q}_1$ and polynomials in A.

7-3-5. Prove the corollary directly by equating the j-th column on each side of $\mathsf{AQ} = \mathsf{QT}$ and then using the fact that the χ_j satisfy a certain three-term recurrence relation to show that $\mathbf{q}_{j+1}$ must be a multiple of $\chi_j(\mathsf{A})\mathbf{q}_1$.

7-3-6. Let $\mathsf{A} = \begin{bmatrix} 4 & 2 & 1 \\ 2 & 6 & 2 \\ 1 & 2 & 8 \end{bmatrix}$ and $\mathbf{q}_1 = \begin{bmatrix} -1 \\ 0 \\ 1 \end{bmatrix} / \sqrt{2}$. Find an upper bound on β_1 and $\beta_1\beta_2$. Use the method of proof of Theorem (7-2-2) to compute, in order, α_1, β_1, $\mathbf{q}_2$, α_2, and β_2. Exhibit χ_1 and χ_2.

7-4 EXPLICIT REDUCTION OF A FULL MATRIX

Any A can be reduced to tridiagonal form T by a sequence of simple orthogonal similarity transformations which introduce zero elements column by column. The first step is typical and can be described most simply by partitioning A as shown below.

$$\mathsf{A} = \mathsf{A}_1 = \begin{bmatrix} \alpha_1 & \mathbf{c}_1^* \\ \mathbf{c}_1 & \mathsf{M}_1 \end{bmatrix}, \alpha_1 = a_{11}.$$

Now consider any orthogonal similarity transformation of A_1 which leaves the (1, 1) element unaltered:

$$\begin{aligned}\hat{\mathsf{A}}_1 &= \begin{bmatrix} 1 & 0^* \\ 0 & \mathsf{P}_1^* \end{bmatrix} \begin{bmatrix} \alpha_1 & c_1^* \\ \mathbf{c}_1 & \mathsf{M}_1 \end{bmatrix} \begin{bmatrix} 1 & 0^* \\ 0 & \mathsf{P}_1 \end{bmatrix} \\ &= \begin{bmatrix} \alpha_1 & \mathbf{c}_1^*\mathsf{P}_1 \\ \mathsf{P}_1^*\mathbf{c}_1 & \mathsf{P}_1^*\mathsf{M}_1\mathsf{P}_1 \end{bmatrix}.\end{aligned}$$

Any orthogonal matrix P_1 such that $\mathsf{P}_1^*\mathbf{c}_1 = \mathbf{e}_1\beta_1$ will cause $\hat{\mathsf{A}}_1$ to be tridiagonal in its first column. Since P_1 is orthogonal

$$|\beta_1| = \|\mathbf{e}_1\beta_1\| = \|\mathsf{P}_1^*\mathbf{c}_1\| = \|\mathbf{c}_1\|.$$

Two practical choices for P_1 are described below. When P_1 has been chosen then $\mathsf{P}_1^*\mathsf{M}_1\mathsf{P}_1$, which we call A_2, is calculated explicitly and, since this step is the heart of the algorithm, its cost is critical for the efficiency of the method. At first glance it appears that two matrix multiplications are involved and this normally requires $2(n-1)^3$ operations. However we can do much better than that.

At the second step the same technique is applied to

$$\mathsf{P}_1^*\mathsf{M}_1\mathsf{P}_1 \equiv \mathsf{A}_2 = \begin{bmatrix} \alpha_2 & \mathbf{c}_2^* \\ \mathbf{c}_2 & \mathsf{M}_2 \end{bmatrix}, \qquad \mathsf{A}_2 \textit{ is } (n-1) \times (n-1).$$

An orthogonal matrix P_2 is chosen so that $\mathsf{P}_2^*\mathbf{c}_2 = \mathbf{e}_1\beta_2$ and then $\mathsf{A}_3 = \mathsf{P}_2^*\mathsf{M}_2\mathsf{P}_2$ must be computed. The crucial observation is that **the similarity transformation at the second step does not destroy the zero elements introduced at the first step.** This can be seen by forming the product of the three matrices shown below.

$$\begin{bmatrix} 1 & 0 & 0^* \\ 0 & 1 & 0^* \\ 0 & 0 & \mathsf{P}_2^* \end{bmatrix} \begin{bmatrix} \alpha_1 & \beta_1 & 0^* \\ \beta_1 & \alpha_2 & \mathbf{c}_2^* \\ 0 & \mathbf{c}_2 & \mathsf{M}_2 \end{bmatrix} \begin{bmatrix} 1 & 0 & 0^* \\ 0 & 1 & 0^* \\ 0 & 0 & \mathsf{P}_2 \end{bmatrix}$$

The process continues, the submatrix which is not yet in tridiagonal form shrinks, and finally after $(n-2)$ steps T is obtained.

There is one more point to notice:

$$\mathsf{T} = \begin{bmatrix} \mathsf{I}_{n-2} & 0^* \\ 0 & \mathsf{P}_{n-2}^* \end{bmatrix} \cdots \begin{bmatrix} 1 & 0^* \\ 0 & \mathsf{P}_1^* \end{bmatrix} \mathsf{A} \begin{bmatrix} 1 & 0^* \\ 0 & \mathsf{P}_1 \end{bmatrix}$$
$$\times \begin{bmatrix} \mathsf{I}_2 & 0^* \\ 0 & \mathsf{P}_2 \end{bmatrix} \cdots \begin{bmatrix} \mathsf{I}_{n-2} & 0^* \\ 0 & \mathsf{P}_{n-2} \end{bmatrix}$$
$$= \mathsf{Q}^*\mathsf{AQ}, \text{ defining } \mathsf{Q},$$

and

$$\mathsf{Qe}_1 = \begin{bmatrix} 1 & 0^* \\ 0 & \mathsf{P}_1 \end{bmatrix} \cdots \begin{bmatrix} \mathsf{I}_{n-2} & 0^* \\ 0 & \mathsf{P}_{n-2} \end{bmatrix} \mathsf{e}_1 = \mathsf{e}_1.$$

By the reduction uniqueness theorem (7-2-2) the **product** of the $n - 2$ orthogonal matrices involving the P_i is uniquely determined despite the varied possibilities for each P_i, $i = 1, \ldots, n - 3$.

7-4-1 EXPLOITING SYMMETRY

When A is full and can be stored in the highspeed memory the preferred choice for P to satisfy $\mathsf{Pc} = \mathsf{e}_1\beta$ is the reflector matrix (see Sec. 6-3)

$$\mathsf{P} = \mathsf{H(w)} = \mathsf{I} - \gamma\mathsf{ww}^*, \qquad \gamma = 2/\mathsf{w}^*\mathsf{w},$$

where

$$\mathsf{w} = \mathsf{c} + \beta\mathsf{e}_1, \qquad \beta = \|\mathsf{c}\| \operatorname{sign}(\mathsf{e}_1^*\mathsf{c}).$$

Two valuable assets accrue from this choice. In the first place, P is not needed as an explicit two-dimensional array; it is determined by w and w differs from c only in its first element. So no extra n by n arrays are required. Second, the similarity transformation $\mathsf{H(w)MH(w)}$ can be carried out with great efficiency, as explained below. It would be ridiculous to fill out a j by j array H and then execute two matrix multiplications at a cost of $2j^3$ scalar multiplications. Instead we use the fact that H is elementary and write

$$\begin{aligned} \mathsf{HMH} &= (\mathsf{I} - \gamma\mathsf{ww}^*)\mathsf{M}(\mathsf{I} - \gamma\mathsf{ww}^*) \\ &= \mathsf{M} - \gamma\mathsf{w}(\mathsf{w}^*\mathsf{M}) - \gamma(\mathsf{Mw})\mathsf{w}^* + \gamma^2\mathsf{w}(\mathsf{w}^*\mathsf{Mw})\mathsf{w}, \\ &= \mathsf{M} - \mathsf{wp}^* - \mathsf{pw}^* + (\gamma\mathsf{w}^*\mathsf{p})\mathsf{ww}^*, \end{aligned} \tag{7-4-1}$$

where

$$\mathsf{p} = \gamma\mathsf{Mw}.$$

This use of p is good and reduces the operation count to $3j^2$ multiplications. Yet there is an extra trick (due to Wilkinson) which is a nice example of the **art** of writing programs to implement numerical methods. The quantity $\gamma \mathbf{w}^*\mathbf{p} = \gamma^2 \mathbf{w}^*\mathbf{M}\mathbf{w}$ is real (even when $\mathbf{M}$ is complex Hermitian) and so the fourth term in (7-4-1) may be shared between the second and third terms as follows. Compute

$$\mathbf{q} = \mathbf{p} - \mathbf{w}(\gamma \mathbf{w}^*\mathbf{p})/2,$$

and then

$$\mathbf{HMH} = \mathbf{M} - \mathbf{w}\mathbf{q}^* - \mathbf{q}\mathbf{w}^*.$$

The order of computation and operation count are shown below:

Quantity	$\mathbf{w}$	γ	$\mathbf{p}$	$\mathbf{w}^*\mathbf{p}$	$\mathbf{q}$	$\mathbf{M}-\mathbf{w}\mathbf{q}^*-\mathbf{q}\mathbf{w}^*$	Total
Multiplications	0	j	$j(j+1)$	j	j	j^2	$2j(j+2)$

At each step of the reduction to tridiagonal form the order j shrinks by one. The total number of ops is $\sum_{j=2}^{n-1} 2j(j+2) = \frac{2}{3}n^3 + n^2 + 0(n)$ while the most naive implementation would require $n^2(n+1)^2/2$ ops. The number of square roots is $(n-2)$.

More details can be found in Contribution II/2 in the Handbook. The method was originally proposed by Householder in 1958.

7-4-2 PLANE ROTATIONS

Another way to reduce $\mathbf{c}$ to $\pm \mathbf{e}_1\|\mathbf{c}\|$ is by a sequence of $(j-1)$ plane rotations. These elementary orthogonal matrices are described in Sec. 6-4. Each rotation affects two rows (and two columns) of the matrix and creates one new zero in the vector which started as $\mathbf{c}$.

The elements can be annihilated in the natural order $2, 3, \ldots, j$ or the reverse. Moreover there are two popular ways of annihilating element k: Either rotate in the $(1, k)$ plane or in the $(k-1, k)$ plane.

Each of the $(j-1)$ rotations requires $4j$ multiplications and 1 square root. This yields a total multiplication count

$$\Sigma 4j(j-1) = \frac{4}{3}n^3 + 0(n)$$

and $n(n-1)/2$ square roots. This method and the whole idea of reducing $\mathbf{A}$ to $\mathbf{T}$ was introduced by Givens in 1954.

For full stored matrices the rotations cannot compete with reflections but they are useful in the treatment of sparse matrices, particularly when used in the scaled, or fast, form which halves the number of operations. See Sec. 6-8. In addition it is easy to skip unnecessary rotations.

Exercises on Sec. 7-4

$$\mathsf{M} = \begin{bmatrix} 0 & & & \text{sym} \\ 1 & -1 & & \\ 2 & 4 & 2 & \\ 4 & -2 & 0 & 4 \end{bmatrix}$$

7-4-1. Carry out the first step of the reduction of M to tridiagonal form by using a reflection. Exhibit the vectors $\mathbf{w}$, $\mathbf{p}$, and $\mathbf{q}$.

7-4-2. Reduce the first column of M above to tridiagonal form using plane rotations in the (2, 3) and (2, 4) planes.

7-4-3. Use the fast Givens transformations described in Sec. 6-8 to reduce M to factored tridiagonal form $\Delta\hat{\mathsf{T}}\Delta$ where Δ is diagonal.

7-5 REDUCTION OF A BANDED MATRIX

A =

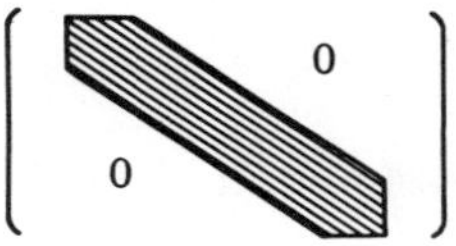

$a_{ij} = 0$ for $|i-j| > m$,
A's bandwidth is $2m + 1$.

A matrix of order 500 and bandwidth 51 requires 13,000 words of storage and can be processed in fast memory **provided that** the bandwidth is not increased during the algorithm. By annihilating pairs of elements a_{ij} and a_{ji} in an ingenious order, banded matrices can be reduced to tridiagonal form without enlarging the bandwidth in the process.

Recall that by use of a suitable plane rotation, any off-diagonal element of A in column j can be annihilated by forming $\mathsf{R^*AR}$ with $\mathsf{R}(i, j, \theta)$. When the doomed element is a_{ij} this is often called a **Jacobi rotation**; otherwise it is called a **Givens rotation.** The choice of the angle θ is different in the two cases. In this algorithm $i = j - 1$ and the doomed element is $a_{\ell j}$ with $\ell < j - 1$.

The idea of the pattern of elements to be annihilated is revealed in Fig. 7-5-1 for the case $n = 10$, $m = 3$.

7-5-1 THE METHOD

The first row is reduced to tridiagonal form, then the second, and so on. The elements in each row are annihilated one by one, from the outside in. But it takes several rotations to eliminate a single element **and** to

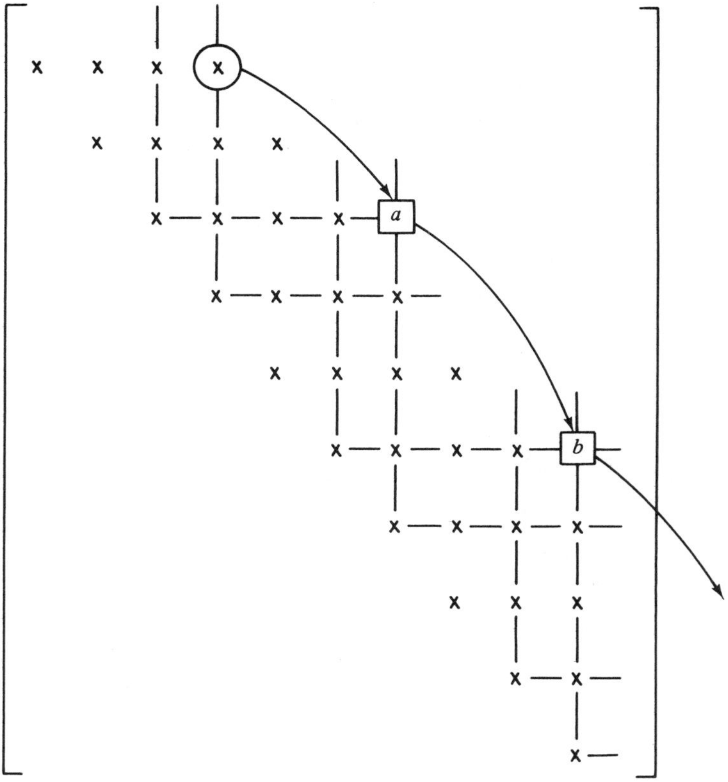

Figure 7-5-1 Reduction of bandwidth

preserve bandwidth. The treatment of the first row is typical.

1. Rotate in plane $(m, m + 1)$ to annihilate element $(1, m + 1)$. This creates a nonzero element (a) at $(m, 2m + 1)$ **outside the band.**
2. Rotate in plane $(2m, 2m + 1)$ to annihilate element $(m, 2m + 1)$. This creates a nonzero element (b) at $(2m, 3m + 1)$ **outside the band.**
3. Rotate in plane $(3m, 3m + 1)$ to annihilate element $(2m, 3m + 1)$. In this case $(n = 10)$; the bulge has been chased off the bottom of the matrix. In general more of these rotations (to a total of $[(n - 1)/m]$) would be required to restore the original bandwidth.

The next element, within the band, to be annihilated is $(1, m)$ and the resulting bulge has to be chased off the end of the matrix, and so on. After the $(1, 3)$ element is treated the second row can be processed.

7-5-2 OPERATION COUNT

The total number of rotations, $N_{\text{rotations}}$, satisfies

$$N_{\text{rotations}} \leqslant n^2(m-1)/(2m).$$

Each rotation requires one square root and, usually, $8m + 13$ ops. Thus

$$N_{\text{ops}} \leqslant n^2(m-1)(4 + 13/2m).$$

Operation counts show that the standard reduction is faster unless $m < n/6$. However it is storage considerations which usually dictate the use of banded reduction.

7-5-3 STORAGE

It is customary to use a rectangular array, n by $(m + 1)$, whose columns hold the successive diagonals of the banded matrix; the main diagonal is in the first column.

7-5-4 REFERENCE

An explicit ALGOL program called **bandrd** is given in [Contribution II/8 by H. R. Schwarz, in Wilkinson and Reinsch, 1971]. This method is in EISPACK.

The fast Givens rotations can be used to halve the work.

Exercises on Sec. 7-5

Let

$$\mathsf{M} = \begin{bmatrix} 6 & & & & \\ -4 & 6 & & \text{sym} & \\ 1 & -4 & 6 & & \\ 0 & 1 & -4 & 6 & \\ 0 & 0 & 1 & -4 & 6 \end{bmatrix}$$

7-5-1. Using the algorithms of the previous section, reduce M to tridiagonal form by means of 4 plane rotations.

7-5-2. Reduce M to tridiagonal form by means of fast Givens rotations to obtain a factored result $\Delta\hat{\mathsf{T}}\Delta$ where Δ is diagonal. A hand held calculator will be useful. You should compute Δ^2 and $\hat{\mathsf{T}}$. See Sec. 6-8.

7-6 IRRELEVANT INSTABILITY

We are not going to present detailed analyses of the influence of roundoff errors on the preceding algorithms. In fact the general analysis given in

Sec. 6-5 covers these methods. In each case the computed tridiagonal matrix will be exactly similar to a perturbation $A + W$ of the original matrix A and $\|W\|/\|A\|$ is a modest multiple of the unit roundoff. Even if the pessimistic bounds on the ratio were attained that would still be quite satisfactory for most calculations.

Instead we present one example (Example 7-6-1) of an interesting phenomenon which appears, at first glance, to give the lie to the paragraph above. Examples of this type have generated a lot of unnecessary worry. In Example 7-6-1 the same reduction program was run on a nice 24 by 24 banded matrix on two different machines and produced quite different results; check α_6 and α_7 for example in Table 7-6-1. Is there a bug in the program, has one (or both) of the computers malfunctioned, or is the algorithm unstable? Surely reproducibility of results is essential to scientific work?

Indeed the two T's are different, but they are each similar to A to working precision and that is what counts in eigenvalue calculations. The **particular sequence** of similar matrices derived from this A by the algorithm is extremely sensitive to the tiny perturbations produced by roundoff. This instability is irrelevant. Indeed it is not always easy to distinguish the means from end and neither is it easy to judge algorithms.

The eigenvalues and eigenvectors of A were computed to as much accuracy as was warranted on each computer.

EXAMPLE 7-6-1 Sensitivity of the Tridiagonal Form to Roundoff Errors in the Reduction

$$A = \begin{bmatrix} M & C_2^* & & \\ C_2 & M & C_3^* & \\ & C_3 & M & C_4^* \\ & & C_4 & M \end{bmatrix}; \qquad M = \begin{bmatrix} H & B^* \\ B & H \end{bmatrix};$$

$$C_i = (e_6 e_1^* - e_1 e_6^*)/10^{2i}$$

$$H = \begin{bmatrix} 1.0 & 2.0 & 3.0 \\ 2.0 & 4.0 & 5.0 \\ 3.0 & 5.0 & 6.0 \end{bmatrix}; \qquad B = \begin{bmatrix} 0.0 & -1.0 & -0.5 \\ 1.0 & 0.0 & -0.33333330 \\ 0.5 & 0.33333330 & 0.0 \end{bmatrix};$$

$$t_{i,i} = \alpha_i; \qquad t_{i,i+1} = t_{i+1,i} = \beta_i.$$

Table 7-6-1

i	β_i		α_i	
	CDC 6400	PDP 11	CDC 6400	PDP 11
1	0.0000	−0.0080	0.4428	0.4428
2	0.0009	−0.0011	−0.8777	−0.8877
3	0.0141	0.1441	0.4429	0.4444
4	0.6169	1.5997	11.4038	11.2215
5	−0.0115	−0.0002	−0.8466	−0.6659
6	−0.0642	−4.3642	0.4431	9.2954
7	0.3520	−0.2957	11.4244	2.5251
8	−0.0007	−0.0002	−0.8676	−0.8205
9	0.5058	−1.3585	11.4115	11.2643
10	0.0114	0.0531	0.4661	0.6113
11	0.0004	−0.0006	−0.8776	−0.8756
12	0.0000	−1.6807	11.4348	11.1716
13	−0.0088	−0.0535	0.4429	0.7040
14	−0.0319	0.0317	−0.8775	−0.8755
15	−0.3579	−0.0118	11.4231	11.4348
16	0.0435	−0.0731	0.5431	−0.8737
17	−0.2052	−1.1474	−0.8728	0.5599
18	0.0001	−0.0000	11.4314	11.3138
19	−0.2107	−0.9605	0.4088	−0.7741
20	1.3226	2.2464	−0.6994	10.8806
21	−0.0000	−0.0000	11.2907	0.8936
22	−0.8784	−0.8784	−0.0453	−0.0453
23	−5.8618	−5.8618	5.0453	5.0453
24	—	—	6.0000	6.0000

(Mantissa: CDC—48 bits; PDP—24 bits.)
Values are given to four decimal places.

$$\max_i |\lambda_i(\mathsf{A})_{\text{CDC}} - \lambda_i(\mathsf{A})_{\text{PDP}}| \leqslant 9.\times 10^{-6} \leqslant \frac{1}{5} n\epsilon_{\text{PDP}}\|\mathsf{A}\|$$

Care was taken so that each machine worked on the same matrix.

7-7 EIGENVALUES ARE SIMPLE

There is little that is special about the eigenvalues of T but:

> LEMMA The eigenvalues of an unreduced T are distinct. (7-7-1)

Proof. For all ξ the minor of the $(1, n)$ element of $\mathsf{T} - \xi$ is $\beta_1\beta_2 \cdots \beta_{n-1} \neq 0$. Consequently $\text{rank}[\mathsf{T} - \xi] \geqslant n - 1$ and the dimension of the null space of $\mathsf{T} - \xi$ is either 0 (when ξ is not an eigenvalue) or 1 (when ξ is an eigenvalue). This means that there is only one linearly independent eigenvector to each eigenvalue. Since T has a full set of orthogonal eigenvectors, the multiplicity of each eigenvalue must therefore be one. □

This result is useful in theoretical work but it can mislead us into the false assumption that if a pair of eigenvalues is very close then some β_i must be small. A well-known counter-example is W_{21}^{+} (discussed in [Wilkinson, 1965]):

$$\begin{array}{llll} \alpha\text{'s:} & 10, 9, 8, \ldots, 1, 0, 1, \ldots, 10; & \beta_i = 1, & \text{all } i. \\ \lambda_{21}[\mathsf{W}^+] = & 10.74619\ 41829\ 0339\ \ldots. & & \\ \lambda_{20}[\mathsf{W}^+] = & 10.74619\ 41829\ 0332\ \ldots. & & \end{array}$$

7-8 ORTHOGONAL POLYNOMIALS

A big advance was made in linear algebra with the realization that the length of a vector is **not** an a priori, intrinsic property of that vector. In fact there are infinitely many legitimate inner product functions and each gives rise to its own version of length and angle. The numerical measure of the angle between two vectors is a matter of convention, not reality. It is the same with functions; there are many inner products.

Important in applied mathematics are real functions $\phi, \psi, \ldots$ of one real variable. We shall not consider the general integral inner products

$$(\phi, \psi) \equiv \int_a^b \omega(x)\phi(x)\psi(x)\,dx$$

but go straight to the discrete case

$$(\phi, \psi) \equiv \sum_{i=1}^{n} \omega_i \phi(\xi_i)\psi(\xi_i). \tag{7-8-1}$$

To each set of n distinct real numbers $\{\xi_1, \ldots, \xi_n\}$ and positive weights $\{\omega_1, \ldots, \omega_n : \omega_i > 0\}$ there corresponds one, and only one, inner product function as defined by (7-8-1). The corresponding norm and angle are given in Chap. 1.

Polynomials are rather special functions and for each inner product there is a **unique** family of monic orthogonal polynomials $\{\phi_0, \phi_1, \ldots, \phi_{n-1}\}$; that is, ϕ_j has degree j, leading coefficient 1, and $(\phi_j, \phi_k) = 0$ for $j \neq k$. This family is the distinguished basis of the inner product (or Hilbert) space of polynomials of degree less than n enriched with the given inner product.

Tridiagonal matrices come into the picture because there is a remarkable three-term recurrence relation among the ϕ's; for $j = 1, 2, \ldots, n-1$, and $\beta_0 = 0$,

$$\phi_j(\xi) = (\xi - \alpha_j)\phi_{j-1}(\xi) - \beta_{j-1}^2\phi_{j-2}(\xi). \qquad (7\text{-}8\text{-}2)$$

Once such a relationship has been guessed it is straightforward to verify what the α's and β's must be (see Ex. 7-8-1). These numbers may be put into an unreduced symmetric tridiagonal matrix T in the obvious way, as in (7-1-1), and then, for $j = 1, \ldots, n$, Ex. 7-8-2 reveals that

$$\phi_j(\xi) = \chi_j(\xi) \equiv \det[\xi - \mathsf{T}_{1,j}]. \qquad (7\text{-}8\text{-}3)$$

Thus the ξ's and the ω's determine a unique T. The question we pose now is how to determine the ξ's and ω's from a given T. In other words, which inner products make the χ_j, $j = 0, 1, \ldots, n$, mutually orthogonal?

> THEOREM Let $\mathsf{T} = \mathsf{S\Theta S^*}$ be the spectral decomposition of an unreduced T. Then the associated inner product of the form (7-8-1) is given by
>
> $$\xi_i = \theta_i, \qquad \omega_i = \gamma s_{1i}^2, \; i = 1, \ldots, n,$$
>
> for any positive γ; $\sum_1^n \omega_i = \gamma$. (7-8-4)

Proof. The essential observation is that S^*, not S, "reduces" Θ to tridiagonal form T and so, by the corollary to Theorem (7-3-1)

$$\mathsf{S^*e}_{j+1}(\beta_j \cdots \beta_1) = \chi_j(\Theta)\mathsf{S^*e}_1.$$

The orthogonality of columns $(j+1)$ and $(k+1)$ of S^* gives

$$0 = \sum_{i=1}^n (\chi_j(\theta_i)s_{1i})(\chi_k(\theta_i)s_{1i}),$$

$$= (\chi_j, \chi_k), \text{ if } \omega_i = s_{1i}^2 \text{ and } \xi_i = \theta_i \text{ in (7-8-1)}.$$

What other values for ω_i will work? The necessary condition $(\chi_0, \chi_k) = 0$ requires that

$$\sum_{i=1}^n \omega_i\chi_k(\theta_i) = 0, \qquad k = 1, 2, \ldots, n-1, \qquad (7\text{-}8\text{-}5)$$

whereas, for $k = 0$,

$$\sum_{i=1}^n \omega_i = \gamma = \text{arbitrary positive number}. \qquad (7\text{-}8\text{-}6)$$

However the Vandermonde-like matrix G whose (i, j) element is $\chi_{j-1}(\theta_i), j = 1, \ldots, n$, is nonsingular (Ex. 7-8-4) and so the system of linear equations specified by (7-8-5) and (7-8-6) has a unique solution $\{\omega_1, \ldots, \omega_n\}$ for each positive γ. □

It is customary to take $\gamma = 1$. In [Golub and Welsh, 1969] it is shown how to adapt the techniques of Chap. 8 to compute the s_{1i}, $i = 1, \ldots, n$ without finding all of S. See also Sec. 7-9.

Exercises on Sec. 7-8

7-8-1. By taking suitable inner products show that if η denotes the identity function $\eta(\xi) \equiv \xi$ then

$$\alpha_{j+1} = (\eta\phi_j, \phi_j)/\ (\phi_j, \phi_j),$$
$$\beta_j^2 = (\eta\phi_{j-1}, \phi_j)/\ (\phi_{j-1}, \phi_{j-1}).$$

7-8-2. Define $\chi_0(\xi) \equiv 1$ and expand $\det[\xi - \mathsf{T}_{1,j}]$ by its last row to discover that the χ_i also obeys (7-8-2). Hence show that $\phi_i = \chi_i$ for all i.

7-8-3. Let F be the matrix with $f_{ij} = \theta_i^{j-1}$. Show that F is nonsingular by using the fact that det[F] is a polynomial in the θ_i, $i = 1, \ldots, n$.

7-8-4. Let $\mathsf{G} = [\chi_{j-1}(\theta_i)]$. Show that G is nonsingular.

7-8-5. What is the inner product corresponding to the second difference matrix $[\cdots -1 \quad 2 \quad -1 \cdots]$ and what is the name of the associated set of polynomials?

7-9 EIGENVECTORS OF T

There is a remarkable formula for the square of any element of any normalized eigenvector of T and there are several beautiful relations among the elements of each eigenvector.

In order to derive these results we recall the notion of the (classical) **adjugate** of a matrix, namely the transpose of the matrix of cofactors. For example, if $|\mathsf{B}| = \det \mathsf{B}$, then

$$\operatorname{adj}\begin{bmatrix} 1 & 0 & 1 \\ 1 & 1 & 1 \\ 2 & -1 & 1 \end{bmatrix} = \begin{bmatrix} \begin{vmatrix} 1 & 1 \\ -1 & 1 \end{vmatrix}, & -\begin{vmatrix} 0 & 1 \\ -1 & 1 \end{vmatrix}, & \begin{vmatrix} 0 & 1 \\ 1 & 1 \end{vmatrix} \\ -\begin{vmatrix} 1 & 1 \\ 2 & 1 \end{vmatrix}, & \begin{vmatrix} 1 & 1 \\ 2 & 1 \end{vmatrix}, & -\begin{vmatrix} 1 & 1 \\ 1 & 1 \end{vmatrix}, \\ \begin{vmatrix} 1 & 1 \\ 2 & -1 \end{vmatrix}, & -\begin{vmatrix} 1 & 0 \\ 2 & -1 \end{vmatrix}, & \begin{vmatrix} 1 & 0 \\ 1 & 1 \end{vmatrix} \end{bmatrix}$$

$$= \begin{bmatrix} 2 & -1 & -1 \\ 1 & -1 & 0 \\ -3 & 1 & 1 \end{bmatrix}.$$

The importance of the adjugate stems from the famous Cauchy-Binet formula

$$\mathsf{B} \cdot \operatorname{adj}[\mathsf{B}] = \det[\mathsf{B}] \cdot \mathsf{I}.$$

We begin with a fact about all symmetric matrices which was proved in [Thompson and McEnteggert, 1968]. It employs the characteristic polynomial χ of A and relates certain adjugates to the spectral projectors $\mathsf{H}_i \equiv \mathsf{z}_i\mathsf{z}_i^*$ described in Chap. 1.

THEOREM Let $\mathsf{A} = \mathsf{Z}\Lambda\mathsf{Z}^*$ be the spectral decomposition of A; $\mathsf{Z} = (\mathsf{z}_1, \ldots, \mathsf{z}_n)$ is orthogonal, $\Lambda = \operatorname{diag}(\lambda_1, \ldots, \lambda_n)$. Then for $i = 1, \ldots, n$,

$$\operatorname{adj}[\lambda_i - \mathsf{A}] = \chi'(\lambda_i)\mathsf{z}_i\mathsf{z}_i^*, \tag{7-9-1}$$

where χ' is the derivative of χ.

Proof. For $\mu \neq \lambda_i$, $\mu - \mathsf{A}$ is invertible and so

$$\begin{aligned}\operatorname{adj}[\mu - \mathsf{A}] &= \det[\mu - \mathsf{A}](\mu - \mathsf{A})^{-1},\\ &= \chi(\mu)\mathsf{Z}(\mu - \Lambda)^{-1}\mathsf{Z}^*,\\ &= \mathsf{Z}\Delta(\mu)\mathsf{Z}^*, \qquad \Delta = \operatorname{diag}(\delta_1, \ldots, \delta_n),\end{aligned} \tag{7-9-2}$$

where

$$\delta_k = \delta_k(\mu) = \chi(\mu)/(\mu - \lambda_k) = \prod_{j\neq k}(\mu - \lambda_j).$$

The elements of adj[B] are sums of products of elements of B and thus continuous functions of them. On letting $\mu \longrightarrow \lambda_i$ both sides of (7-9-2) have limits. In particular

$$\delta_k(\lambda_i) = \begin{cases} 0, & k \neq i,\\ \chi'(\lambda_i), & k = i,\end{cases}$$

and the result follows. □

Before applying this result to T we recall the definition of $\mathsf{T}_{\mu,\nu}$ in (7-1-1) and define

$$\chi_{\mu,\nu}(\tau) = \begin{cases} \det[\tau - \mathsf{T}_{\mu,\nu}], & \mu \leqslant \nu,\\ 1, & \mu > \nu.\end{cases}$$

The following corollary of Theorem (7-9-1) was given in [Paige, 1971].

THEOREM Let $T = S\Theta S^*$ be the spectral decomposition of tridiagonal T; $S = (s_1, \ldots, s_n)$ is orthogonal, $\Theta = \mathrm{diag}(\theta_1, \ldots, \theta_n)$. Then, for $\mu \leq \nu$ and all j, the elements $s_{\mu j} (\equiv e_\mu^* s_j)$ of the normalized eigenvectors of T obey

$$\chi'_{1,n}(\theta_j) s_{\mu j} s_{\nu j} = \chi_{1,\mu-1}(\theta_j)\beta_\mu \cdots \beta_{\nu-1}\chi_{\nu+1,n}(\theta_j). \tag{7-9-3a}$$

In particular, when θ_j is a simple eigenvalue (T unreduced),

$$s_{\mu j}^2 = \chi_{1,\mu-1}(\theta_j)\chi_{\mu+1,n}(\theta_j)/\chi'_{1,n}(\theta_j). \tag{7-9-3b}$$

The proof is left as Ex. 7-9-3.

On giving special values to μ and ν several valuable relations among the elements of S become visible. Recall that $\chi(\theta) \equiv \chi_{1,n}(\theta)$.

COROLLARY For all $j \leq n$,

$$s_{1j}s_{nj}\chi'(\theta_j) = \beta_1\beta_2 \cdots \beta_{n-1} \tag{7-9-4a}$$

$$s_{1j}^2\chi'(\theta_j) = \chi_{2,n}(\theta_j) \tag{7-9-4b}$$

$$s_{nj}^2\chi'(\theta_j) = \chi_{1,n-1}(\theta_j) \tag{7-9-4c}$$

The next result is useful in establishing convergence for various iterative methods for computing eigenvectors.

THEOREM If T is unreduced then its eigenvector matrix has no zero elements in its first and last rows nor in the columns corresponding to the extreme eigenvalues. (7-9-5)

Proof. Since T is unreduced its eigenvalues are distinct by Lemma (7-7-1). Thus $\chi'(\theta_j) \neq 0$. By Corollary (7-9-4a) above, $s_{1j}s_{nj} = \Pi\beta_i/\chi'(\theta_j) \neq 0$. Now consider the two columns. By Cauchy's interlace theorem (Sec. 10-1) the zeros of $\chi_{1,n-1}$ and $\chi_{2,n}$ and **all** $\chi_{\mu,\nu}$ lie strictly between θ_1 and θ_n. By Paige's formula (7-9-3b) the corresponding eigenvectors have no zero elements. □

A picture of S for a random 25 by 25 T is given in Example 13-4-1.

EXAMPLE 7-9-1

An instructive illustration of (7-9-4b) is the case when $n = 100$, $\alpha_i = 0$, $\beta_i = 1$, for all i. The eigenvectors are the same as for the famous second difference matrix ($\alpha_i = -2$, $\beta_i = 1$, for all i) and $\chi_{1,n}(\tau)$ is related to the Tchebychev polynomial T_{n+1} described in Appendix *B*. Let $\omega = \pi/(n+1)$. In this example it is convenient to order the θ_j by $\theta_1 > \theta_2 > \cdots$

$$\text{Eigenvalues:} \qquad \theta_j = 2\cos j\omega, \qquad j = 1, \ldots, n$$

$$\text{Eigenvectors:} \qquad s_{ij} = \sqrt{\kappa}\,\sin ij\omega, \qquad \kappa = 2/(n+1).$$

$$\left[s_{1j} = (-1)^{j+1}s_{nj}; \qquad \text{hence } \chi_{2,n}(\theta_j) = (-1)^{j+1}\beta_1 \cdots \beta_{n-1} = (-1)^{j+1}\right]$$

In fact $\chi(\theta) = \kappa T'_{101}(\theta/2)$.

j	1	2	49	50
θ_j	0.1999×10^1	0.1996×10^1	0.9328×10^{-1}	0.3110×10^{-1}
$\chi'(\theta_j)$	0.5221×10^5	-0.1307×10^5	0.5061×10^2	-0.5051×10^2
$s_{ij}^2 = s_{nj}^2$	0.1915×10^{-4}	0.7654×10^{-4}	0.1976×10^{-1}	0.1980×10^{-1}
$\chi_{2,100}(\theta_j)$	1.0	-1.0	1.0	-1.0

Note the large values for $\chi'(\theta_i)$, $i = 1, 2$, although θ_1 and θ_2 are close.

Exercises on Sec. 7-9

7-9-1. Compute adj$\begin{bmatrix} 2 & 1 & 0 \\ 1 & 2 & 1 \\ 0 & 1 & 2 \end{bmatrix}$ and adj$[\mathsf{T}_{1,3}]$ for a general T.

7-9-2. Show that $\sum_{i=1}^{n} \text{adj}[\lambda_i - \mathsf{A}] = \chi'(\mathsf{A})$.

7-9-3. Prove Paige's theorem (7-9-3) by evaluating the (μ, ν) element of adj $[\theta_j - \mathsf{T}]$.

7-9-4. Give the proper values to μ and ν to establish the corollary to Paige's theorem.

7-9-5. Compute the values of s_{1j} and s_{nj}, for all j, for

$$\begin{bmatrix} 2 & 1 \\ 1 & 1 \end{bmatrix}, \quad \begin{bmatrix} 10 & 1 & 0 \\ 1 & 0 & 1 \\ 0 & 1 & 10 \end{bmatrix}, \quad \begin{bmatrix} 10 & 1 & 0 \\ 1 & 0 & 1 \\ 0 & 1 & -10 \end{bmatrix}.$$

7-9-6. Compute $\chi'(\lambda_1)\mathsf{z}_1\mathsf{z}_1^*$ and adj$[\lambda_1 - \mathsf{A}]$ for $\mathsf{A} = \begin{bmatrix} 2 & 1 \\ 1 & 1 \end{bmatrix}$.

7-10 STURM SEQUENCES

For a given T the sequence of characteristic polynomials $\{\chi_j(\xi), j = 0, 1, \ldots, n\}$ defined in (7-8-3) is a **Sturm sequence**; that is, the zeros of χ_{j-1} interlace those of χ_j. This result is an application of the Cauchy interlace theorem (Sec. 10-1). An interesting property of Sturm sequences is that the number of sign agreements between consecutive terms of the numerical sequence $\{\chi_k(\xi), k = 0, 1, \ldots, n\}$, call it $\boldsymbol{\alpha}(\xi)$, equals the number of zeros of χ_n which are less than ξ. If $\xi < \eta$ then

$$\boldsymbol{\alpha}(\eta) - \boldsymbol{\alpha}(\xi)$$

is the number of eigenvalues of T in the half open interval $[\xi, \eta)$. Fig. 7-10-1 illustrates a Sturm sequence for a simple 3 by 3 matrix T.

The interlacing of the zeros is strict if T is unreduced but the matrix W_{21}^{-} introduced in Sec. 3-6 shows how weak the strict inequality can be. For W_{21}^{-} the largest zero of χ_{20} agrees with the largest zero of χ_{21} through the first 14 decimals.

By using the three-term recurrence, shown in Sec. 7-8, the sequence $\{\chi_k(\xi)\}$ can be evaluated at a cost of $2n$ multiplications. By the method of bisection (Sec. 3-5) eigenvalues can be approximated as accurately, or as crudely, as desired subject to the limitations of roundoff.

This ingenious technique was presented by Givens in 1954 and has been used extensively ever since. However the same information can be obtained by use of triangular factorization and Sylvester's inertia theorem, a method which is both more stable and has wider applicability. It is described in Chap. 3. Hence the computation of the Sturm sequence $\{\chi_j(\xi)\}$ by the three-term recurrence may seek honorable retirement after many years of valuable service.

The observant reader may have noticed that if θ is an eigenvalue of T then

$$[1, \chi_1(\theta)/\beta_1, \ldots, \chi_{n-1}(\theta)/(\beta_1\beta_2\cdots\beta_{n-1})]^* \qquad (7\text{-}10\text{-}1)$$

is an eigenvector of T (Ex. 7-10-2). However, in practice, when θ is merely a very accurate approximation the vector given above can be almost orthogonal to the true eigenvector. The technique is violently unstable. A detailed and illuminating discussion is given in [Wilkinson, 1965, p. 316]. The topic is developed in some of the following exercises. However the algorithm warrants further study because its storage demands are so modest.

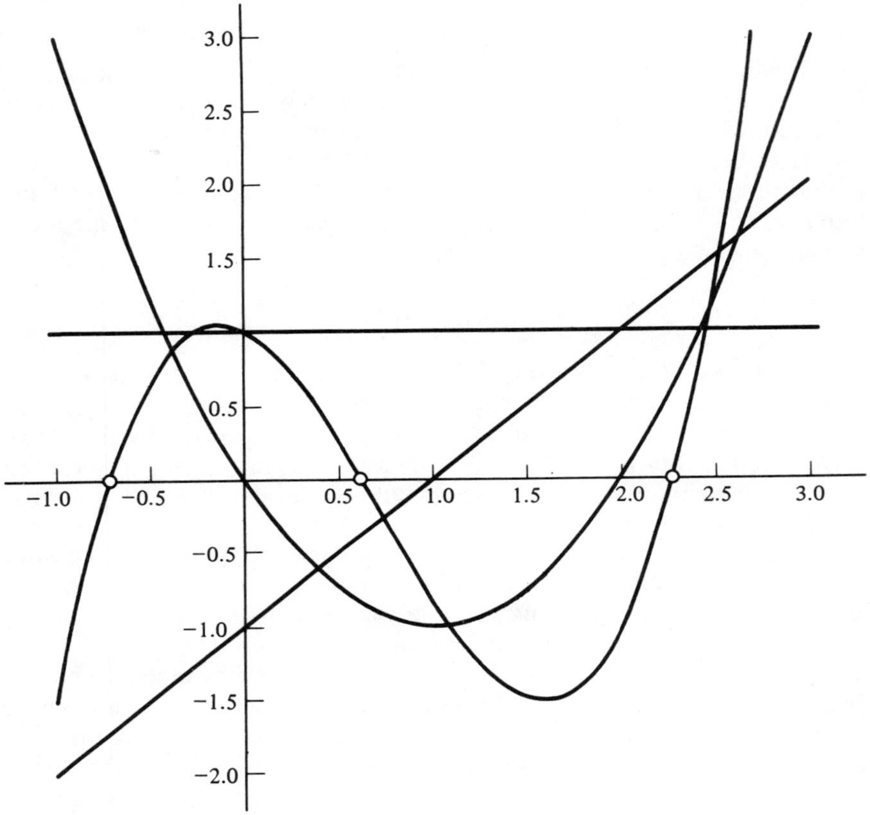

$$A = \begin{bmatrix} 1 & -1 & 0 \\ -1 & 1 & 1 \\ 0 & 1 & -\frac{1}{6} \end{bmatrix}$$

ξ	−1	0	1.5	3
χ_0	1	1	1	1
χ_1	−2	−1	0.5	2
χ_2	3	0	−0.75	3
χ_3	−0.5	1	−1.75	7.5
α	0	1	2	3

Figure 7-10-1 Sturm sequence

Exercises on Sec. 7-10

7-10-1. Show that if $\chi_j(\xi)$ had been defined as $\det[T_{1,j} - \xi]$ then $\alpha(\xi)$ would give the number of eigenvalues of T greater than ξ. However χ_j would not be monic for odd j.

7-10-2. By examining the three-term recurrence for $\{\chi_j(\theta)\}$ show that (7-10-1) gives an eigenvector of T belonging to the eigenvalue θ.

7-10-3. Show that when θ is not an eigenvalue then the vector v given in (7-10-1) satisfies

$$(T - \theta)v = e_n\chi_n(\theta)/(-\beta_1 \cdots \beta_{n-1}).$$

7-10-4. Replace θ by $\theta + \delta\theta$ and expand each $\chi_j(\theta + \delta\theta)$ about θ to obtain an expression for the effect of perturbations in θ on (7-10-1).

7-10-5. Consider the tridiagonal matrix $(\cdots -1 \quad 2 \quad -1 \cdots)$ whose eigenvalues are given in Example 7-9-1 and hence find the change in the approximate vector for various eigenvalues (inner and outer) when $\delta\theta = 10^{-10}\theta$ and $n = 100$.

7-11 WHEN TO NEGLECT AN OFF-DIAGONAL ELEMENT

If $\beta_i = 0$ then T is completely reduced to $\mathrm{diag}(T_1, T_2)$ and the two submatrices T_1 and T_2 can be processed independently with some small gain in efficiency. Of more importance is the fact that some processes (the *QL* algorithm, for example) **require** that any such reductions by recognized; the process will be marred, if not ruined, by an unexpected zero value for a β.

In practice, therefore we must recognize when T is reduced "to within working accuracy." The treatment of this problem is the least satisfactory aspect of the computation of eigenvalues of symmetric tridiagonal matrices. It is useful to distinguish some distinct but related questions.

7-11-1 THE MATHEMATICAL PROBLEM

Find an upper bound, in terms of T's elements, on the change $\delta\theta$ induced in any eigenvalue θ when β_i is replaced by zero.

Simple Answer: $|\delta\theta_j| \leqslant |\beta_i|$, for all j (Ex. 7-11-1).

Wrong Answer: $|\delta\theta_j| \leqslant \beta_i^2/\min|\alpha_\mu - \alpha_\nu|$ over all $\alpha_\mu \neq \alpha_\nu$.

Example:

$$\mathsf{T} = \begin{bmatrix} 1 & \sqrt{2} & 0 \\ \sqrt{2} & 2 & \beta \\ 0 & \beta & 0 \end{bmatrix}$$

Eigenvalues

$3, \pm\beta/\sqrt{3} + O(\beta^2)$,
when β is tiny,
$3, \pm 0$, when $\beta = 0$.

Complicated Answer: Take $\beta_0 = \beta_n = 0$ and define

$$\delta_i \equiv (\alpha_{i+1} - \alpha_i)/2, \quad \rho_i^2 \equiv (1 - 1/\sqrt{2})(\beta_{i-1}^2 + \beta_{i+1}^2). \qquad (7\text{-}11\text{-}1)$$

Then

$$\sum_{j=1}^{n} (\delta\theta_j)^2 \leqslant \frac{\beta_i^2}{\delta_i^2 + \rho_i^2}\left[2\rho_i^2 + \frac{\delta_i^2\beta_i^2}{\delta_i^2 + \rho_i^2}\right] \equiv \omega_i^2. \qquad (7\text{-}11\text{-}2)$$

The proof is given in [Kahan, 1966]. He applies the Wielandt-Hoffman theorem (see Sec. 1-4) to the result of performing a rotation in the $(i, i+1)$ plane. The bound is then minimized as a function of the angle of rotation.

There are several ways to weaken this bound in the interests of simplicity. See Ex. 7-11-3.

The simple bound $|\beta_i|$ is far too crude in most applications.

EXAMPLE 7-11-1

For $\mathsf{T} = \begin{bmatrix} 3 & 10^{-5} & 0 \\ 10^{-5} & 2 & 10^{-5} \\ 0 & 10^{-5} & 1 \end{bmatrix}$ the wrong answer given above is correct.

7-11-2 ALGORITHMIC PROBLEM NO. 1

Find an inexpensive criterion for neglecting β_i given a tolerance on the permitted disturbance to an eigenvalue. If ϵ is the relative precision of the basic arithmetic operations then the tolerance may be absolute, $\epsilon\|\mathsf{A}\|$, or relative $\epsilon(|\alpha_i| + |\alpha_{i+1}|)$.

7-11-3 ALGORITHMIC PROBLEM NO. 2

For those T which are derived from full A by similarity transformations, find an inexpensive criterion to decide when suppression of β_i is "equivalent to" the roundoff errors already made in obtaining T. In other words, ensure that T is exactly similar to a matrix which is acceptably close to A.

The desire for a simple test arises from the fact that the *QL* algorithm can transform T into a more nearly diagonal $\hat{T}$ at a cost of approximately $11n$ multiplications and n square roots. Our test will be applied to all $(n - 1)$ β's and will nearly always fail.

Here is the quandary: The simple test $\beta_i < \epsilon\|T\|$ will miss β_i which should be neglected and will occasionally degrade the *QL* algorithm; yet a test which costs as little as 5 multiplications will increase execution time of this part of the computation significantly.

The programmer's escape from this quandary is a two-level test:

0. . . .
1. If $|\beta_i| \geqslant \sqrt{\epsilon}\,\|T\|$ then go to 4.
2. If $|\beta_i| \geqslant$ favorite test then go to 4.
3. Set $\beta_i = 0$.
4. . . . .

Several practical tests are suggested in the exercises.

7-11-4 THE LAST OFF-DIAGONAL ELEMENT

In the *QL* algorithm the last β to be calculated is β_1 and the case for using a refined test seems strong because the goal of the algorithm is to make β_1 negligible.

In this section we have pointed to some problems but have offered no clear-cut solution and we must report that the 1977 versions of the EISPACK programs use a simple test exclusively; $|\beta_i| < \epsilon(|\alpha_i| + |\alpha_{i+1}|)$.

Exercises on Sec. 7-11

7-11-1. Apply the Weyl monotonicity theorem (Sec. 10-3) to show that the change $\delta\theta$ in any eigenvalue, when β_i is replaced by 0, satisfies

$$|\delta\theta| < |\beta_i|.$$

Hint: A special rank two matrix is subtracted from T.

7-11-2. Find the eigenvalues of

$$\begin{bmatrix} 1 & \sqrt{2} & 0 \\ \sqrt{2} & 2 & \beta \\ 0 & \beta & 0 \end{bmatrix}$$

as functions of β.

7-11-3. Using the notation of (7-11-1) and (7-11-2) show that

$$\omega_i^2 < (\beta_{i-1}^2 + \beta_i^2 + \beta_{i+1}^2)\beta_i^2/\delta_i^2.$$

7-11-4. Apply a Jacobi rotation to annihilate β_i. This introduces $-s\beta_{i-1}$ into element $(i+1, i-1)$ and $s\beta_{i+1}$ into $(i+2, i)$, where $s = \sin\theta$. Hence show that $|\beta_i|(|\beta_{i-1}| + |\beta_{i+1}|) < \epsilon|\delta_i|(|\alpha_i| + |\alpha_{i+1}|)$ is a reasonable criterion for neglecting β_i.

7-12 INVERSE EIGENVALUE PROBLEMS

The fundamental note of a uniform vibrating string, clamped at each end, is related to the smallest eigenvalue λ of the differential equation $-\phi'' + \sigma\phi = \lambda\phi$ with boundary conditions $\phi(a) = \phi(b) = 0$. The higher harmonics (overtones) are related to the other eigenvalues. If the string is defective and uneven then the constant σ must be replaced by a function $\sigma(\xi)$ which is related to the density of the string. The eigenvalues of the new problem will depend on σ and the inverse problem asks whether it is possible to locate the defect, i.e., find $\sigma(\xi)$, when all the eigenvalues are known. The answer is that, under certain conditions and with enough extra information, it can be done. One ambitious project in the same vein hopes to determine properties of the earth's core from extensive monitoring of seismic activity at the surface.

When the differential equation is discretized it yields a tridiagonal matrix and the discrete inverse problem is to find T when given its eigenvalues $\theta_1, \theta_2, \ldots, \theta_n$ and some extra information. The previous sections have given enough understanding to solve this problem in a number of important cases.

We have written the spectral decomposition of T as

$$\mathsf{T} = \mathsf{S}\Theta\mathsf{S}^* = (\mathsf{S}^*)^*\Theta\mathsf{S}^* \tag{7-12-1}$$

where S is the matrix of normalized eigenvectors. If Θ is given then the inverse problem can be regarded as the reduction (perhaps we should say expansion) of Θ to tridiagonal form. By the uniqueness of the reduction [Theorem (7-2-2)] T and S^* are determined by $\mathsf{S}^*\mathbf{e}_1$ and Θ. Sections 7-4 and 7-5 as well as the proof of Theorem (7-2-2), describe ways to carry out the reduction once $\mathsf{S}^*\mathbf{e}_1$ is in hand; namely the methods of Secs. 7-4 and 7-5 may be applied to $\mathsf{P}^*\Theta\mathsf{P}$ where P is any orthogonal matrix whose first column is $\mathsf{S}^*\mathbf{e}_1$. All that remains is to find the $s_{1j}, j = 1, \ldots, n$.

Case 1: Symmetry about the Midpoint. Suppose that T must be symmetric about the secondary diagonal (top right-bottom left) as well as the main diagonal, i.e.,

$$\alpha_j = \alpha_{n+1-j}, \qquad \beta_j = \beta_{n-j} > 0, \qquad j = 1, \ldots, [n/2].$$

This condition corresponds to a string and associated function $\sigma(\xi)$ which are symmetric about the middle of the interval. The double symmetry of T implies (Ex. 7-12-2) that $s_{1j} = (-1)^{n-j} s_{nj}, j = 1, \ldots, n$. Now use this in formula [7-9-4(a)].

$$s_{1j} s_{nj} \chi'(\theta_j) = \beta_1 \beta_2 \cdots \beta_{n-1} \equiv \beta, \qquad j = 1, \ldots, n, \quad (7\text{-}12\text{-}2)$$

where χ is the characteristic polynomial of both T and Θ. Thus $\chi'(\theta_j) = \prod_{i=1}^{n} (\theta_j - \theta_i), i \neq j$, and is determined when Θ is given. Moreover β is determined by the normalization condition,

$$1 = \sum_{j=1}^{n} s_{1j}^2 = \beta \sum_{j=1}^{n} 1/|\chi'(\theta_j)|, \quad (7\text{-}12\text{-}3)$$

and so the starting vector $\mathbf{e}_1^* \mathbf{S}$ is the vector $(|\chi'(\theta_1)|^{-1/2}, |\chi'(\theta_2)|^{-1/2}, \ldots, |\chi'(\theta_n)|^{-1/2})$ after normalization. There is no loss of generality in making the elements of $\mathbf{S}^*\mathbf{e}_1$ positive.

Case 2: Submatrix Spectrum. Suppose that, in addition to Θ, the eigenvalues of the $T_{2,n}$ are given. These values $\mu_i, i = 1, \ldots, n-1$ cannot be arbitrary but, by the Cauchy interlace theorem (Sec. 10-1), must interlace the θ's, i.e.,

$$\theta_1 < \mu_1 < \theta_2 < \cdots < \mu_{n-1} < \theta_n.$$

Formula (7-9-4b) gives

$$s_{1j}^2 \chi'(\theta_j) = \chi_{2,n}(\theta_j), \qquad j = 1, \ldots, n,$$

where $\chi_{2,n}(\theta_j) = \prod_{i=1}^{n-1} (\theta_j - \mu_i)$ and is computable. As before $\chi'(\theta_j) = \prod_{i=1}^{n} (\theta_j - \theta_i), i \neq j$, and the interlace condition guarantees that $\chi_{2,n}/\chi'$ is positive at the θ's. Consequently $|s_{1j}|$ is determined for $j = 1, \ldots, n$, and a unique T can be constructed with positive β's.

Case 2 corresponds to the physical problem in which the discretized string is clamped at the first interior mesh point and the new frequencies then computed.

Example 7-12-1 gives the result for Case 2 on a small and easy problem.

EXAMPLE 7-12-1

θ's $\{\frac{1}{5}, \frac{1}{4}, \frac{1}{3}, \frac{1}{2}, 1\}$.

μ's $\{\frac{9}{40}, \frac{7}{24}, \frac{5}{12}, \frac{3}{4}\}$.

s_{1j} $\{0.633, 0.437, 0.356, 0.333, 0.413\}$.

α's $\{0.6, 0.66326, 0.42933, 0.32508, 0.26566\}$.

β's $\{0.34046, 0.16216, 0.085823, 0.044658\}$.

The extent to which the continuous function $\sigma(\xi)$ is determined by the eigenvalues of the equation $-\phi'' + \sigma\phi = \lambda\phi$, where $\phi(a) = \phi(b) = 0$, is a much more subtle and interesting problem. The reader is referred to [Hald, 1977] and [Hochstadt, 1975]. Both authors also study the matrix cases given above but they use less stable algorithms for computing T.

Our technique of reducing the matrix $P^*\Theta P$ to tridiagonal form is both simple in concept and stable in action, but it is not very efficient for large problems. The latest and best techniques are described in [De Boor and Golub, 1979].

Exercises on Sec. 7-12

7-12-1. Show that $\chi'(\theta_j) = \prod_i (\theta_j - \theta_i)$, $i \neq j$, for each zero θ_j of the polynomial χ.

7-12-2. Let $\tilde{\mathsf{I}} = (\mathbf{e}_n, \ldots, \mathbf{e}_1)$ and suppose $\mathsf{T} = \tilde{\mathsf{I}}\mathsf{T}\tilde{\mathsf{I}}$. Use the fact that eigenvectors are determined up to a scalar multiple to show that $s_{1j} = \pm s_{nj}$, $j = 1, \ldots, n$. Show that if $\beta_i > 0$ for all i then $s_{1j} = s_{nj}(-1)^{n-j}$.

7-12-3. Project. Consider the storage requirements for the various ways of computing T when Θ and $\mathsf{S}^*\mathbf{e}_1$ are given. Is it necessary to keep an n by n array in fast storage?

7-12-4. Project. Write a program to compute $\mathsf{H}\Theta\mathsf{H}$ where $\mathsf{H} = \mathsf{H}(\mathbf{w})$, $\mathbf{w} = \mathbf{f} \pm \mathbf{e}_1$, where $\mathbf{f} = \mathsf{S}^*\mathbf{e}_1$. Then call the local library subroutines to reduce $\mathsf{H}\Theta\mathsf{H}$ to T and compute T's eigenvalues and compare them with the given θ_i. Try various sets of interlacing μ's for each set of θ's.

Notes and References

The goal of reducing A to tridiagonal form T instead of finding the coefficients of χ_A is found in two seminal reports [Lanczos, 1950] and [Givens, 1954]. The transformation is described in several books: [Wilkinson, 1965], [Householder, 1964], [Stewart, 1973], [Fox, 1964] to name a few.

The formulas for the eigenvectors of T in terms of the eigenvalues were used in [Paige, 1971] but are not well known. On the other hand the connection of T with orthogonal polynomials is standard material in analysis [Szegö, 1939] and [Golub and Welsch, 1969]. The criteria for neglecting off-diagonal terms come from the unpublished report [Kahan, 1966].

There is an extensive literature on inverse problems for differential equations. Only since 1975 have there been stable and efficient methods for computing T's with given spectral properties.

8

The QL and QR Algorithms

8-1 INTRODUCTION

Although the QL and QR algorithms were not conceived before 1958 they have emerged as the most effective way of finding all the eigenvalues of a small symmetric matrix. A full matrix is first reduced to tridiagonal form by a sequence of reflections (Sec. 7-4) and then the QL algorithm swiftly reduces the off-diagonal elements until they are negligible. The algorithm repeatedly applies a complicated similarity transformation to the result of the previous transformation, thereby producing a sequence of matrices that converges to a diagonal matrix. What is more, the tridiagonal form is preserved.

The transformation is based on the orthogonal-triangular decomposition of a matrix B say, and this decomposition is the matrix formulation of the Gram-Schmidt orthonormalization process (Sec. 6-6) applied to the columns of B. If the columns are taken in the order $\mathsf{b}_1, \mathsf{b}_2, \ldots, \mathsf{b}_n$ then the factorization is $\mathsf{B} = \mathsf{Q}_R\mathsf{R}$ where R is right (or upper) triangular and $\mathsf{Q}_R^* = \mathsf{Q}_R^{-1}$. If the columns are taken in the reverse order $\mathsf{b}_n, \mathsf{b}_{n-1}, \ldots, \mathsf{b}_1$ then the result is $\mathsf{B} = \mathsf{Q}_L\mathsf{L}$ where L is left (or lower) triangular and $\mathsf{Q}_L^* = \mathsf{Q}_L^{-1}$.

A strong psychological deterrent to the discovery of the algorithms must have been the apparently high cost of this factorization. However the orthonormal matrices Q_L and Q_R which loom large in the rest of this chapter will never be formed explicitly. The QL and QR transformations are defined for any square matrix and so we forsake symmetric matrices for those sections which give the formal definitions and basic properties. Next comes the relation to other, simpler methods together with the very satisfactory convergence theory. Finally there is a discussion of various implementations.

8-2 THE QL TRANSFORMATION

Given B and a scalar σ, called the **origin shift**, consider the orthogonal-lower triangular factorization

$$\mathsf{B} - \sigma = \mathsf{QL}. \tag{8-2-1}$$

From Q, L, and σ define $\hat{\mathsf{B}}$, the **QL transform** of B, by

$$\begin{aligned} \hat{\mathsf{B}} &\equiv \mathsf{LQ} + \sigma, \\ &= \mathsf{Q}^*(\mathsf{B} - \sigma)\mathsf{Q} + \sigma, \qquad \text{using (8-2-1)}, \\ &= \mathsf{Q}^*\mathsf{BQ}, \qquad \text{since } \mathsf{Q}^* = \mathsf{Q}^{-1}. \end{aligned} \tag{8-2-2}$$

It is not clear that $\hat{\mathsf{B}}$ is any improvement over B; no zero elements have been created; nevertheless the shifted QL algorithm doggedly iterates the transformation $\mathsf{B} \longrightarrow \hat{\mathsf{B}}$.

8-2-1 QL ALGORITHM

For $k = 1, 2, \ldots$ and given $\{\sigma_k\}$, to be discussed later,

1. Let $\mathsf{Q}_k\mathsf{L}_k$ be the QL factorization of $\mathsf{B}_k - \sigma_k$.
2. Define $\mathsf{B}_{k+1} \equiv \mathsf{Q}_k^*\mathsf{B}_k\mathsf{Q}_k$.

The matrices B_k are all orthogonally similar to each other. The relation of B_{k+1} to B_1 is

$$\begin{aligned} \mathsf{B}_{k+1} &= \mathsf{Q}_k^*\mathsf{B}_k\mathsf{Q}_k \\ &= \mathsf{Q}_k^*(\mathsf{Q}_{k-1}^*\mathsf{B}_{k-1}\mathsf{Q}_{k-1})\mathsf{Q}_k \\ &= \cdots \\ &= \mathsf{P}_k^*\mathsf{B}_1\mathsf{P}_k, \end{aligned} \tag{8-2-3}$$

where

$$\mathsf{P}_k \equiv \mathsf{Q}_1\mathsf{Q}_2 \cdots \mathsf{Q}_k \tag{8-2-4}$$

$$
\mathsf{B}_1 - \sigma_1 = \begin{bmatrix} 6.0 & 0.6 & 0 \\ 1.0 & 4.0 & 2.0 \\ -3.0 & -0.8 & 0 \end{bmatrix} = \begin{bmatrix} 0.8 & 0.6 & 0 \\ 0 & 0 & 1.0 \\ 0.6 & -0.8 & 0 \end{bmatrix} \begin{bmatrix} 3.0 & & \\ 6.0 & 1.0 & \\ 1.0 & 4.0 & 2.0 \end{bmatrix}
$$

$$
\mathsf{B}_2 - \sigma_1 = \begin{bmatrix} 2.4 & 1.8 & 0 \\ 4.8 & 3.6 & 1.0 \\ 2.0 & -1.0 & 4.0 \end{bmatrix} = \begin{bmatrix} 3.0 & & \\ 6.0 & 1.0 & \\ 1.0 & 4.0 & 2.0 \end{bmatrix} \begin{bmatrix} 0.8 & 0.6 & 0 \\ 0 & 0 & 1.0 \\ 0.6 & -0.8 & 0 \end{bmatrix}
$$

Figure 8-2-1 One step of QL

is a product of orthonormal matrices and hence orthonormal.

The QR transformation is similar but QR replaces QL in (8-2-1). Of course, the new Q and the new $\hat{\mathsf{B}}$ will differ from the ones given in (8-2-2).

Exercises on Sec. 8-2

8-2-1. Show that $\mathsf{P}_k\mathsf{L}_k = (\mathsf{B}_1 - \sigma)\mathsf{P}_{k-1}$ and hence that $\mathsf{P}_k\mathsf{e}_n$ is a normalized version of $(\mathsf{B}_1 - \sigma)\mathsf{P}_{k-1}\mathsf{e}_n$.

8-2-2. Show that if $\sigma_k = 0$ for all k then

$$\mathsf{P}_k(\mathsf{L}_k \cdots \mathsf{L}_2\mathsf{L}_1) = \mathsf{B}_1^k.$$

8-2-3. Perform one step of the QL algorithm with $\sigma_1 = 0$ on

$$\mathsf{B}_1 = \begin{bmatrix} 2 & 2 \\ 1 & 1 \end{bmatrix}.$$

8-2-4. Perform one step of the QR algorithm on the example of Fig. 8-2-1.

8-3 PRESERVATION OF BANDWIDTH

If $\mathsf{B} = \mathsf{A} = \mathsf{A}^*$ then symmetry is preserved because the transformations are congruences as well as similarities. The fact that bandwidth is also preserved can be seen from Fig. 8-3-1.

The zero elements **above** the diagonal of $\hat{\mathsf{A}}$ are preserved simply by the shape of Q and L; the elements **below** the diagonal must be zero, through cancellation, since $\hat{\mathsf{A}}$ is symmetric.

The preservation of tridiagonal form is very valuable. The fact that such a complicated and far-from-sparse Q_k can preserve the tridiagonal form is quite surprising.

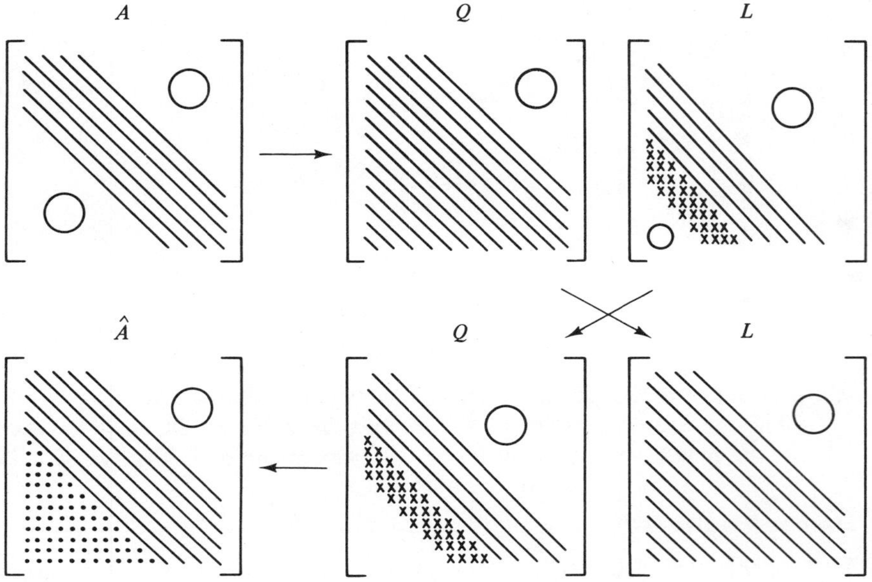

Key: x Nonzero elements

• Elements which vanish through cancellation

Figure 8-3-1 Preservation of bandwidth

Exercises on Sec. 8-3

8-3-1. Perform one step of QL with $\sigma_1 = 0$ on

$$\mathsf{A} = \begin{bmatrix} 0 & 1 & 0 \\ 1 & 0 & 1 \\ 0 & 1 & 1 \end{bmatrix}.$$

Exhibit the cancellation in the (3, 1) element of $\hat{\mathsf{A}}$.

8-3-2. Show that if $\mathsf{b}_{ij} = 0$ for $i < j - 1$ then, under the QL transform $\hat{b}_{ij} = 0$ for $i < j - 1$.

8-4 RELATION BETWEEN QL AND QR

The QL algorithm is simply a reorganization of the original QR algorithm and the two methods need not be distinguished (from each other) for theoretical purposes.

To give the formal relationship it is convenient to denote by $\mathsf{Q}_R[\mathsf{B}]$, the result of orthonormalizing the columns of any square B in the order $\mathsf{b}_1, \mathsf{b}_2, \ldots, \mathsf{b}_n$, while $\mathsf{Q}_L[\mathsf{B}]$ comes from the reverse order $\mathsf{b}_n, \mathsf{b}_{n-1}, \ldots, \mathsf{b}_1$. When there is no ambiguity concerning the matrix B we write simply Q_R

or Q_L. Also needed is the matrix $\tilde{\mathsf{I}}$ obtained by reversing the order of the columns of the identity matrix. Note that $\tilde{\mathsf{I}}^* = \tilde{\mathsf{I}}^{-1} = \tilde{\mathsf{I}}$.

There is no trivial relationship between the orthonormal matrices $\mathsf{Q}_R[\mathsf{B}]$ and $\mathsf{Q}_L[\mathsf{B}]$. On the other hand, when B is invertible,

LEMMA $\quad \mathsf{Q}_L[\tilde{\mathsf{I}}\mathsf{B}\tilde{\mathsf{I}}] = \tilde{\mathsf{I}}\mathsf{Q}_R[\mathsf{B}]\tilde{\mathsf{I}}. \qquad$ (8-4-1)

Proof.

$$\tilde{\mathsf{I}}\mathsf{B}\tilde{\mathsf{I}} = \tilde{\mathsf{I}}\mathsf{Q}_R[\mathsf{B}]\mathsf{R}_B\tilde{\mathsf{I}},$$

$$= (\tilde{\mathsf{I}}\mathsf{Q}_R[\mathsf{B}]\tilde{\mathsf{I}})(\tilde{\mathsf{I}}\mathsf{R}_B\tilde{\mathsf{I}}). \qquad (8\text{-}4\text{-}2)$$

Since R_B is upper triangular the second line of (8-4-2) gives the unique QL factorization of $\tilde{\mathsf{I}}\mathsf{B}\tilde{\mathsf{I}}$ as claimed. Uniqueness is proved in Sec. 6-6. □

Let us return to the two transformations

$$\text{QL:} \quad \mathsf{B} \longrightarrow \mathsf{Q}_L^*\mathsf{B}\mathsf{Q}_L, \qquad \mathsf{Q}_L = \mathsf{Q}_L[\mathsf{B} - \sigma],$$

$$\text{QR:} \quad \mathsf{B} \longrightarrow \mathsf{Q}_R^*\mathsf{B}\mathsf{Q}_R, \qquad \mathsf{Q}_R = \mathsf{Q}_R[\mathsf{B} - \sigma].$$

LEMMA Let $\{\mathsf{B}_k^L\}$ and $\{\mathsf{B}_k^R\}$ be sequences generated by the QL and QR algorithms using the same sequence of origin shifts. If $\mathsf{B}_1^R = \tilde{\mathsf{I}}\mathsf{B}_1^L\tilde{\mathsf{I}}$ then

$$\mathsf{B}_k^R = \tilde{\mathsf{I}}\mathsf{B}_k^L\tilde{\mathsf{I}}, \qquad \text{for all } k > 1. \qquad (8\text{-}4\text{-}3)$$

The proof is left as Ex. 8-4-1.

If B_k^L tends to lower-triangular form, as $k \longrightarrow \infty$, with the smaller eigenvalues (in absolute value) at the top then B_k^R tends to upper-triangular form with the smaller eigenvalues at the bottom. For symmetric matrices both the QL and the QR algorithm converge to diagonal form. The sole reason for introducing the QL algorithm is that certain types of problem yield **graded** matrices in which the smaller elements are already at the top of the matrix and the bigger ones at the bottom as indicated in Sec. 8-13. Matrices of this form can arise from the generalized problem $\mathsf{A} - \lambda\mathsf{M}$ when M is ill conditioned but positive definite. The lower-triangular Choleski factor C satisfying $\mathsf{C}\mathsf{C}^* = \mathsf{M}$ often has columns which shrink steadily,

$$\|\mathsf{c}_1\| > \|\mathsf{c}_2\| > \cdots > \|\mathsf{c}_n\|$$

and the reduced matrix $\mathsf{C}^{-1}\mathsf{A}\mathsf{C}^{-*}$ will be graded like the one in Sec. 8-13.

Exercise on Sec 8-4

8-4-1. Prove Lemma 8-4-3.

8-5 QL, THE POWER METHOD, AND INVERSE ITERATION

Now we link QL to the more familiar processes described in Chap. 4. Given A and any sequence $\{\sigma_k, k = 1, 2, \cdots\}$ of origin shifts the three algorithms of the section heading produce three different sequences.

QL (with $\mathsf{A}_1 = \mathsf{A}$): $\{\mathsf{A}_k\}$ where $\mathsf{A}_{k+1} = \mathsf{Q}_k^*\mathsf{A}_k\mathsf{Q}_k$, and $\mathsf{A}_k - \sigma_k = \mathsf{Q}_k\mathsf{L}_k$.

PM (v_1 any unit vector): $\{\mathsf{v}_k\}$ where $\mathsf{v}_{k+1} = (\mathsf{A} - \sigma_k)\mathsf{v}_k/\nu_k$, and $\|\mathsf{v}_k\| = 1$.

INVIT (u_1 any unit vector): $\{\mathsf{u}_k\}$ where $(\mathsf{A} - \sigma_k)\mathsf{u}_{k+1} = \mathsf{u}_k\tau_k$, and $\|\mathsf{u}_k\| = 1$.

It is interesting that if v_1 and u_1 are chosen appropriately then the three sequences are intimately related. What connects them is the matrix introduced in Sec. 8-2 to connect A_{k+1} to A, namely

$$\mathsf{P}_k = \mathsf{Q}_1\mathsf{Q}_2\cdots\mathsf{Q}_k, \qquad \mathsf{P}_0 = \mathsf{I}, \tag{8-5-1}$$

Recall, from (8-2-3), that $\mathsf{A}_{k+1} = \mathsf{P}_k^*\mathsf{A}\mathsf{P}_k$.

THEOREM Assume that no shift σ_k is an eigenvalue of A. If $\mathsf{u}_1 = \mathsf{e}_1$ and $\mathsf{v}_1 = \mathsf{e}_n$ then, for $k \geqslant 1$, $\mathsf{u}_k = \mathsf{P}_{k-1}\mathsf{e}_1$ and $\mathsf{v}_k = \mathsf{P}_{k-1}\mathsf{e}_n$. (8-5-2)

Recall that if A is tridiagonal then so are all the A_k's and moreover A_{k+1} is uniquely determined by A_k together with either $\mathsf{P}_k\mathsf{e}_1$ or $\mathsf{P}_k\mathsf{e}_n$ (Sec. 7-2).

Proof. Rewrite $\mathsf{A}_{k+1} = \mathsf{P}_k^*\mathsf{A}\mathsf{P}_k$ as

$$\mathsf{P}_{k-1}\mathsf{A}_k = \mathsf{A}\mathsf{P}_{k-1}. \tag{8-5-3}$$

Now

$$\begin{aligned}\mathsf{P}_k\mathsf{L}_k &= \mathsf{P}_{k-1}\mathsf{Q}_k\mathsf{L}_k, \quad \text{by (8-5-1)}\\ &= \mathsf{P}_{k-1}(\mathsf{A}_k - \sigma_k), \quad \text{by QL,}\\ &= (\mathsf{A} - \sigma_k)\mathsf{P}_{k-1}, \quad \text{by (8-5-3).}\end{aligned} \tag{8-5-4}$$

Since L_k is lower triangular, equate the last columns of (8-5-4):

$$\mathsf{P}_k\mathsf{e}_n = (\mathsf{A} - \sigma_k)\mathsf{P}_{k-1}\mathsf{e}_n/\nu_k, \qquad \nu_k = \mathsf{e}_n^*\mathsf{L}_k\mathsf{e}_n. \tag{8-5-5}$$

Observe that (8-5-5) defines the power sequence, with shifts, generated by A and $\mathsf{P}_0\mathsf{e}_n$. So if, in algorithm PM above, $\mathsf{v}_1 = \mathsf{P}_0\mathsf{e}_n$ $(= \mathsf{e}_n)$ then $\mathsf{v}_{k+1} = \mathsf{P}_k\mathsf{e}_n$.

To obtain the analogous relation for u_k a little manipulation is needed to turn rows into columns. Transpose (8-5-4) to obtain

$$\mathsf{L}_k^*\mathsf{P}_k^* = \mathsf{P}_{k-1}^*(\mathsf{A} - \sigma_k).$$

Pre-multiply this by P_{k-1} and post-multiply by P_k to get

$$\mathsf{P}_{k-1}\mathsf{L}_k^* = (\mathsf{A} - \sigma_k)\mathsf{P}_k. \tag{8-5-6}$$

Since L_k^* is upper triangular,

$$\mathsf{P}_{k-1}\mathsf{e}_1\tau_k = (\mathsf{A} - \sigma_k)\mathsf{P}_k\mathsf{e}_1, \qquad \tau_k = \mathsf{e}_1^*\mathsf{L}_k^*\mathsf{e}_1. \tag{8-5-7}$$

Since $\mathsf{P}_0 = \mathsf{I}$ (8-5-7) defines inverse iteration, with shifts, started from e_1. So if, in INVIT above, $\mathsf{u}_1 = \mathsf{P}_0\mathsf{e}_1 = \mathsf{e}_1$ then $\mathsf{P}_k\mathsf{e}_1 = \mathsf{u}_{k+1}$. □

This result raises the question of how these different methods could ever come up with the same shifts σ_k in practice. That question is answered in Sec. 8-7.

8-6 CONVERGENCE OF THE BASIC QL ALGORITHM

When no origin shifts are used, i.e., $\sigma_k = 0$ for all k, then QL is called **basic** and is simultaneously linked to both direct and inverse iteration, without shifts. The convergence of basic QL can be seen as a corollary of the convergence properties of these two simple iterations.

> THEOREM Suppose that A's eigenvalues satisfy
>
> $$0 < |\lambda_1| < |\lambda_2| \leq \cdots \leq |\lambda_{n-1}| < |\lambda_n| = \|\mathsf{A}\|.$$
>
> Let z_i be the normalized eigenvector of λ_i, $i = 1, n$, and let $\{\mathsf{A}_k\}$ be the basic QL sequence derived from $\mathsf{A}_1 \equiv \mathsf{A}$. If $\mathsf{e}_i^*\mathsf{z}_i \neq 0$, $i = 1, n$, then as $k \longrightarrow \infty$.
>
> $$\mathsf{A}_k\mathsf{e}_1 \longrightarrow \mathsf{e}_1\lambda_1, \qquad \mathsf{A}_k\mathsf{e}_n \longrightarrow \mathsf{e}_n\lambda_n. \tag{8-6-1}$$

Proof. Let $\{\mathsf{u}_k\}$ be the inverse iteration sequence derived from $\mathsf{u}_1 = \mathsf{e}_1$ with no shifts. The hypotheses, by Theorem 4-2-3 (Corollary), ensure that $\mathsf{u}_k = \mathsf{z}_1 + 0(|\lambda_1/\lambda_2|^k)$ as $k \to \infty$. By Theorem (8-5-2)

$$\mathsf{e}_1 = \mathsf{P}_{k-1}^*(\mathsf{P}_{k-1}\mathsf{e}_1) = \mathsf{P}_{k-1}^*\mathsf{u}_k = \mathsf{P}_{k-1}^*\mathsf{z}_1 + O(|\lambda_1/\lambda_2|^k),$$

and, after using (8-2-3),

$$\begin{aligned} A_k e_1 &= P_{k-1}^* A u_k \\ &= P_{k-1}^* [z_1 \lambda_1 + O(|\lambda_1/\lambda_2|^k)], \\ &= e_1 \lambda_1 + O(|\lambda_1/\lambda_2|^k), \qquad k \longrightarrow \infty. \end{aligned}$$

Similarly, using the power sequence $\{v_k\}$, $A_k e_n \longrightarrow e_n \lambda_n$. □

There are two notions of convergence associated with the QL and QR algorithms. Strictly speaking, convergence would be taken to mean the convergence of the matrix sequence $\{A_k\}$ to some limit matrix. In practice, however, as soon as $\|A_k e_1 - e_1 a_{11}^{(k)}\|$ is negligible $a_{11}^{(k)}$ is taken as an eigenvalue and the computation then proceeds on the **submatrix** obtained by ignoring the first row and column. In other words, a stable form of deflation is built into the QL and QR algorithms, as discussed in Sec. 5-3. Hence the second notion simply concerns the convergence of the vector sequence $\{A_k e_1\}$ to a limit $e_1 \lambda$.

We shall employ the second meaning exclusively.

Theorem (8-6-1) is important but not exciting. For a full matrix each QL step is expensive (n^3 ops) and convergence is linear with unknown and often very poor convergence factors. The power of the practical algorithm comes from (a) preservation of bandwidth (reducing the cost of a step to $O(n)$ ops for tridiagonals) and (b) the use of origin shifts to reduce the number of steps.

As shown in Sec. 7-9 the hypotheses $e_1^* z_1 \neq 0$ and $e_n^* z_n \neq 0$ in Theorem (8-6-1) do hold for all unreduced definite tridiagonal matrices.

We now turn to the choice of origin shifts σ_k.

8-7 THE RAYLEIGH QUOTIENT SHIFT

The basic QL algorithm on an unreduced tridiagonal A converges in one step if A is singular. Thus shifts are chosen to approximate eigenvalues. The convergence of the basic algorithm ensures that $a_{11}^{(k)} = e_1^* A_k e_1 \longrightarrow \lambda_1$ as $k \longrightarrow \infty$, and so it seems reasonable to use $a_{11}^{(k)}$ as a shift **after** this element has settled down to some extent. However the shift strategy discussed below casts caution aside and uses $a_{11}^{(k)}$ as the shift right from the start. Formally,

$$\sigma_k = a_{11}^{(k)} = e_1^* A_k e_1, \qquad k = 1, 2, \cdots. \qquad (8\text{-}7\text{-}1)$$

Surprisingly it turns out that σ_k is the same Rayleigh quotient ρ_k used in RQI (Sec. 4-6). To distinguish RQI from INVIT, of which it is a special case, we write x_k instead of u_k.

> THEOREM If RQI is started with $\mathsf{x}_1 = \mathsf{e}_1$ and QL uses (8-7-1) then ρ_k $(\equiv \mathsf{x}_k^* \mathsf{A} \mathsf{x}_k) = \sigma_k$ for all k. (8-7-2)

Proof. Initially,

$$\sigma_1 = \mathsf{e}_1^* \mathsf{A}_1 \mathsf{e}_1 = \mathsf{x}_1^* \mathsf{A} \mathsf{x}_1 = \rho_1.$$

Assume, as induction hypothesis, that $\sigma_k = \rho_k$ for $k = 1, \ldots, j$. Then Theorem (8-5-2) shows that $\mathsf{P}_k \mathsf{e}_1 = \mathsf{x}_{k+1}$ for $k = 1, \ldots, j$. Hence

$$\sigma_{j+1} = \mathsf{e}_1^* \mathsf{A}_{j+1} \mathsf{e}_1 = \mathsf{e}_1^* \mathsf{P}_j^* \mathsf{A} \mathsf{P}_j \mathsf{e}_1 = \mathsf{x}_{j+1}^* \mathsf{A} \mathsf{x}_{j+1} = \rho_{j+1}.$$

By the principle of induction the result holds for all k. □

All the convergence properties of RQI (Secs. 4-7, 4-8, and 4-9) can now be translated into statements about QL with this particular shift.

In particular let $\mathsf{r}_k \equiv (\mathsf{A} - \rho_k)\mathsf{x}_k$ denote the residual vector of x_k and let QL, using (8-7-1), produce

$$\mathsf{A}_k = \begin{bmatrix} \alpha_k & \mathsf{b}_k^* \\ \mathsf{b}_k & \mathsf{M}_k \end{bmatrix}.$$

Then (Ex. 8-7-5), $\|\mathsf{b}_k\| = \|\mathsf{r}_k\|$. Further let $\phi_k = \angle(\mathsf{x}_k, \mathsf{z}_1)$, the error angle. If $\phi_k \longrightarrow 0$ as $k \longrightarrow \infty$, then (Ex. 8-7-3)

$$\|\mathsf{r}_k\| / |\sin \phi_k| \longrightarrow |\lambda_2 - \lambda_1|. \tag{8-7-3}$$

The ultimate cubic convergence of ϕ_k to zero (Theorem 4-7-3) entails the same behavior for $\|\mathsf{r}_k\|$ $(= \|\mathsf{b}_k\|)$. The usual proofs that $\|\mathsf{b}_{k+1}\| / \|\mathsf{b}_k\|^3 \longrightarrow$ constant as $k \longrightarrow \infty$ (assuming that $\|\mathsf{b}_k\| \longrightarrow 0$) are rather complicated.

Turning to the global behavior of QL we conclude that $\|\mathsf{b}_k\| \leqslant \|\mathsf{b}_{k-1}\|$ for **all** k (from Sec. 4-8) and $\|\mathsf{b}_k\| \longrightarrow 0$ for almost all A (Sec. 4-9). However there do exist nondiagonal matrices which are invariant under QL with the Rayleigh quotient shift. The simplest example is

$$\mathsf{A} = \begin{bmatrix} 0 & 1 \\ 1 & 0 \end{bmatrix}.$$

However the slightest perturbation of this A leads to convergence.

Although such matrices must exist no one has exhibited an A_1 such that, in exact arithmetic, the QL sequence $\{\mathsf{A}_k\}$, using (8-7-1), neither converges nor is stationary. The simple Rayleigh quotient shift strategy makes QL a powerful tool for diagonalizing matrices. In Sec. 8-9 we turn to a strategy which is slightly more complicated and even more satisfactory.

Exercises on Sec. 8-7

8-7-1. State and prove the analogue of Theorem (8-7-2) for the QR algorithm.

8-7-2. Compute A_2 when $A_1 = \begin{bmatrix} 10 & 1 & 0 \\ 1 & 20 & 2 \\ 0 & 2 & 30 \end{bmatrix}$ and $\sigma_1 = 10$.

8-7-3. Using the expansion $x_k = z_1 \cos\phi_k + w_k \sin\phi_k$, $z_1^* w_k = 0$, $\|w_k\| = 1$, show that

$$\|r_k\|^2 = \sin^2\phi_k\left[\|(A - \rho_k)w_k\|^2 + \tfrac{1}{4}(\lambda_1 - \rho(w_k))^2 \sin^2 2\phi_k\right]$$

and hence establish (8-7-3).

8-7-4. Prove that, as $k \longrightarrow \infty$ and $x_k \longrightarrow z_1$,

$$|\lambda_1 - \rho_{k+1}|/|\lambda_1 - \rho_k|^3 \longrightarrow 1.$$

8-7-5. Show that although $b_k \neq r_k$ we have $\|b_k\| = \|r_k\|$ using the partition of A_k shown above.

8-7-6. Show that the basic QL algorithm applied to a singular unreduced tridiagonal matrix must converge in one step.

8-8 THE OFF-DIAGONAL ELEMENTS

In the context of inverse iteration the only vector on hand at the k-th step is the current eigenvector approximation u_k and it is natural to choose its Rayleigh quotient as shift. However in the context of the QL algorithm A_k is available and so more accurate approximations to λ_1 can be computed. Moreover when A_1 is tridiagonal these improvements can be obtained at negligible cost. Before discussing such shifts in detail we look at the result of a single QL transformation with any shift σ when A is tridiagonal. The formulas are

$$A = T = \begin{bmatrix} \alpha_1 & \beta_1 & & \\ \beta_1 & \alpha_2 & \cdot & \\ & \cdot & \cdot & \beta_{n-1} \\ & & \beta_{n-1} & \alpha_n \end{bmatrix},$$

$$T - \sigma = QL, \qquad \hat{T} = Q^*TQ. \qquad (8\text{-}8\text{-}1)$$

By the invariance of bandwidth $\hat{T}$ is also tridiagonal and so, in a formal sense, $\hat{T}$ is the result of "reducing" T to tridiagonal form using Q. By Theorem (7-1-3) $\hat{T}$ is completely determined by T and either q_1 or q_n. Consider the last column on each side of $T - \sigma = QL$ to see that q_n is a multiple of $(0, \ldots, 0, \beta_{n-1}, \alpha_n - \sigma)^*$ and this fact is used in **implementing** the transformation as discussed in Secs. 8-12 and 8-13. It turns out that the description of $\hat{\mathsf{T}}$ in terms of $\mathbf{q}_1$ rather than $\mathbf{q}_n$ keeps the **analysis** simple. If σ is an eigenvalue the algorithm will converge immediately; therefore, we need only consider the case when σ is not an eigenvalue.

The relation with inverse iteration [Theorem (8-5-2), $k = 2$] gives

$$\mathbf{q}_1 \;(= \mathsf{P}_1\mathbf{e}_1) = (\mathsf{T} - \sigma)^{-1}\mathbf{e}_1\tau,$$

or

$$\begin{aligned}(\mathsf{T} - \sigma)\mathbf{q}_1 &= \mathbf{e}_1\tau,\\ \tau &= 1/\|(\mathsf{T} - \sigma)^{-1}\mathbf{e}_1\| = \|(\mathsf{T} - \sigma)\mathbf{q}_1\|. \qquad (8\text{-}8\text{-}2)\end{aligned}$$

The normalizing factor τ plays a central role in what follows. The results of Sec. 7-3 characterize the new off-diagonal elements $\hat{\beta}_i$ in terms of certain monic polynomials. In particular,

$$\begin{aligned}|\hat{\beta}_1| &= \min\|\phi(\mathsf{T})\mathbf{q}_1\| \text{ over all } \phi(\xi) = \xi - \mu,\\ &\leqslant \|(\mathsf{T} - \sigma)\mathbf{q}_1\| = \tau, \quad \text{by (8-8-2)}, \qquad (8\text{-}8\text{-}3)\end{aligned}$$

and

$$\begin{aligned}|\hat{\beta}_1\hat{\beta}_2| &= \min\|\phi(\mathsf{T})\mathbf{q}_1\| \text{ over all monic } \phi \text{ of degree 2},\\ &\leqslant \|(\mathsf{T} - \alpha_1)(\mathsf{T} - \sigma)\mathbf{q}_1\|, \quad \text{the artful choice},\\ &= \|(\mathsf{T} - \alpha_1)\mathbf{e}_1\tau\|, \quad \text{by (8-8-2)},\\ &= \|\mathbf{e}_2\beta_1\tau\| = |\beta_1|\tau. \qquad (8\text{-}8\text{-}4)\end{aligned}$$

These relations (8-8-3) and (8-8-4), which hold for all $\sigma \neq \lambda_j[\mathsf{T}]$ show the potential usefulness of the normalizing factor τ given in (8-8-2).

Although exact expressions for $\hat{\beta}_1$ and $\hat{\beta}_1\hat{\beta}_2$ are available (Sec. 8-11) they are more complicated than upper bounds on τ and for our purposes the bounds are adequate.

8-9 RESIDUAL BOUNDS USING WILKINSON'S SHIFT

Given T, as in (7-1-1), then Wilkinson's shift ω is that eigenvalue of $\begin{bmatrix}\alpha_1 & \beta_1\\ \beta_1 & \alpha_2\end{bmatrix}$ which is closer to α_1. In case of a tie ($\alpha_1 = \alpha_2$) we choose the smaller, namely $\alpha_1 - |\beta_1|$. This shift is obviously better than α_1 when either β_1 or β_2 is small. One formula for ω is

$$\omega = (\alpha_1 + \alpha_2)/2 - \text{sign}(\delta)\sqrt{\delta^2 + \beta_1^2}$$

where $\delta = (\alpha_2 - \alpha_1)/2$, but a better one (Ex. 8-9-1) is

$$\omega = \alpha_1 - \text{sign}(\delta)\beta_1^2/\left(|\delta| + \sqrt{\delta^2 + \beta_1^2}\right).$$

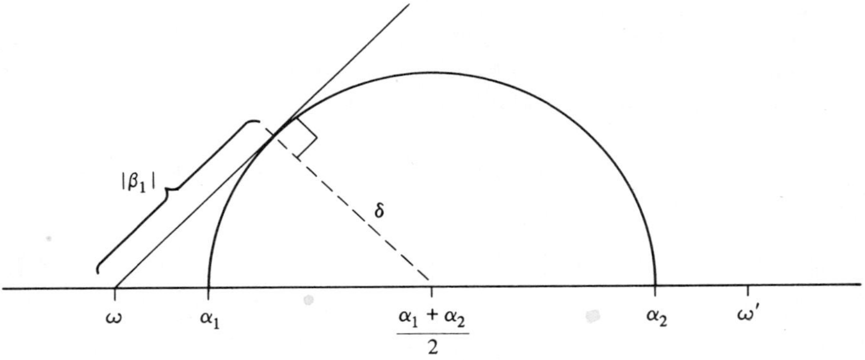

Figure 8-9-1 Wilkinson's shift ω

A glance at Fig. 8-9-1 shows that

$$|\alpha_1 - \omega| \leqslant |\alpha_2 - \omega|, \tag{8-9-1}$$

with equality if, and only if, $\delta = 0$. By noting that $|\beta_1|$ is the geometric mean of $|\alpha_1 - \omega|$ and $|\alpha_2 - \omega|$ we have

$$\frac{|\alpha_1 - \omega|}{|\beta_1|} = \frac{|\beta_1|}{|\alpha_2 - \omega|} = \sqrt{\left|\frac{\alpha_1 - \omega}{\alpha_2 - \omega}\right|} \leqslant 1, \tag{8-9-2}$$

with equality if, and only if, $\delta = 0$.

Our goal, following (8-8-4), is to bound $\tau = \|(\mathsf{T} - \omega)\mathbf{q}_1\| = 1/\|\mathbf{p}\|$ where $\mathbf{p}$ is defined by

$$(\mathsf{A} - \omega)\mathbf{p} = \mathbf{e}_1. \tag{8-9-3}$$

LEMMA If Wilkinson's shift ω is not an eigenvalue of T then the unit vector $\mathbf{q}_1$ is determined by $(\mathsf{T} - \omega)\mathbf{q}_1 = \mathbf{e}_1\tau$ and

$$\|(\mathsf{T} - \omega)\mathbf{q}_1\|^2 = \tau^2 \leqslant \min\{2\beta_1^2, \beta_2^2, |\beta_1\beta_2|/\sqrt{2}\} \tag{8-9-4}$$

Proof. Let $\mathbf{p}$ in (8-9-3) have elements $\pi_1, \pi_2, \ldots, \pi_n$. Then

$$\tau^2 = 1/\|\mathbf{p}\|^2 \leqslant 1/\left(\pi_1^2 + \pi_2^2 + \pi_3^2\right). \tag{8-9-5}$$

For convenience let $\bar{\alpha}_i = \alpha_i - \omega$. Then the first two equations in (8-9-3) may be written

$$\bar{\alpha}_1\pi_1 + \beta_1\pi_2 = 1, \tag{8-9-6}$$

$$\beta_1\pi_1 + \bar{\alpha}_2\pi_2 + \beta_2\pi_3 = 0. \tag{8-9-7}$$

Eliminate π_1 and use $\bar{\alpha}_1\bar{\alpha}_2 = \beta_1^2$ to find

$$0 + 0 + \beta_2\pi_3 = -\beta_1/\bar{\alpha}_1. \tag{8-9-8}$$

It is remarkable that π_3 depends only on $\bar{\alpha}_1$, β_1, β_2. In contrast π_1 and π_2 depend on all the elements of T − ω. However a simple bound on $\pi_1^2 + \pi_2^2$ comes from the linear equation (8-9-6) and elementary geometry (Ex. 8-9-2),

$$\pi_1^2 + \pi_2^2 \geqslant 1/\left(\bar{\alpha}_1^2 + \beta_1^2\right). \tag{8-9-9}$$

Substitute (8-9-8) and (8-9-9) into (8-9-5) to get

$$\tau^2 \leqslant \left\{\frac{1}{\bar{\alpha}_1^2 + \beta_1^2} + \frac{\beta_1^2}{\bar{\alpha}_1^2\beta_2^2}\right\}^{-1}. \tag{8-9-10}$$

This is more complicated than necessary. By (8-9-2)

$$\tau^2 \leqslant \left(1/2\beta_1^2 + 1/\beta_2^2\right)^{-1} \leqslant \min\left\{2\beta_1^2, \beta_2^2\right\}. \tag{8-9-11}$$

Finally, because the arithmetic mean exceeds the geometric mean, (8-9-11) can also yield

$$\tau^2 \leqslant \frac{1}{2}\sqrt{(2\beta_1^2)\beta_2^2} \tag{8-9-12}$$

and the lemma is proved. □

The middle expression in (8-9-11) is half the harmonic mean of $2\beta_1^2$ and β_2^2.

Exercises on Sec. 8-9

8-9-1. Show that both formulas for ω are the same in exact arithmetic. Find an example in which the first formula loses half the digits and show that the second formula cannot suffer such a loss.

8-9-2. Find the point (ξ, η) on the line $\lambda\xi + \mu\eta = 1$ which is closest to the origin.

8-10 TRIDIAGONAL QL ALWAYS CONVERGES

For any unreduced tridiagonal matrix T_1 the QL algorithm produces a sequence of unreduced tridiagonal matrices $\{\mathsf{T}_k, k = 1, 2, \ldots\}$ and the glorious fact is that, with Wilkinson's shift, $\beta_1^{(k)} \longrightarrow 0$ rapidly as $k \longrightarrow \infty$. The only assumption is that the arithmetic is done exactly.

Several practical advantages accrue from this theory: (1) There is no need to test for the right moment to switch from one shift strategy to another; (2) there is no need to add statements to the programs to check for, and cope with, rare special cases; and (3) an upper limit, such as 30, can be put on the number of iterations allowed to find any eigenvalue. This last feature guarantees quick execution in finite precision without essentially restricting the algorithm's applicability. See Ex. 8-10-1.

In contrast to the Rayleigh quotient shift strategy where $\sigma_k = \alpha_1^{(k)}$, the off-diagonal element $\beta_1^{(k)}$ need not decrease at each step. Monotonicity has been sacrificed to win guaranteed convergence to zero.

One difficulty in analyzing a nonmonotonic process is to known how many consecutive steps must be considered in order to capture the essential pattern. Fortunately one step suffices in our case because the quantity $\beta_1^2\beta_2$ does decline monotonically to zero and also dominates $\hat{\beta}_1^3$.

> THEOREM The tridiagonal QL algorithm using Wilkinson's shift always converges, i.e., $\beta_1^{(k)} \longrightarrow 0$ as $k \longrightarrow \infty$. (8-10-1)

Proof. Let T and $\hat{\text{T}}$ be consecutive terms in the tridiagonal QL algorithm. By lemma 8-9-4,

$$\hat{\beta}_1^2 < \tau^2 \leq \min\{2\beta_1^2, \beta_2^2, |\beta_1\beta_2|/\sqrt{2}\}.$$

Now combine the first and third candidates for this minimum;

$$|\hat{\beta}_1|^3 = |\hat{\beta}_1| \cdot |\hat{\beta}_1^2| < (\sqrt{2}\,|\beta_1|)(|\beta_1\beta_2|/\sqrt{2}) = |\beta_1^2\beta_2|. \quad (8\text{-}10\text{-}2)$$

Next consider the sequence $\{(\beta_1^{(k)})^2\beta_2^{(k)}\}$. The product of (8-8-3), $|\hat{\beta}_1| < \tau$, and (8-8-4), $|\hat{\beta}_1\hat{\beta}_2| < |\beta_1|\tau$, yields

$$|\hat{\beta}_1^2\hat{\beta}_2| \leq |\beta_1|\tau^2 \leq |\beta_1| \cdot |\beta_1\beta_2|/\sqrt{2}\,. \quad (8\text{-}10\text{-}3)$$

Consequently, as $k \longrightarrow \infty$,

$$|\beta_1^{(k+1)}|^3 < |(\beta_1^{(k)})^2\beta_2^{(k)}| < |(\beta_1^{(1)})^2\beta_2^{(1)}|/(\sqrt{2})^{k-1} \longrightarrow 0. \quad \square$$

The next section shows that for large enough k, which often means $k > 2$, the actual convergence rate is much better than indicated in the proof given above. The importance of Theorem (8-10-1) is that convergence is guaranteed and is at least linear right from the start.

It is most satisfactory that the modest change from the Rayleigh quotient shift ($\sigma = a_{11}$) to Wilkinson's shift ($\sigma = \omega$) transforms the convergence from almost always to always. However with the Rayleigh quotient shift there is no restriction to tridiagonal matrices; the theory in Chap. 4 is coordinate-free. One might ask if there is a shift strategy which guarantees convergence of inverse iteration from any starting vector. The answer is yes because the tridiagonal assumption is not a real restriction, but rather a convenient normalization. The only task is to formulate Wilkinson's shift in geometric form and this is quickly done.

Let u be the current vector in inverse iteration for a given symmetric matrix A. By the uniqueness of reduction [Theorem (7-2-2)] there is a unique orthogonal transformation Q such that $Q^*u = e_1$ and $Q^*AQ = T$ is tridiagonal. Now we pull back the shift ω from the tridiagonal form to the original setting.

Definition. Given A and u, with $\|u\| = 1$, define

$$\alpha_1 = u^*Au, \qquad r = Au - u\alpha_1, \qquad \beta_1 = \|r\|, \qquad \alpha_2 = r^*Ar/\beta_1^2,$$

$$W_2 = \begin{bmatrix} \alpha_1 & \beta_1 \\ \beta_1 & \alpha_2 \end{bmatrix}.$$

Then the eigenvalue ω of W_2 which is closer to α_1 is **Wilkinson's shift** for inverse iteration:

$$(A - \omega)\hat{u} = u\tau.$$

When A is full the computation of ω requires the formation of Ar as well as Au, a heavy price. With the QL algorithm $u = e_1$ and it is only the computation of Ar which discourages the use of ω for general matrices. However when A has bandwidth $2m + 1$ the cost of Ar is m^2 ops and when $m << n$ it seems a pity not to use ω and enjoy the fruits of guaranteed convergence.

Exercises on 8-10

8-10-1. Let T_0 be your favorite 10 by 10 tridiagonal matrix. Let $T(m, \delta)$ be the tridiagonal matrix obtained by linking m copies of T_0 as indicated.

$T(3, \delta) =$

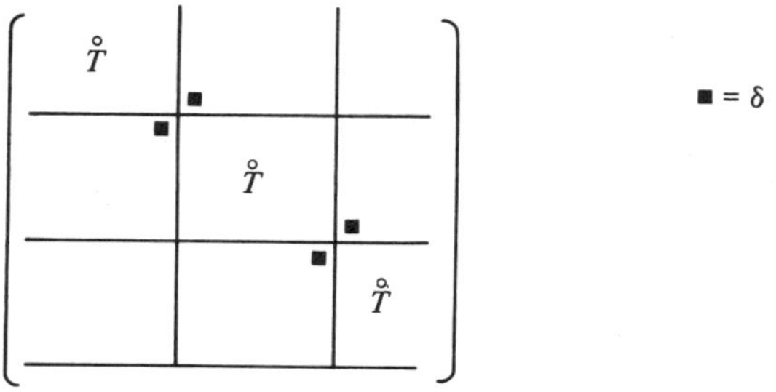

Use a local QL or QR program to compute the eigenvalues of $T(m, \delta)$ and plot the maximum number of iterations needed for any eigenvalue against small values of δ ($\varepsilon/10, \varepsilon, 10\varepsilon, \ldots, 10^5\varepsilon$), for various values of m

(5, 10, 15, 20). We find that it is possible to have QL converge slowly. It is not clear how to choose the shift to make just one of the δ's decrease.

8-10-2. Justify the given definition of Wilkinson's shift for A's which are not tridiagonal.

8-11 ASYMPTOTIC CONVERGENCE RATES

With the Rayleigh quotient shift QL converges almost always and when it does so the asymptotic convergence rate is cubic; in the tridiagonal case successive values of β_1 might be 10^{-1}, 10^{-3}, 10^{-9}, 10^{-27}. With Wilkinson's shift the asymptotic rate is better than cubic, almost always, but no one has been able to rule out the possibility of (mere) quadratic convergence to a certain very special limit matrix. In discussing the asymptotic regime it is convenient to suppress the iteration count k on which all quantities depend.

A typical step of QL transforms tridiagonal T into $\hat{\mathsf{T}} = \mathsf{Q}^*\mathsf{TQ}$ where

$$|\hat{\beta}_1| < \tau = 1/\|(\mathsf{T} - \sigma)^{-1}\mathsf{e}_1\|, \tag{8-11-1}$$

and σ is the shift. The bound on τ in Lemma (8-9-4), which was crucial for establishing global convergence, is too crude for the asymptotic regime.

First we obtain an exact expression for $|\hat{\beta}_1|$.

LEMMA Let $\hat{\mathsf{T}} = \mathsf{Q}^*\mathsf{TQ}$ be the QL transform of T with shift σ, i.e.,

$$(\mathsf{T} - \sigma)\mathsf{q}_1 = \mathsf{e}_1\tau. \tag{8-11-2}$$

Then

$$|\hat{\beta}_1| = \tau|\sin \angle(\mathsf{q}_1, \mathsf{e}_1)|. \tag{8-11-3}$$

Proof. Let $\theta = \angle(\mathsf{q}_1, \mathsf{e}_1)$. Rearrange $\mathsf{Q}\hat{\mathsf{T}}\mathsf{e}_1 = \mathsf{Tq}_1$ to find

$$\begin{aligned}
\mathsf{q}_2\hat{\beta}_1 &= \mathsf{Tq}_1 - \mathsf{q}_1\hat{\alpha}_1, \\
&= (\mathsf{I} - \mathsf{q}_1\mathsf{q}_1^*)\mathsf{Tq}_1, \quad \text{since } \hat{\alpha}_1 = \mathsf{q}_1^*\mathsf{Tq}_1, \\
&= (\mathsf{I} - \mathsf{q}_1\mathsf{q}_1^*)(\mathsf{q}_1\sigma + \mathsf{e}_1\tau), \quad \text{by (8-11-2)}, \\
&= \tau(\mathsf{e}_1 - \mathsf{q}_1 \cos\theta).
\end{aligned}$$

Since q_2 is a unit vector

$$\hat{\beta}_1^2 = \tau^2(1 - 2\mathsf{q}_1^*\mathsf{e}_1 \cos\theta + \cos^2\theta) = \tau^2 \sin^2\theta. \qquad \square$$

Normally as $\beta_1 \longrightarrow 0$ so do β_2 and β_3 but much more slowly. Consequently α_1, α_2, and α_3 all approach eigenvalues, but precisely which eigenvalues we cannot say in general. The outcome depends strongly on the initial shift. However when the eigenvalues are found in the natural order we can make a precise statement about the asymptotic rate of convergence. As above we suppress the iteration index k.

The element $\hat{\beta}_1$ is completely determined by q_1. A convenient multiple of q_1 is the vector $\mathsf{p} = (\pi_1, \pi_2, \ldots, \pi_n)^*$ defined in (8-9-3) by

$$(\mathsf{T} - \sigma)\mathsf{p} = \mathsf{e}_1. \tag{8-11-4}$$

THEOREM Let the QL algorithm with Wilkinson's shift be applied to an unreduced tridiagonal matrix T. Then, as $k \longrightarrow \infty$, $\beta_1 \longrightarrow 0$. If, in addition,

$$\beta_2 \longrightarrow 0, \beta_3 \longrightarrow 0, \quad \text{and} \quad \alpha_i \longrightarrow \lambda_i[\mathsf{T}], i = 1, 2, 3, \tag{8-11-5}$$

then, as $k \longrightarrow \infty$,

$$|\hat{\beta}_1/\beta_1^3\beta_2^2| \longrightarrow 1/|\lambda_2 - \lambda_1|^3|\lambda_3 - \lambda_1| \neq 0. \tag{8-11-6}$$

It is convenient to let $\bar{\alpha}_i \equiv \alpha_i - \omega$. Then, by (8-9-2), $\bar{\alpha}_1\bar{\alpha}_2 = \beta_1^2$.

Proof. When $\sigma = \omega$ the first three equations in (8-11-4) may be solved in terms of π_4 to obtain (Ex. 8-11-1),

$$\pi_1 = -\frac{\bar{\alpha}_2^2\bar{\alpha}_3}{\beta_1^2\beta_2^2} + \frac{\bar{\alpha}_2\beta_3\pi_4}{\beta_1\beta_2} + \frac{\bar{\alpha}_2}{\beta_1^2},$$

$$\pi_2 = \frac{\bar{\alpha}_2\bar{\alpha}_3}{\beta_1\beta_2^2} - \frac{\beta_3\pi_4}{\beta_2},$$

$$\pi_3 = -\frac{\bar{\alpha}_2}{\beta_1\beta_2}. \tag{8-11-7}$$

The guaranteed convergence of QL ensures that $\beta_1 \longrightarrow 0$, $\bar{\alpha}_1 \longrightarrow 0$, as $k \longrightarrow \infty$, but the fate of β_2 is not certain. Assumption (8-11-5) gives $\beta_2 \longrightarrow 0$ and, what is crucial, also guarantees that all the other elements of p, namely $\pi_4, \ldots, \pi_n$, are $O(|\beta_3\pi_3|)$ as $k \longrightarrow \infty$. The demonstration of this is left in Ex. 8-11-2. Eventually π_1 and π_2 dominate p and the first terms on the right in (8-11-7) dominate π_1

and π_2. So, as $k \longrightarrow \infty$,

$$\tau = \frac{1}{\|\mathbf{p}\|} \sim \frac{\beta_1^2\beta_2^2}{\bar{\alpha}_2^2|\bar{\alpha}_3|},$$

$$|\sin\theta| = \left\{\frac{\pi_2^2 + \cdots \pi_n^2}{\pi_1^2 + \pi_2^2 + \cdots + \pi_n^2}\right\}^{\frac{1}{2}} \sim \left|\frac{\pi_2}{\pi_1}\right| \sim \left|\frac{\beta_1}{\bar{\alpha}_2}\right|.$$

The result follows from Lemma (8-11-3). □

In practice the convergence of β_1 is so rapid that calculations with only 14 decimal digits rarely let β_2 and β_3 enter the asymptotic regime before deflation occurs. Example 8-11-1 illustrates this phenomenon.

EXAMPLE 8-11-1 Local Convergence of QL

$$A = (a_{ij}), \quad i, j = 1, \ldots, 10, \quad \text{where } a_{ij} = \begin{cases} 2i - 1 & \text{for } i = j \\ 1 & \text{for } i - j = 1 \\ 0 & \text{otherwise} \end{cases}$$

i	1	2	3	4
$\lambda_i[A]$	0.549129	2.95307	4.99785	6.99995

Table 8-11-1

Step	*Shift*	$\lvert\hat{\beta}_1/\beta_1^3\beta_2^2\rvert$	β_1	β_2	β_3	β_4
0	———	———	1.0	1.0	1.0	1.0
1	0.585786	0.01724	−0.01724	0.5427	0.6825	0.7589
2	0.549132	0.01434	2×10^{-8}	0.2991	0.4736	0.5801
3	0.549129	———	10^{-21}†	0.1617	0.3280	0.4437

$0.016181 = 1/61.801 = 1/[(\lambda_2 - \lambda_1)^3(\lambda_3 - \lambda_1)]$

Step	*Shift*	$\lvert\hat{\beta}_2/\beta_2^3\beta_3^2\rvert$	β_1	β_2	β_3	β_4
3	———	———	10^{-21}	0.1617	0.3280	0.4437
4	2.95323	0.02778	10^{-21}	-1×10^{-5}	0.1663	0.2976
5	2.95307	———	10^{-21}	10^{-18}†	0.08393	0.1994

$0.028902 = 1/34.599 = 1/(\lambda_3 - \lambda_2)^3(\lambda_4 - \lambda_2)$

†Noise.

It remains to indicate why assumption (8-11-5) might fail. Consider the matrix

$$\mathsf{T}_\infty = \begin{bmatrix} 0 & 0 & 0 & 0 \\ 0 & 0 & 1 & 0 \\ 0 & 1 & 0 & * \\ 0 & 0 & * & * \end{bmatrix}$$

If a QL program failed to recognize that $\beta_1 = 0$ then QL with Wilkinson's shift would leave T_∞ invariant. It seems possible that some delicate perturbation of T_∞ might cause QL to converge to T_∞. In such a case $\beta_2 \not\longrightarrow 0$ while $\bar{\alpha}_2 \longrightarrow 0$. However π_1 still dominates $\mathbf{p}$ but it is the third term in the expression which brings this about. Then, as $k \longrightarrow \infty$,

$$\tau = 0(\beta_1^2/|\bar{\alpha}_2|) = 0(|\bar{\alpha}_1|),$$

$$|\sin\theta| = 0(|\beta_1|),$$

$$|\hat{\beta}_1| = 0(|\bar{\alpha}_1\beta_1|) = 0(\beta_1^2).$$

Thus convergence is quadratic even in these circumstances.

Here ends our discussion of the convergence of the QL algorithm. The remaining sections turn to the implementation problems.

Exercises on Sec. 8-11

8-11-1. Solve (8-11-4) to obtain (8-11-7).

8-11-2. Let $\mathbf{p}^* = (\pi_1, \pi_2, \pi_3, \mathbf{s}^*)$. Show that $\mathbf{e}_1\pi_3\beta_3 + (\mathsf{T}_{4,n} - \omega)\mathbf{s} = \mathbf{o}$. Use the Cauchy interlace theorem (8-10-1) to show that $\|(\mathsf{T}_{4,n} - \omega)^{-1}\|$ is bounded as $\omega \longrightarrow \lambda_1$. Conclude that $|\pi_n/\pi_3| = 0(\beta_3)$ as $k \longrightarrow \infty$. Assume that (8-11-5) holds.

8-11-3. Investigate the asymptotic convergence rate under the assumption that $\beta_2 \longrightarrow 0$, $\bar{\alpha}_3 \longrightarrow 0$, $\beta_3 \not\longrightarrow 0$.

8-12 TRIDIAGONAL QL WITH EXPLICIT SHIFT

The tridiagonal form is preserved under the QL transformation (see Sec. 8-3) and this section shows how economically T can be turned into $\hat{\mathsf{T}} \equiv \mathsf{Q}^*\mathsf{TQ}$. If σ is the shift then $\mathsf{T} - \sigma = \mathsf{QL}$ and the orthogonal matrix Q must be a so-called **lower Hessenberg matrix** (i.e., $q_{ij} = 0$ whenever $i < j - 1$), as a glance at Fig. 8-3-1 reveals. Fortunately Q need never be formed explicitly as we now show.

Imagine the reduction of $\mathsf{T} - \sigma$ to L by a sequence of $n - 1$ plane rotations,

$$\mathsf{Q}^*(\mathsf{T} - \sigma) = \mathsf{R}_1 \cdots \mathsf{R}_{n-1}(\mathsf{T} - \sigma) = \mathsf{L} \tag{8-12-1}$$

where each $\mathsf{R}_j = \mathsf{R}(j, j + 1, \theta_j)$ is chosen to annihilate the $(j, j + 1)$ element of the matrix on hand. The plane rotations are described in Sec. 6-4. Let us take a few snapshots of the process. It is convenient to let c_j denote $\cos \theta_j$ and s_j denote $\sin \theta_j$.

We start with the two rows at the bottom and let $\bar{\alpha}_i = \alpha_i - \sigma$.

$$\begin{bmatrix} c_{n-1} & -s_{n-1} \\ s_{n-1} & c_{n-1} \end{bmatrix} \begin{bmatrix} \cdot & \cdot & 0 & \beta_{n-2} & \bar{\alpha}_{n-1} & \beta_{n-1} \\ \cdot & \cdot & \cdot & 0 & \beta_{n-1} & \bar{\alpha}_n \end{bmatrix} =$$

$$\begin{bmatrix} \cdot & \cdot & 0 & c_{n-1}\beta_{n-2} & \pi_{n-1} & 0 \\ \cdot & \cdot & \cdot & s_{n-1}\beta_{n-2} & * & \zeta_n \end{bmatrix} \tag{8-12-2}$$

where $\zeta_n^2 = \beta_{n-1}^2 + \bar{\alpha}_n^2$, $\pi_{n-1} = c_{n-1}\bar{\alpha}_{n-1} - s_{n-1}\beta_{n-1}$, $c_{n-1}\beta_{n-1} - s_{n-1}\bar{\alpha}_n = 0$. The last equation determines θ_{n-1} but of course we only need $s_{n-1} = \beta_{n-1}/\zeta_n$ and $c_{n-1} = \bar{\alpha}_n/\zeta_n$. Next we apply R_{n-2}; rows $n - 2$ and $n - 1$ become

$$\begin{bmatrix} c_{n-2} & -s_{n-2} \\ s_{n-2} & c_{n-2} \end{bmatrix} \begin{bmatrix} \cdot & \cdot & 0 & \beta_{n-3} & \bar{\alpha}_{n-3} & \beta_{n-2} & 0 \\ \cdot & \cdot & \cdot & 0 & c_{n-1}\beta_{n-2} & \pi_{n-1} & 0 \end{bmatrix} =$$

$$\begin{bmatrix} \cdot & \cdot & 0 & c_{n-2}\beta_{n-3} & \pi_{n-2} & 0 & 0 \\ \cdot & \cdot & \cdot & s_{n-2}\beta_{n-3} & * & \zeta_{n-1} & 0 \end{bmatrix},$$

and so on until L is obtained.

The second stage builds up $\hat{\mathsf{T}}$ from

$$\hat{\mathsf{T}} - \sigma = \mathsf{LQ} = \mathsf{LR}_{n-1}^* \cdots \mathsf{R}_1^*. \tag{8-12-3}$$

The trouble with such a straightforward approach is the need to store all the c_i and s_i, $i = n - 1, n - 2, \ldots, 2$, from the first stage given in (8-12-1). However when $\mathsf{R}_{n-2}\mathsf{R}_{n-1}\mathsf{T}$ has been computed the last two **columns** of L are already in their final form. Consequently R_{n-1}^* can be applied on the right at once, without waiting for L to be formed completely. After that is done c_{n-1} and s_{n-1} are no longer needed. The calculation actually proceeds in an even more intertwined fashion than we have suggested.

Because $\hat{\mathsf{T}}$ is tridiagonal the $(j, j - 2)$ elements of L must be annihilated eventually. They do not affect any other quantities in the calculation and can be ignored. In fact, by ingenious algebraic manipulations $\hat{\mathsf{T}} - \sigma$

can be written over T with the aid of only six temporary words of storage, $(n-1)$ square roots, $3(n-1)$ divisions, $10(n-1)$ multiplications, and $10(n-1)$ additions. The fluctuation in the ratios of the execution times of the basic arithmetic operations makes it difficult to summarize the operation count both neatly and accurately.

Full details concerning the algorithm are given in Contribution II/3 in the Handbook and the corresponding sections of the EISPACK guide (see Sec. 2-8).

The shift σ is not restored; it is $\hat{\mathsf{T}} - \sigma$ which is computed. Consequently the running sum of the shifts must be maintained during the iteration.

Occasionally an unfortunate early choice of shift can wipe out relevant information in the α's and so we turn to an alternative implementation.

8-13 CHASING THE BULGE

There is a way to effect the transformation $\mathsf{T} \longrightarrow \hat{\mathsf{T}} = \mathsf{Q^*TQ}$ so that the **only** modifications to T are plane rotations and there is no need to subtract σ from the diagonal elements.

The first rotation R_{n-1} is chosen exactly as in the previous section, namely $\tan\theta_{n-1} = \beta_{n-1}/(\alpha_n - \sigma)$, and T is pre-multiplied by R_{n-1} as shown in (8-12-2). Next, instead of pre-multiplying by R_{n-2} the matrix is post-multiplied by R^*_{n-1} thus completing a similarity transformation of T. Applying R^*_{n-1} to (8-12-2) gives

$$\begin{bmatrix} \cdot & \cdot & \alpha_{n-2} & c_{n-1}\beta_{n-2} & \boxed{s_{n-1}\beta_{n-2}} \\ \cdot & \cdot & c_{n-1}\beta_{n-2} & \gamma_{n-1} & s_{n-1}\pi_{n-1} \\ \cdot & \cdot & \boxed{s_{n-1}\beta_{n-2}} & s_{n-1}\pi_{n-1} & \hat{\alpha}_n \end{bmatrix} \qquad (8\text{-}13\text{-}1)$$

where $\gamma_{n-1} = c_{n-1}\pi_{n-1}$ and $\hat{\alpha}_n$ can be expressed in terms of c_{n-1} and s_{n-1}. However because the trace is preserved we have $\hat{\alpha}_n + \gamma_{n-1} = \alpha_{n-1} + \alpha_n$.

The tridiagonal form has been spoilt in positions $(n, n-2)$ and $(n-2, n)$. This is the bulge. All the remaining rotations are devoted to restoring the matrix in (8-13-1) to tridiagonal form by the method of Givens described in Sec. 7-5. The quantity $s_{n-1}\beta_{n-2}$ is annihilated by a rotation $\mathring{\mathsf{R}}_{n-2} \equiv \mathsf{R}(n-2, n-1, \mathring{\theta}_{n-2})$ with $\tan\mathring{\theta}_{n-2} = s_{n-1}\beta_{n-2}/s_{n-1}\pi_{n-1} =$

β_{n-2}/π_{n-1}. This similarity yields

$$\mathring{R}_{n-2}R_{n-1}TR^*_{n-1}\mathring{R}^*_{n-2} =$$

$$\begin{bmatrix} & & \cdot & 0 & 0 & 0 \\ \cdot & \cdot & \alpha_{n-3} & c_{n-2}\beta_{n-3} & \boxed{s_{n-2}\beta_{n-3}} & 0 \\ \cdot & \cdot & c_{n-2}\beta_{n-3} & \gamma_{n-2} & s_{n-2}\pi_{n-2} & 0 \\ \cdot & \cdot & \boxed{s_{n-2}\beta_{n-3}} & s_{n-2}\pi_{n-2} & \mathring{\alpha}_{n-1} & \mathring{\beta}_{n-1} \\ \cdot & \cdot & 0 & 0 & \mathring{\beta}_{n-1} & \hat{\alpha}_n \end{bmatrix} \qquad (8\text{-}13\text{-}2)$$

where $\gamma_{n-2} = c_{n-2}\pi_{n-2}$ and $\mathring{\alpha}_{n-1} + \gamma_{n-2} = \alpha_{n-2} + \gamma_{n-1}$.

The bulge has been chased from $(n, n-2)$ to $(n-1, n-3)$ and (8-13-2) reveals the essential pattern. The plane rotations are continued until the bulge goes to (2, 0) and vanishes. The dot on $\mathring{R}_j$ is to emphasize that, in principle, there is no reason why $\mathring{R}_j$ should equal the R_j of Sec. 8-12. The final matrix is $\mathring{T}$ and it is not clear at this point how the transformation $T \longrightarrow \mathring{T}$ is related to QL (See Sec. 8-15 for more details concerning the transformation $T \longrightarrow \hat{T}$.)

By the reduction uniqueness theorem (Sec. 7-2) $\mathring{T}$ is completely determined by T and the first (or last) column of $\mathring{Q} = R^*_{n-1}\mathring{R}^*_{n-2} \cdots \mathring{R}^*_1$. Note that there is no dot over R_{n-1}. By the special form of plane rotations

$$\begin{aligned} \mathring{Q}e_n &= R^*_{n-1}\mathring{R}^*_{n-2} \cdots \mathring{R}^*_1 e_n = R^*_{n-1}\mathring{R}^*_{n-2} \cdots \mathring{R}^*_2 e_n \\ &= \cdots \\ &= R^*_{n-1}e_n. \end{aligned} \qquad (8\text{-}13\text{-}3)$$

Similarly for the transformation matrix of the previous section, and so

$$Qe_n = R^*_{n-1}R^*_{n-2} \cdots R^*_1 e_n = R^*_{n-1}e_n = \mathring{Q}e_n. \qquad (8\text{-}13\text{-}4)$$

By (8-13-4) and Theorem (7-1-2) if either $\hat{T}$ or $\mathring{T}$ is unreduced then they must be identical (up to signs of the off-diagonal elements). All that is necessary in the new formulation is that the first rotation angle θ_{n-1} be determined by the QL factorization of $T - \sigma$.

By forming $\mathring{T}$, instead of $\hat{T} - \sigma$, the transformation has been effected **implicitly**.

To complete the theory behind this indirect formulation of QL one more result is needed.

LEMMA Let $\hat{T} = Q^*TQ$ where QL is the QL factorization of $T - \sigma$. If T is unreduced and σ is not an eigenvalue then $\hat{T}$ is unreduced too. (8-13-5)

Proof. Let $\hat{\pi}_j \equiv \hat{\beta}_j \cdots \hat{\beta}_{n-1}, j = n-1, \ldots, 1$. We must show that $\hat{\pi}_j \neq 0$ for $j \geqslant 1$. An easy variation on the results of Sec. 7-3 yields

$$|\hat{\pi}_{n-j}| = \|\tilde{\chi}_j(\hat{\mathsf{T}})\mathbf{e}_n\|, \qquad 1 \leqslant j \leqslant n,$$

where $\tilde{\chi}_j$ are monic polynomials of degree j. So

$$\begin{aligned} |\hat{\pi}_{n-j}| &= \|\mathsf{Q}^*\tilde{\chi}_j(\mathsf{T})\mathsf{Q}\mathbf{e}_n\|, \quad \text{since } \hat{\mathsf{T}} = \mathsf{Q}^*\mathsf{T}\mathsf{Q}, \\ &= \|\tilde{\chi}_j(\mathsf{T})\mathbf{q}_n\|, \quad \text{by orthogonal invariance of } \|\cdot\|, \\ &= \|\tilde{\chi}_j(\mathsf{T})(\mathsf{T}-\sigma)\mathbf{e}_n\|/\ell_{nn}, \quad \text{by the QL factorization,} \end{aligned}$$

and $\ell_{nn} > 0$, since σ is not an eigenvalue. Moreover polynomials in a matrix commute; so if $\hat{\pi}_{n-j} = 0$, then

$$(\mathsf{T}-\sigma)\tilde{\chi}_j(\mathsf{T})\mathbf{e}_n = \mathbf{o},$$

and, since $\mathsf{T} - \sigma$ is invertible,

$$\tilde{\chi}_j(\mathsf{T})\mathbf{e}_n = \mathbf{o}.$$

The $(n-j)$-th element of the vector on the left is $\pi_{n-j} \equiv \beta_{n-j} \cdots \beta_{n-1}$. In other words $\hat{\pi}_{n-j} = 0$ if, and only if, $\pi_{n-j} = 0$. □

8-13-1 *COMPARISON OF EXPLICIT AND IMPLICIT SHIFTS*

The only defect of the explicit shift is that if a big σ is subtracted from a small α_j then information in α_j is irretrievably lost.

More details and a complete algorithm are given in Contribution II/4 of the Handbook and in EISPACK. The operation count is essentially the same as for the explicit shift, $9n$ mults, $2n$ divisions, and $(n-1)$ square roots. Usually the subtraction of σ does **no more** damage to the eigenvalues than some of the preceding arithmetic operations. In general small eigenvalues are not determined to as high relative accuracy as are the large ones. However for some "graded," or nearly diagonal, matrices—as shown below—the small eigenvalues can be determined to high relative accuracy, provided that an appropriate algorithm is used. The implicit shift delivers this bonus always. The explicit shift also yields comparable accuracy provided that the eigenvalues are found in order of increasing absolute value.

EXAMPLE 8-13-1 A "Graded" Matrix

$$\mathsf{T} = \begin{bmatrix} 1 & 1 & & & \\ 1 & 10^3 & 10^2 & & \\ & 10^2 & 10^6 & 10^5 & \\ & & 10^5 & 10^9 & 10^8 \\ & & & 10^8 & 10^{12} \end{bmatrix}$$

The small eigenvalues can be determined to high **relative** accuracy.

This section must end with a warning. The theorems invoked to equate the implicit shift algorithm and QL are based on exact arithmetic and the unreduced property of T. In practice it is appropriate to test whether T is unreduced **to working precision**. This is not an easy decision. See Sec. 7-11 for a discussion of various tests.

The penalty for not spotting negligible β's is not loss of accuracy but a wash-out of the shift and a resulting slow down in convergence. See [Stewart, 1970] for more details.

8-14 SHIFTS FOR ALL SEASONS

Experience shows that all the eigenvalues of T can be computed with an average of approximately 1.7 QL transforms per eigenvalue when Wilkinson's shift is used. A rule of thumb (Sec. 8-15) says that $9n^2$ ops suffice to find all the eigenvalues and consequently it is difficult for any modification to make a **significant** improvement in this computation. Recall that reduction of a full A to T already requires $(2/3)n^3$ ops.

The situation changes markedly when P, the product of all the Q_k, is to be accumulated during the algorithm. This way of computing P ensures that the computed eigenvectors are orthogonal to working accuracy always, even when some eigenvalues are very close to each other. The computation of P_k $(= \mathsf{P}_{k-1}\mathsf{Q}_k)$ requires $4n(n-1)$ ops and raises the cost of one QL iteration from $24n$ ops (Sec. 8-13) to $4n^2 + 21n$ ops. The spectral decomposition of T by this method is an $O(n^3)$ process and there is some incentive to reduce the number of QL transformations.

We will now give brief comments on some strategies which have been considered.

8-14-1 NO SHIFTS

Eigenvalues will be found in monotonic order by absolute value—in exact arithmetic. It is too slow for general use.

8-14-2 RAYLEIGH-QUOTIENT SHIFT

This is very simple to program and there is very fast convergence but troublesome cases could arise.

8-14-3 WILKINSON'S SHIFT

This is simple to program for tridiagonals. It is feasible to compute for all A, there is very fast convergence. Eigenvalues are computed in a loosely monotonic order.

8-14-4 NEWTON'S SHIFT

This is designed to produce the eigenvalues of tridiagonals in monotone order without losing second-order convergence. T must be definite (positive or negative) initially; then $T + \chi_T(0)/\chi_T'(0)$ is still definite. Bauer found a clever way to implement QL so that $-\chi_T(0)/\chi_T'(0)$ is produced as a byproduct and can be used as a shift at the next iteration. Details are given in Contribution II/6 in the Handbook and in EISPACK program RATQR. Difficulties with underflow have been reported.

Convergence can be rather slow in difficult cases and the caution built into the Newton shift is only needed for the stability of the implementation. It is quicker to risk finding a few more eigenvalues than are wanted, using Wilkinson's or some more powerful shift, and then check which eigenvalues have been found by slicing the spectrum as described in Chap. 3.

8-14-5 SAAD'S SHIFTS

Since each update of P costs $4n^2$ ops it is not unreasonable to spend $O(n)$ ops to find a better shift than Wilkinson's. [Saad, 1974] makes some specific suggestions. It is possible to evaluate $\chi(\omega)$ and $\chi'(\omega)$ together in $4n$ ops without modifying T and then compute one QL transform with shift $\omega - \chi(\omega)/\chi'(\omega)$ and update P. Of course there is no need to restrict the shift to the first Newton iterate of ω. In fact Saad proposes using Newton's method to compute an eigenvalue to the desired accuracy and only then to perform a QL transform using the eigenvalue as shift. Such techniques bring the total number of QL iterations close to n and permit the computation of $P = \prod_{k=1}^{n} Q_k$ in approximately $\sum_{1}^{n} 4k(k-1) = 4(n^3 - n)/3$ ops, a 40% reduction in cost.

8-14-6 THE ULTIMATE SHIFTS

If two n-vectors of extra storage are available then the root-free QL algorithm of Sec. 8-15 can be applied to a copy of T to compute all the eigenvalues in perhaps $9n^2 + 42n$ ops. The QL transform is then applied to T by using the eigenvalues in monotonic order as shifts and accumulating the product of the Q's.

The advantage of using QL instead of Newton's method to find the eigenvalues is not so much in the operation count as in the guaranteed convergence of QL with Wilkinson's shift. More care is needed with Newton's method to make it converge in all situations.

*8-15 CASTING OUT SQUARE ROOTS

The implicit QL transformation of T into $\hat{T}$ requires $9n$ multiplications, $2n$ divisions, and $(n - 1)$ square roots. Usually $(1.7)n$ transformations suffice to compute all the eigenvalues and the algorithm is rightly regarded as both efficient and stable. It lies at the heart of current eigenvalue programs. Nevertheless it can be quickened when eigenvectors are not wanted.

In 1963 Ortega and Kaiser observed that the QR transformation can be reorganized to eliminate all the square roots. Unfortunately their program occasionally gives inaccurate results. This can happen because the new version no longer literally carries out orthogonal similarity transformations. That paper led to a fascinating sequence of investigations devoted to casting square roots out of the QR algorithm [Rutishauser, Wilkinson, Welsch, Stewart, Glauz, Pereyra, and Sack, to name a few]. Each contributor seemed to cure a defect in his predecessor's program only to introduce a subtle blemish of his own. The sequence appeared to terminate with [Reinsch, 1971] who gave a streamlined, stable algorithm called TQLRAT. However Reinsch's code has a feature which occasionally prevents the calculation of small eigenvalues to their maximum relative accuracy. The quest for an optimal implementation was not over.

This section presents an unpublished algorithm developed by Pal and Walker (in 1968–1969!) as a project in a graduate course at the University of Toronto taught by Kahan. We call it the PWK algorithm. It avoids tolerances without sacrificing either stability or elegance.

> Using either the PWK or TQLRAT it is usually possible to compute the p smallest or p largest eigenvalues of T with approximately $20\,pn$ multiplications.

The best shift strategies yield the eigenvalues in a loosely monotonic order. After p eigenvalues have been computed it is easy to slice the spectrum (Sec. 3-3) and see whether or not any wanted eigenvalues have been missed. If so the algorithm proceeds until all p have been found. A simple estimate is 2 iterations per eigenvalue. This is too low if $p = 1$ and too high if $p \geqslant n/3$. If a division is equivalent to 2 multiplications then each root-free QL transform requires only $10n$ multiplications.

All eigenvalues of T can be computed in approximately $9n^2$ ops.

Justification: Assume an average of 1.8 transforms for each eigenvalue, the order dropping by one each time.

8-15-1 DERIVATION OF THE ALGORITHM

Our object is to exhibit the QL transform of Sec. 8-13 in such a way that it is clear how to avoid the square root and yet preserve accuracy in all circumstances. We start with $\mathsf{T} = \mathsf{T}_n$ and gradually build up $\hat{\mathsf{T}} = \mathsf{T}_1$ by means of a sequence of plane rotations $\mathsf{T}_i = \mathsf{R}_i \mathsf{T}_{i+1} \mathsf{R}_i^*$, $i = n - 1, \ldots, 1$, where $\mathsf{R}_i = \mathsf{R}(i, i+1, \theta_i)$ and T_i bulges in positions $(i \pm 1, i \mp 1)$.

An important ingredient in this undertaking is to see the structure of the active elements of T_{i+1}. This pattern emerged in Sec. 8-13 and it is shown, in general position, in Fig. 8-15-1. It is helpful to ignore the shift σ now and it is easy to put it in place at the end.

In order to move the bulge in T_{i+1} the variables c_i and s_i, which determine R_i, must satisfy

$$c_i(s_{i+1}\beta_i) - s_i(s_{i+1}\pi_{i+1}) = 0. \tag{8-15-1}$$

This dictates

$$\zeta_i = \sqrt{\pi_{i+1}^2 + \beta_i^2}, \qquad c_i = \pi_{i+1}/\zeta_i, \qquad s_i = \beta_i/\zeta_i. \tag{8-15-2}$$

Now

$$\hat{\beta}_{i+1} = c_i(s_{i+1}\pi_{i+1}) + s_i(s_{i+1}\beta_i) = s_{i+1}\zeta_i. \tag{8-15-3}$$

The interesting question is how to express the new values of elements (i, i), $(i, i+1)$, and $(i+1, i+1)$. It helps to lay out the active rows of $\mathsf{R}_i\mathsf{T}_{i+1}$,

$$\begin{bmatrix} \cdot & \alpha_{i-1} & \beta_{i-1} & 0 & 0 & \cdot \\ \cdot & c_i\beta_{i-1} & \pi_i & 0 & 0 & \cdot \\ \cdot & s_i\beta_{i-1} & \kappa_i & c_{i+1}\zeta_i & s_{i+1}\zeta_i & \cdot \\ \cdot & 0 & s_{i+1}\beta_i & s_{i+1}\pi_{i+1} & \hat{\alpha}_{i+2} & \cdot \end{bmatrix} \tag{8-15-4}$$

$$
\mathsf{T}_{i+1} = \begin{bmatrix}
\alpha_1 & \beta_1 & & & & & & \\
\beta_1 & \cdot & \cdot & & & & & \\
 & \cdot & \alpha_{i-1} & \beta_{i-1} & & & & \\
 & & \beta_{i-1} & \alpha_i & c_{i+1}\beta_i & \boxed{s_{i+1}\beta_i} & & \\
 & & & c_{i+1}\beta_i & c_{i+1}\pi_{i+1} & s_{i+1}\pi_{i+1} & & \\
 & & & \boxed{s_{i+1}\beta_i} & s_{i+1}\pi_{i+1} & \hat{\alpha}_{i+2} & \hat{\beta}_{i+2} & \\
 & & & & & \hat{\beta}_{i+2} & \cdot & \cdot \\
 & & & & & & \cdot & \hat{\alpha}_n
\end{bmatrix}
$$

Transformation of the (i, i) element:

$$\alpha_i \longrightarrow \gamma_i \longrightarrow \hat{\alpha}_i$$

$$\mathsf{T}_{i+1} \longrightarrow \mathsf{T}_i \longrightarrow \mathsf{T}_{i-1}$$

$$\gamma_i = c_i\pi_i, \qquad \text{from Sec. 8-13}$$

Figure 8-15-1 The QL transform showing the bulge

where

$$\pi_i = c_i\alpha_i - s_i(c_{i+1}\beta_i). \tag{8-15-5}$$

When R_i^* is applied on the right the (i, i) element becomes

$$\begin{aligned}
\gamma_i &= c_i\pi_i - s_i0, \\
&= c_i^2\alpha_i - c_is_ic_{i+1}\beta_i, && \text{by (8-15-5)} \\
&= c_i^2\alpha_i - s_i^2c_{i+1}\pi_{i+1}, && \text{by (8-15-1)} \\
&= c_i^2\alpha_i - s_i^2\gamma_{i+1}. && \text{(8-15-6)}
\end{aligned}$$

Considerable efforts have been made to get rid of either c_i^2 or s_i^2 but PWK prefer to keep them both for the sake of stability. The constant trace condition gives a nice way to update element $(i + 1, i + 1)$, namely

$$\hat{\alpha}_{i+1} = \gamma_{i+1} + \alpha_i - \gamma_i. \tag{8-15-7}$$

Having updated γ by (8-15-6) we are free to update π by

$$\pi_i = \gamma_i/c_i. \tag{8-15-8}$$

This will not do when $c_i = 0$ and it looks as though (8-15-8) might be unreliable when c_i is tiny. The error analysis shows that the redundancy in (8-15-6) keeps (8-15-8) stable in the presence of tiny values of c_i. It is left as Ex. 8-15-1 to show that the case $c_i = 0$ is a harmless interchange: $s_i = 1$, $\hat{\alpha}_{i+1} = \alpha_i$, but $\pi_i = -\beta_i c_{i+1}$. Thus (8-15-8) becomes

$$\pi_i = \begin{cases} \gamma_i / c_i, & c_i \neq 0, \\ -\beta_i c_{i+1}, & c_i = 0. \end{cases} \tag{8-15-9}$$

Formulas (8-15-2), (8-15-3), (8-15-6), (8-15-7), and (8-15-9) yield a slightly unorthodox implementation of the inner loop of the standard QL algorithm. To get rid of the square root in (8-15-2) it is only necessary to square formulas (8-15-2), (8-15-3), and (8-15-9). No essential information has been lost since (8-15-6) uses c_i^2 and s_i^2. The inner loop is laid out at the end of the section.

The nice feature is that (8-15-9) embodies a simple zero test to switch between two formulas. The shift appears in (8-15-6) only since (8-15-7) is invariant. Before examining stability we look at some other versions.

8-15-2 ORTEGA-KAISER VERSION

A multiplication is saved by writing

$$\gamma_i = \alpha_i - s_i^2(\alpha_i + \gamma_{i+1}) \tag{8-15-10}$$

instead of $\gamma_i = c_i^2 \alpha_i - s_i^2 \gamma_{i+1}$. When $s_i^2 \doteqdot 1$ and $\gamma_{i+1} \ll \alpha_i$ then γ_i should be very close to $-\gamma_{i+1}$ but the computed value will have a very large relative error. Of itself this is not serious. The damage is done in the next step (8-15-9) wherein $\pi_i^2 = \gamma_i^2 / c_i^2$. When $s_i^2 \doteqdot 1$ the error in π_i^2 will be large relative to $\|\mathsf{T} - \sigma\|$.

8-15-3 REINSCH'S ALGORITHM

Reinsch also wished to avoid the redundancy of using both c_i^2 and s_i^2 and so used (8-15-10) to update γ. This necessitated rewriting (8-15-9) in a clever way, as follows:

$$\pi_i^2 = \gamma_i^2 / c_i^2 = \gamma_i \cdot \eta_i \tag{8-15-11}$$

where

$$\begin{aligned} \eta_i &\equiv \gamma_i / c_i^2 \\ &= \alpha_i - s_i^2 \gamma_{i+1} / c_i^2, \quad \text{by (8-15-6)} \\ &= \alpha_i - \gamma_{i+1} \beta_i^2 / \pi_{i+1}^2, \quad \text{by (8-15-1)}, \\ &= \alpha_i - \beta_i^2 c_{i+1} / \gamma_{i+1}, \quad \text{by (8-15-9)}, \\ &= \alpha_i - \beta_i^2 / \eta_{i+1}. \end{aligned} \tag{8-15-12}$$

In fact η_i is the i-th diagonal element in the triangular factorization of T from the bottom upwards. To modify (8-15-12) for $\eta_{i+1} = 0$ Reinsch replaces 0 by $\varepsilon \max_k |\alpha_k - \sigma|$. This can be justified in terms of backward error analysis as being equivalent to perturbing α_i by no more than $\varepsilon \max|\alpha_k - \sigma|$ which is a small perturbation relative to $\|\mathsf{T} - \sigma\|$. Nevertheless it is nice to have an algorithm in which such a rigid replacement is not needed.

8-15-4 *SLICING THE SPECTRUM*

Reinsch observed that ν, the number of negative η_i, is equal to the number of eigenvalues of T that are less than the shift σ. The reasons are given in Sec. 3-3. Since $\gamma_i = \eta_i c_i^2$ the same information can be obtained from the Pal-Walker-Kahan scheme. When σ has converged to working accuracy then $\nu + 1$ will give its index. This is valuable information when only a few eigenvalues of T are wanted.

On the other hand it is slightly quicker to test the index outside the handcrafted inner loop of the QL scheme.

8-15-5 *STABILITY*

There is a place for a formal error analysis of the PWK algorithm but not in this book. The treatment of most of the formulas is straightforward. We will focus on the way γ_i is formed in (8-15-6). We show that, in practice, the division in (8-15-8) is not to be feared.

Let overbars denote computed quantities and assume that

$$\begin{aligned} fl(\xi + \eta) &= \xi(1 + \epsilon) + \eta(1 + \epsilon), \\ fl(\xi\eta) &= \xi\eta(1 + \epsilon). \end{aligned} \tag{8-15-13}$$

The quantity ϵ represents a possibly different, tiny value at each appearance.

Despite errors the algorithm maintains the following relationships:

$$\bar{\pi}_{i+1}\bar{s}_i = \beta_i\bar{c}_i(1 + \epsilon), \quad \text{by (8-15-1)}, \tag{8-15-14}$$

$$\bar{\gamma}_i = \bar{c}_i\bar{\pi}_i(1 + \epsilon), \quad \text{by (8-15-9)}. \tag{8-15-15}$$

Then, in (8-15-6),

$$\begin{aligned} \bar{\gamma}_i &= fl\left[(\alpha_i - \sigma)\bar{c}_i^2 - \bar{\gamma}_{i+1}\bar{s}_i^2\right] \\ &= \left[\alpha_i(1 + \epsilon) - \sigma(1 + \epsilon)\right]\bar{c}_i^2(1 + \epsilon)^2 - \bar{\gamma}_{i+1}\bar{s}_i^2(1 + \epsilon)^2, \quad \text{by (8-15-13)}, \\ &= \gamma_i + 3\epsilon\alpha_i\bar{c}_i^2 + 3\epsilon\sigma\bar{c}_i^2 + 2\epsilon\bar{\gamma}_{i+1}\bar{s}_i^2, \end{aligned} \tag{8-15-16}$$

neglecting all terms in ϵ^2. The last term in (8-15-16) appears to invite disaster when $\bar{\gamma}_i$ is divided by a small $\bar{c}_i$. This is not the case because

$$\begin{aligned} \bar{\gamma}_{i+1}\bar{s}_i^2 &= \bar{c}_{i+1}\bar{\pi}_{i+1}(1+\epsilon)\bar{s}_i^2, \quad \text{by (8-15-15)}, \\ &= (1+\epsilon)\bar{c}_{i+1}\bar{s}_i\beta_i\bar{c}_i(1+\epsilon), \quad \text{by (8-15-14)}. \end{aligned} \tag{8-15-17}$$

It follows that, after (8-15-8) is executed,

$$|\bar{\pi}_i - \pi_i| \leqslant \epsilon\{3(|\alpha_i| + |\sigma|)\bar{c}_i + 2|\bar{c}_{i+1}\beta_i| \cdot |\bar{s}_i| + |\pi_i|\} \tag{8-15-18}$$

and so the error is always tiny relative to neighboring elements in the matrix.

We now display the inner loop of the algorithm.

Table 8-15-1 PWK Inner Loops

	QR (*alters* $B_{k-1}^{\ddagger}$)	*QL* (*alters* $B^{\ddagger}_m$)
Initialize	$C \leftarrow 1,\ S \leftarrow 0,$ $\gamma \leftarrow \alpha_k - \sigma,\ P \leftarrow \gamma\cdot\gamma$	$C \leftarrow 1,\ S \leftarrow 0,$ $\gamma \leftarrow \alpha_m - \sigma,\ P \leftarrow \gamma\cdot\gamma$
Loop	$i \leftarrow k, k+1, \ldots, m-1$: $BB \leftarrow B_i$ $R \leftarrow P + BB$ $B_{i-1} \leftarrow S\cdot R$ $OLDC \leftarrow C$ $C \leftarrow P/R$ $S \leftarrow BB/R$ $old\gamma \leftarrow \gamma$ $\alpha \leftarrow \alpha_{i+1}$ $\gamma \leftarrow C\cdot(\alpha - \sigma) - S\cdot old\gamma$ $\alpha_i \leftarrow old\gamma + (\alpha - \gamma)$ $P \leftarrow \begin{cases} (\gamma\cdot\gamma)/C, & C \neq 0, \\ OLDC\cdot BB, & C = 0. \end{cases}$	$i \leftarrow m-1, \ldots, k+1, k$: $BB \leftarrow B_i$ $R \leftarrow P + BB$ $B_{i+1} \leftarrow S\cdot R$ $OLDC \leftarrow C$ $C \leftarrow P/R$ $S \leftarrow BB/R$ $old\gamma \leftarrow \gamma$ $\alpha \leftarrow \alpha_i$ $\gamma \leftarrow C\cdot(\alpha - \sigma) - S\cdot old\gamma$ $\alpha_{i+1} \leftarrow old\gamma + (\alpha - \gamma)$ $P \leftarrow \begin{cases} (\gamma\cdot\gamma)/C, & C \neq 0, \\ OLDC\cdot BB, & C = 0. \end{cases}$
Termination	$B_{m-1} \leftarrow S\cdot P$ $\alpha_m \leftarrow \sigma + \gamma$	$B_k \leftarrow S\cdot P$ $\alpha_k \leftarrow \sigma + \gamma$
Cost	3 divisions, 4 multiplications, 5 adds per iteration	
Storage	$BB, R, P, S, C, OLDC, \gamma, old\gamma$	
Scaling	The matrix should be scaled up, by a power of the radix of the arithmetic unit, before starting in order to avoid unnecessary underflows in the QL transformations.	

‡Notation: Capital letters denote squared quantities, $B_i = \beta_i^2$, $i = k, \ldots, m-1$.

Exercise on Sec. 8-15

8-15-1. Examine the case when $c_{i+1} \neq 0$ but $c_i = 0$ and show how the interchange leads to $\pi_i = -\beta_i c_{i+1}$.

8-16 QL FOR BANDED MATRICES

If 4 eigenvalues of a matrix A of order 400 and bandwidth 41 are wanted then it is somewhat inefficient to reduce A to tridiagonal form as described in Sec. 7-5. Because bandwidth is preserved the shifted QL algorithm can be implemented with economy in both storage requirements and operation count. The algorithm is presented as Contribution II/7 in the Handbook. It is a clever piece of mathematical software that is far from a blind realization of the transformation defined in Sec. 8-2. We will not go into those details here but we will indicate the main ideas.

With a given shift σ the matrix $\bar{A} = A - \sigma$ is, in principle, pre-multiplied by a sequence of reflectors $H_n, H_{n-1}, \ldots$ to reduce it, column by column, to lower triangular form L. Reflectors are elementary orthogonal matrices described in Sec. 6-3. Thus

$$H_2 \cdots H_n\bar{A} = L, \tag{8-16-1}$$

where $H_n = H(w_n)$ is chosen to put the n-th column into triangular form, H_{n-1} puts the $(n-1)$-st column of $H_n\bar{A}$ into triangular form, and so on. The proper choice of w_i is straightforward and is given in Sec. 6-3.

The second phase of the transformation requires the post-multiplication of L by all the H_i and a difficulty becomes apparent. The vectors w_i defining the H_i must be saved and this requires $n(m+1)$ storage locations. Here $(m+1)$ is the half-bandwidth. The problem goes away when it is observed that there is no need to find L before beginning the second phase. Inspection of Fig. 8-16-1 shows that after m H_i have been applied on the left the second phase can be started. This suggests that one whole QL transformation can be accomplished by $m + n$ minor steps of the following form:

$$H_i\bar{A}H_{i+m}, \qquad i = n, n-1, \ldots, -m \tag{8-16-2}$$

where $\bar{A}$ is the current matrix and $H_j = I$ for $j \leq 1$ and $j > n$. Further inspection of Fig. 8-16-1 shows that w_{i+2m} can be discarded at this step.

In fact the algorithm does not implement (8-16-2) directly. Instead

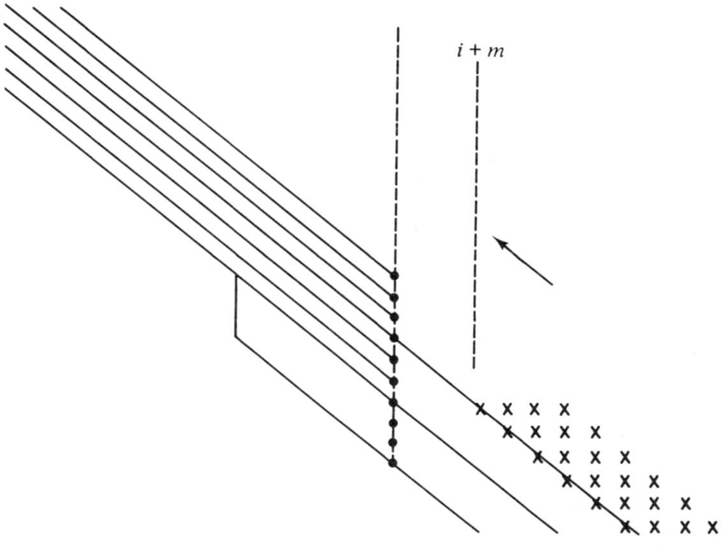

Key: —— Elements of A (untouched)

• • • Active column

x x x Elements of $\hat{A}$

Figure 8-16-1 The banded algorithm

$(\mathsf{A} - \sigma)\mathbf{e}_i$ is taken straight into the corresponding column of L via

$$\mathsf{H}_i \cdots \mathsf{H}_{i+2m}(\mathbf{a}_i - \sigma \mathbf{e}_i),$$

$\mathbf{w}_i$ is stored away, other columns of A are **not** touched, and the $(i + m)$-th row of the final matrix $\hat{\mathsf{A}}$ is computed and stored. Full advantage is taken of symmetry and storage is handled very nicely.

8-16-1 ORIGIN SHIFT

The algorithm will only be used when a small number of eigenvalues are needed and so the order in which eigenvalues are found is important. The Handbook uses a two-stage strategy; the origin shift σ is 0 initially; then after the first off-diagonal row passes a test Wilkinson's tridiagonal shift is used. This is the only inelegant feature of this program.

At the cost of one triangular factorization ($m^2 n/2$ ops) one can find the exact position of any computed eigenvalue in the spectrum. This makes it less important to shift with great caution.

8-16-2 OPERATION COUNT

$3(m+1)(2m+1)$ ops per minor step.

$(m+n)$ minor steps per iteration.

5 iterations (average) for the first eigenvalue

1 iteration for each later eigenvalue

8-16-3 STORAGE

$n \times (m+1)$	for $\mathbf{A}$ (by diagonals)
$(m+1) \times (2m+1)$	for the $\mathbf{w}_i$
$2 \times (2m+1)$	for work space

This method is in EISPACK, release 2 under the name BQR.

Notes and References

For practical purposes the QL algorithm solves the problem of computing eigenvalues of small matrices. Yet it was not invented until 1958–1959 and was not appreciated until the mid-1960s. The key idea came from Rutishauser with his construction of a related algorithm called LR in 1958.

Credit also goes to a young systems programmer in England, J. G. F. Francis, who made the three observations required to make the method successful, to wit (1) the basic QR transformation, (2) the invariance of the Hessenberg form, and (3) the use of origin shifts to hasten convergence. It should be added that Francis received some assistance from Strachey and Wilkinson. The original papers are [Francis, 1961–1962]. Independently, and in the same period, Kublanovskaja discovered (1) and (3), but without the improvements induced by (2) the algorithm is very slow. See [Kublanovskaja, 1961].

The intimate connection between the Rayleigh quotient iteration (RQI) and QR was noticed by both Kahan and Wilkinson. In fact, the monotonic decay of the last off-diagonal element in the tridiagonal QR algorithm led Kahan to the global convergence analysis of RQI presented in Chap. 4.

Wilkinson and Reinsch encouraged us to switch to the QL formulation by using it in the influential Handbook.

In 1968 Wilkinson proved that the tridiagonal QL algorithm can never fail when his shift is used. His proof is based on the monotonic decline of the product $|\beta_1 \beta_2|$. However the analysis is simpler when $\beta_1^2 \beta_2$ is used in place of $\beta_1 \beta_2$ and this recent approach is presented in Secs. 8-12 and 8-13 which follow [Hoffman and Parlett, 1978] and use some insights buried in [Dekker and Traub, 1971].

The PWK version of the root-free QL algorithm (Sec. 8-15) has not appeared in the open literature. It seems to be the best way to compute all the eigenvalues of a symmetric tridiagonal matrix. Moreover the fastest way to compute a full set of orthonormal eigenvectors is to execute the QL algorithm with the fast Givens transformation using the previously computed eigenvalues as origin shifts. This combination of techniques seems to have been overlooked so far.

Some of the ideas which led to the discovery of the QR algorithm are described in [Parlett, 1964].

9

Jacobi Methods

9-1 ROTATION IN THE PLANE

In two dimensions a rotation through an angle θ is accomplished by pre-multiplying vectors by

$$\mathsf{R}(\theta) = \begin{bmatrix} c & -s \\ s & c \end{bmatrix}.$$

Throughout this chapter we use c, s, and t for the scalar quantities $\cos\theta$, $\sin\theta$, and $\tan\theta$. The associated similarity transformation on a matrix A is

$$\begin{aligned}\mathsf{RAR}^* &= \begin{bmatrix} c & -s \\ s & c \end{bmatrix}\begin{bmatrix} \alpha & \gamma \\ \gamma & \beta \end{bmatrix}\begin{bmatrix} c & s \\ -s & c \end{bmatrix} \\ &= \begin{bmatrix} \alpha c^2 - 2\gamma sc + \beta s^2, & (c^2 - s^2)\gamma + (\alpha - \beta)sc \\ (c^2 - s^2)\gamma + (\alpha - \beta)sc, & \alpha s^2 + 2\gamma sc + \beta c^2 \end{bmatrix}.\end{aligned} \tag{9-1-1}$$

The new matrix will be diagonal if

$$\tan 2\theta = \frac{\sin 2\theta}{\cos 2\theta} = \frac{2sc}{c^2 - s^2} = \frac{2\gamma}{\beta - \alpha}. \tag{9-1-2}$$

There is no need to find θ explicitly. Let

$$\delta = |\beta - \alpha|/2, \qquad \nu = \sqrt{\gamma^2 + \delta^2}\,. \tag{9-1-3}$$

Using standard trigonometric identities (Ex. 9-1-1) we find that

$$c^2 = \frac{1}{2}(1 + \delta/\nu), \qquad s^2 = \frac{1}{2}(1 - \delta/\nu). \tag{9-1-4}$$

The expression for s^2 is a classic example of a treacherous formula for finite precision calculation. Whenever θ is small then δ/ν is close to 1 and the formula would be fine were it easy to calculate δ/ν to extra precision. With fixed length storage for numbers the remedy is to use better formulas.

There are several stable ways of computing $\cos\theta$ and $\sin\theta$ from the data. Perhaps the nicest is found in Rutishauser's Jacobi program in the Handbook. He uses the trigonometric identities for $\tan 2\theta$ in terms of $t = \tan\theta$,

$$\frac{1 - t^2}{2t} = \cot 2\theta = \frac{\beta - \alpha}{2\gamma} \equiv \zeta. \tag{9-1-5}$$

Thus t is the smaller root, in magnitude, of

$$t^2 + 2\zeta t - 1 = 0 \tag{9-1-6}$$

and so

$$t = \operatorname{sign}(\zeta)/\left(|\zeta| + \sqrt{1 + \zeta^2}\right). \tag{9-1-7}$$

Then

$$c = 1/\sqrt{1 + t^2}, \qquad s = ct. \tag{9-1-8}$$

By forcing c to be positive we obtain a rotation with $|\theta| \leq \pi/4$. There is one other solution to $\cot 2\theta = \zeta$ and that is the angle $(\pi/2 + \theta)$. The reason for insisting on the small angle is given at the end of Sec. 9-3 on convergence.

Further use of trigonometry (Ex. 9-1-2) shows that the new matrix is

$$\begin{bmatrix} \alpha - \gamma t & 0 \\ 0 & \beta + \gamma t \end{bmatrix}. \tag{9-1-9}$$

Exercises on Sec. 9-1

9-1-1. Use the relations between trigonometric functions of θ and 2θ to prove (9-1-4).

9-1-2. Show that the diagonal elements of $\mathbf{RAR^*}$ can be written $\alpha - \gamma t$ and $\beta + \gamma t$.

9-2 JACOBI ROTATIONS

The traditional formulas defining the rotation matrix $\mathsf{R}(\theta)$ which diagonalizes a 2 by 2 matrix A are given above in (9-1-5), (9-1-7), and (9-1-8). The idea behind Jacobi methods is to apply that technique to the case of an n by n matrix A using the plane rotations $\mathsf{R}(i, j, \theta)$ described in Sec. 6-4. By identifying $\alpha = a_{ii}$, $\beta = a_{jj}$, $\gamma = a_{ij}$, the formula (9-1-5) given above for c and s implicitly defines the angle θ such that the (i, j) and (j, i) elements of $\mathsf{A}' = \mathsf{R}(i, j, \theta)\mathsf{A}\mathsf{R}(i, j, \theta)^*$ are zero, namely

$$\tan 2\theta = 2a_{ij} / (a_{jj} - a_{ii}), \qquad i < j.$$

With this choice $\mathsf{R}(i, j, \theta)$ is called a **Jacobi rotation matrix** and the associated similarity transformation is a **Jacobi rotation.** It annihilates the (i, j) element and, since θ is fixed by (9-1-5), we can write R_{ij} for $\mathsf{R}(i, j, \theta)$ without ambiguity.

What is new when $n > 2$ is that there are other off-diagonal elements in rows i and j which are transformed. Thus, for $k \neq i, j$,

$$\begin{aligned} a'_{ik} &= c \cdot a_{ik} - s \cdot a_{jk}, \\ a'_{jk} &= s \cdot a_{ik} + c \cdot a_{jk}. \end{aligned} \qquad (9\text{-}2\text{-}1)$$

Note also that for $k \neq i, j$,

$$(a'_{ik})^2 + (a'_{jk})^2 = a_{ik}^2 + a_{jk}^2. \qquad (9\text{-}2\text{-}2)$$

Cost. By taking advantage of symmetry and working with either the lower or the upper triangle of A a traditional Jacobi rotation can be made with $4n$ multiplications and 2 square roots (Ex. 9-2-1).

Alternatives. Before leaving this topic it is well to emphasize that there are other possible choices for θ when rotating the (i, j) plane.

The Jacobi rotation is distinguished by causing the greatest decrease in the sum of squares of the off-diagonal elements. It is locally optimal but not necessarily best for the total computation. A Givens rotation in the (i, j) plane chooses θ to annihilate some element other than a_{ij}.

EXAMPLE 9-2-1 Annihilation of a_{13} by Rotations in Two Different Planes

$$\mathsf{A} = \begin{bmatrix} 3.0 & -12.0 & 10.0 \\ -12.0 & 64.0 & -60.0 \\ 10.0 & -60.0 & 60.0 \end{bmatrix}.$$

Trace $(\mathsf{A}) = 127.0$, $\omega(\mathsf{A}) \left(= \sum_{i<j} a_{ij}^2\right) = 3844.0.$

JACOBI

$$\mathsf{A}'_J = \mathsf{R}^*(1, 3, \theta)\mathsf{A}\mathsf{R}(1, 3, \theta) = \begin{bmatrix} 61.70 & -61.16 & 0. \\ -61.16 & 64.0 & 1.75 \\ 0. & 1.75 & 1.30 \end{bmatrix}.$$

$$\theta = 9.67°, \qquad \tan(\theta) = 0.17, \qquad \omega(\mathsf{A}'_J) = 3744.0.$$

GIVENS

$$\mathsf{A}'_G = \mathsf{R}^*(2, 3, \theta)\mathsf{A}\mathsf{R}(2, 3, \theta) = \begin{bmatrix} 3.0 & -15.62 & 0. \\ -15.62 & 121.38 & -8.85 \\ 0. & -8.85 & 2.62 \end{bmatrix}.$$

$$\theta = -39.81°, \qquad \tan(\theta) = -0.83, \qquad \omega(\mathsf{A}'_G) = 322.37.$$

The values of ω appear to contradict the assertion above that a Jacobi rotation produces the maximal decrease in ω. The resolution of this quandary constitutes Ex. 9-2-3.

9-2-1 RUTISHAUSER'S MODIFICATIONS

In the application we are about to describe the angles θ will often be small and there are alternative expressions to (9-2-1) which cost no more and have better roundoff properties. Define

$$\tau \equiv \tan \theta/2 = s/\,(1 + c) \tag{9-2-3}$$

and then (Ex. 9-2-2) for $k \neq i, j$,

$$\begin{aligned} a'_{ik} &= a_{ik} - s(a_{jk} + \tau \cdot a_{ik}), \\ a'_{jk} &= a_{jk} + s(a_{ik} - \tau \cdot a_{jk}). \end{aligned} \tag{9-2-4}$$

We close this section by mentioning another effective device introduced by Rutishauser. It is simple but no one thought to use it before 1965. The modifications ($\pm\, ta_{ij}$) to the diagonal elements are accumulated in a separate array of length n, for a whole sweep through all the off-diagonal elements and only then are the totals of these (usually) small quantities added to the diagonal elements, which are also stored in a separate array. Such devices are the essence of good mathematical software.

Exercises on 9-2

9-2-1. Do the operation count for one Jacobi rotation using (9-1-5), (9-1-7), and (9-1-8) and modifying only the upper triangular part of A.

9-2-2. Derive (9-2-3) and (9-2-4) using standard trigonometric identities.

9-2-3. The Givens rotation used to annihilate (1, 3) in Example 9-2-1 causes a far greater reduction in ω than does the Jacobi rotation in the (1, 3) plane. Does this contradict the assertion that Jacobi rotations produce the greatest decrease in ω?

9-3 CONVERGENCE

Jacobi methods seek to diagonalize a given A by a sequence of Jacobi rotations. Zero elements created at one step will be filled in later and any diagonalizing sequence must be, in principle, infinite. Jacobi methods vary soley in their strategies for choosing the next doomed element.

Before discussing particular strategies we present the notions on which their analysis rests. Recall the Frobenius matrix norm

$$\|\mathsf{A}\|_F = \left(\sum_i \sum_j a_{ij}^2\right)^{\frac{1}{2}}$$

and Fact 1-10: If Q is orthogonal then $\|\mathsf{QAQ}^*\|_F = \|\mathsf{A}\|_F$. It is helpful to split $\|\mathsf{A}\|_F^2$ into two parts,

$$\delta(\mathsf{A}) \equiv \sum a_{ii}^2,$$

$$2\omega(\mathsf{A}) = \sum_{i \neq j}\sum a_{ij}^2. \tag{9-3-1}$$

For **any** plane rotation $\mathsf{R}(p, q, \theta)$ it happens that

$$\delta(\mathsf{RAR}^*) = \delta(\mathsf{A}) + 2\left[a_{pq}^2 - (\text{new } a_{pq})^2\right]. \tag{9-3-2}$$

In particular, for a Jacobi rotation R_{ij} (Ex. 9-3-1),

$$\omega(\mathsf{R}_{ij}\mathsf{A}\mathsf{R}_{ij}^*) = \omega(\mathsf{A}) - a_{ij}^2. \tag{9-3-3}$$

In whatever order the off-diagonal elements are annihilated the quantity ω is monotonic decreasing and the only question is whether the limit is zero or not. Provided only that, at each step, the pair (i, j) is chosen so that

$$a_{ij}^2 \geqslant \text{the average of } \{a_{pq}^2 : p < q\} = 2\omega(\mathsf{A})/n(n-1) \tag{9-3-4}$$

then, using (9-3-3), we get

$$\omega(\mathsf{R}_{ij}\mathsf{A}\mathsf{R}_{ij}^*) \leqslant \left(1 - \frac{2}{n(n-1)}\right)\omega(\mathsf{A}) \tag{9-3-5}$$

and, by (9-3-5), ω must converge to zero.

Condition (9-3-4) ensures convergence to diagonal **form** and that is sufficient for practical purposes. Nevertheless it is legitimate to wonder whether the diagonal elements either wander about or converge to specific eigenvalues in the limit.

Fact 1-11, the Wielandt-Hoffman theorem, shows (Ex. 9-3-4) that there is **some** ordering π of the eigenvalues so that

$$|\lambda_{\pi(i)} - a_{ii}|^2 \leqslant \sum_{\nu} |\lambda_{\pi(\nu)} - a_{\nu\nu}|^2 \leqslant 2\omega(\mathsf{A}). \tag{9-3-6}$$

Our fear is that this ordering might change at each step. We may apply (9-3-6) at a late stage in the Jacobi iteration at which

$$\omega(\mathsf{A}) \leqslant \frac{1}{4}\min|\lambda_p - \lambda_q|^2 \equiv \sigma^2, \tag{9-3-7}$$

where the minimum is over **distinct** pairs of eigenvalues. Whenever (9-3-7) holds there is a diagonal element in the interval $[\lambda - \sigma, \lambda + \sigma]$ for each distinct eigenvalue λ and the number of a_{ii} in this interval is precisely the multiplicity of λ (Ex. 9-3-2).

Can any a_{ii} escape to another interval at a subsequent step? Yes, indeed, because there are **two** angles θ which satisfy the Jacobi condition

$$\tan 2\theta = 2a_{ij}/\,(a_{jj} - a_{ii}). \tag{9-3-8}$$

The larger angle forces both a_{ii} and a_{jj} to leave their intervals and the smaller angle prevents their escape.

EXAMPLE 9-3-1

$$\begin{bmatrix} 0 & -1 \\ 1 & 0 \end{bmatrix}\begin{bmatrix} 2 & 0 \\ 0 & 1 \end{bmatrix}\begin{bmatrix} 0 & 1 \\ -1 & 0 \end{bmatrix} = \begin{bmatrix} 1 & 0 \\ 0 & 2 \end{bmatrix}.$$

THEOREM If a Jacobi method satisfies (9-3-4) at each step and chooses θ in the half open interval $(-\pi/4, \pi/4]$ to satisfy (9-3-8), then the sequence of matrices converges to a diagonal matrix.

(9-3-9)

The proof is left as Ex. 9-3-3.

Exercises on 9-3

9-3-1. Prove (9-3-3) using (9-2-2) and the invariance of $2\omega + \delta$.

9-3-2. Use (9-3-6) and (9-3-7) to show that each interval $[\lambda - \sigma, \lambda + \sigma]$ contains at least one diagonal element.

9-3-3. Prove theorem 9-3-9.

9-3-4. Show that (9-3-6) is a corollary of the Wielandt-Hoffman theorem for suitable choices of A and M.

9-4 VARIOUS STRATEGIES

9-4-1 CLASSICAL JACOBI

At each step the maximal off-diagonal element is annihilated. Convergence follows from (9-3-5).

On some computers it is relatively expensive to search for the largest off-diagonal element. Considerable effort went into the means for reducing this cost, but inequality (9-3-5) shows that the choice of any reasonably large element will suffice for convergence and this fact suggests other strategies.

9-4-2 THE CYCLIC JACOBI METHODS

The simplest scheme is to annihilate elements, regardless of size, in the order

$$(1, 2), (1, 3), \ldots, (1, n), (2, 3), \ldots, (2, n), (3, 4), \ldots, (n-1, n)$$

and then begin again for another sweep through the matrix.

In [Henrici, 1958] it was shown that if the angles are suitably restricted the method does indeed converge. The difficulty in the analysis is the possibility that the larger off-diagonal elements are always moving around **ahead** of the current element (i, j) in the sequence. If this were to happen then significant reductions in ω would never occur.

In principle the off-diagonal elements could be annihilated in any order at each sweep, provided that no element was missed, but convergence has never been proved.

The cyclic methods waste time annihilating small elements in the early sweeps.

9-4-3 THRESHOLD METHODS

The preferred strategy is to use the simple cyclic pattern but to skip rotations when a_{ij} is less than some threshold value τ which can either be fixed or be varied at each rotation. The fixed threshold is changed, of course, when all off-diagonal elements are below it.

Rutishauser chooses to calculate, for each of the first 4 sweeps, a fixed τ which is an approximation to ω, namely

$$\tau = \frac{1}{5}\left(\sum_{p<q}\sum |a_{pq}|/n\right)^2.$$

Perhaps the fastest of the known techniques is the **variable** threshold strategy of Kahan and Corneil [Corneil, 1965]: Initially one computes $\omega = \sum\sum_{p<q} a_{pq}^2$ at the cost of $N = \frac{n(n-1)}{2}$ multiplications, and then $\tau = \sqrt{\omega/N}$, the true root mean square (RMS) of the off-diagonal elements. At each actual rotation ω is reduced by a_{ij}^2 and τ is recomputed; the cost is 1 multiplication, 1 division, and 1 square root **per rotation**. The square root is justified when the number of actual rotations per sweep is well below $N/3$. The alternative is to change the test to

$$Na_{ij}^2 > \omega$$

which costs 2 multiplications **per test**, i.e., $2N$ multiplications per sweep regardless of the number of rotations. As the cost of square roots continues to tumble it becomes more and more attractive to recompute τ at each rotation.

9-5 ULTIMATE QUADRATIC CONVERGENCE

After 3 or 4 initial sweeps through all the off-diagonal elements convergence to diagonal form is usually very rapid. If at the end of some sweep $\omega/\sigma^2 \leqslant 2^{-j}$ we can expect $\omega/\sigma^2 \leqslant 2^{-2j}$ at the end of the next sweep.

Rigorous proofs of this quadratic convergence property [Schönhage 1961; Wilkinson, 1962] are too involved to present here. However the central idea is simple and goes as follows.

Consider a specific position in the matrix, say the $(1, n)$ element, after several sweeps in cyclic order by rows, namely $(1, 2), (1, 3), \ldots, (n-1, n)$. After being annihilated $(1, n)$ stays zero until the Jacobi rotation involving R_{2n}. Its new value, by (9-2-1), is

$$\begin{aligned} a'_{1n} &= ca_{1n} - sa_{2n} \\ &= -sa_{2n}, \quad \text{since } a_{1n} = 0. \end{aligned} \tag{9-5-1}$$

Moreover, since $|\theta| < \pi/4$,

$$|s| = |\sin\theta| \leqslant |\theta| \leqslant |\tan\theta| \leqslant \frac{1}{2}|\tan 2\theta| = |a_{2n}|/|a_{nn} - a_{22}|. \tag{9-5-2}$$

Combining (9-5-1) and (9-5-2) yields the key observation

$$|a'_{1n}| \leqslant |a_{2n}|^2/|a_{nn} - a_{22}|. \tag{9-5-3}$$

Now $a_{ii} \to \lambda_i$ as the sweeps continue. After **some stage**

$$|a_{kk} - a_{\ell\ell}| > \sigma \equiv \min_{i\neq j}|\lambda_i - \lambda_j|/2, \qquad \text{for all } k \neq \ell, \tag{9-5-4}$$

and we suppose for the moment that all the λ_i are simple. By the same argument we see that the next time the $(1, n)$ element is changed its new value is

$$\begin{aligned} |a''_{1n}| &= |ca'_{1n} - sa_{3n}|, \\ &\leq |a'_{1n}| + |sa_{3n}|, \\ &\leq (a_{2n}^2 + a_{3n}^2)/\sigma, \quad \text{since } |s| < |a_{3n}|/\sigma. \end{aligned} \tag{9-5-5}$$

At the end of some later sweep (9-5-4) will hold and, in addition, $\max_{i \neq j} |a_{ij}| \leq \eta$ for some η smaller than σ. At the end of the following sweep (9-5-3) and (9-5-5) suggest that

$$\max_{i \neq j} |\text{new } a_{ij}| \leq \eta^2(n-1)/\sigma = 0(\eta^2) \qquad \text{as } \eta \to 0. \tag{9-5-6}$$

This behavior is called **(asymptotic) quadratic convergence.**

9-5-1 MULTIPLE EIGENVALUES

The foregoing analysis suggests that the onset of rapid convergence to diagonal form will be greatly delayed when eigenvalues are close (i.e., when σ is very small) and may not even be ultimately quadratic when multiple eigenvalues are present. However these fears are groundless as the following surprising observation shows.

There is no loss of generality in supposing that all the a_{ii} which converge to a multiple eigenvalue λ are at the bottom of A's diagonal. Now partition A accordingly as

$$\mathsf{A} = \begin{bmatrix} \mathsf{A}_1 & \mathsf{B} \\ \mathsf{B}^* & \mathsf{A}_2 \end{bmatrix}, \qquad \mathsf{A}_1 \text{ is } (n-m) \text{ by } (n-m), \mathsf{A}_2 \text{ is } m \text{ by } m. \tag{9-5-7}$$

Our assumption is that $\mathsf{A}_2 \to \lambda \mathsf{I}_m$ where m is the multiplicity of λ and that all other eigenvalues of A are separated from λ by 2δ. After a certain number of sweeps all the eigenvalues of A_1 will be separated from λ by δ or more; in other words,

$$\|(\mathsf{A}_1 - \lambda)^{-1}\| = 1/\min|\lambda_i - \lambda| \leq 1/\delta.$$

The argument for quadratic convergence was based on the assumption that the angles of rotation, $|a_{ij}|/|a_{ii} - a_{jj}|$, become small. This assumption **fails** for the off-diagonal elements of A_2, where each $a_{kk} \to \lambda$ but **holds** for B. Fortunately the angles for A_2 do not matter because all the off-diagonal elements **are already tiny**, as the next theorem shows.

THEOREM Let A be partitioned as in (9-5-7). If $\|(\mathsf{A}_1 - \lambda)^{-1}\| < 1/\delta$ and if λ is an eigenvalue of multiplicity m then

$$\|\mathsf{A}_2 - \lambda\| \leq \|\mathsf{B}\|^2/\delta. \tag{9-5-8}$$

The proof is an application of block triangular factorization.

Proof.

$$\mathsf{A} - \lambda = \begin{bmatrix} \mathsf{A}_1 - \lambda & \mathsf{B} \\ \mathsf{B}^* & \mathsf{A}_2 - \lambda \end{bmatrix},$$

$$= \begin{bmatrix} \mathsf{I} & \mathsf{O} \\ \mathsf{B}^*(\mathsf{A}_1 - \lambda)^{-1} & \mathsf{I}_m \end{bmatrix} \begin{bmatrix} \mathsf{A}_1 - \lambda & \mathsf{O} \\ \mathsf{O}^* & \mathsf{X}_\lambda \end{bmatrix} \begin{bmatrix} \mathsf{I} & (\mathsf{A}_1 - \lambda)^{-1}\mathsf{B} \\ \mathsf{O}^* & \mathsf{I}_m \end{bmatrix},$$

where

$$\mathsf{X}_\lambda = (\mathsf{A}_2 - \lambda) - \mathsf{B}^*(\mathsf{A}_1 - \lambda)^{-1}\mathsf{B}.$$

Now observe that the rank of $\mathsf{A} - \lambda$ is $n - m$. By Sylvester's law of inertia the rank of the block diagonal factor must also by $n - m$, and so $\mathsf{X}_\lambda = \mathsf{O}$. Thus

$$\|\mathsf{A}_2 - \lambda\| = \|\mathsf{B}^*(\mathsf{A}_1 - \lambda)^{-1}\mathsf{B}\| \leqslant \|\mathsf{B}\|^2/\delta. \qquad \square$$

A corollary of this theorem is that, for $j > n - m$,

$$|a_{jj} - \lambda| \leqslant \|\mathsf{B}\|^2/\delta, \tag{9-5-9}$$

so the diagonal elements converge faster to the multiple eigenvalues than to the well-separated ones.

9-6 ASSESSMENT OF JACOBI METHODS

To reject all plane rotations except Jacobi rotations in a procedure to diagonalize A is like going into a boxing ring with one hand tied behind your back. What happens if we "generalize" Jacobi methods very slightly to allow the use of Givens rotations to annihilate elements?

We can, first of all, use Givens' original method to reduce A to tridiagonal form T (see Chap. 7). These rotations may not reduce ω^2 but they do preserve the zeros that have been so carefully created. Next we can do a Jacobi rotation $\mathsf{R}_{n-1,n}$. This fills in elements $(n, n-2)$ and $(n-2, n)$ and creates a bulge in the tridiagonal form. This bulge can be chased up the matrix and off the top by a sequence of Givens rotations as described in Sec. 8-6. At this point we have executed one tridiagonal QL transformation with zero shift. This can be repeated until the off-diagonal elements are negligible. Lo and behold, Jacobi has turned into Givens followed by QL without shifts.

If we relax the habit of zeroing a matrix element at **every** step we can do a plane rotation in the $(n-1, n)$ plane through an angle other than the Jacobi angle. This permits us to incorporate shifts into the QL algorithm and thus accelerate convergence dramatically.

Are there reasons for artificially restricting ourselves to rotating through the Jacobi angle instead of permitting other choices of θ? First we must say that the best Jacobi programs are only about three times slower than the tridiagonal QL methods, but Jacobi's claim to continued attention has been **simplicity** rather than efficiency. This may be very important in special circumstances (hand held calculators or computations in space vehicles).

Let us examine the claim to simplicity. Given below are counts of the executable operations (at the assembly language level) of various codes in the Handbook. None of the codes tried to minimize these op counts.

All eigenvalues and eigenvectors:

$$\text{Jacobi 121 versus} \quad \begin{cases} \text{Tred 2 (tridiagonal reduction)} & 81 \\ \text{Tql 2 (QL algorithm)} & 89 \end{cases}$$

All eigenvalues, no eigenvectors:

$$\text{Jacobi 99 versus} \quad \begin{cases} \text{Tred 1} & 58 \\ \text{Tql 1} & 87 \end{cases}$$

Fetches and stores were ignored when they occurred as part of arithmetic expressions.

The Jacobi advantage is not great and it must be said that there is some advantage, when fast storage is tight, to having the computation split into two separate stages.

Notes and References

In [Jacobi, 1846] plane rotations were used to diagonalize a real 7 by 7 symmetric matrix. One hundred years later the method was rediscovered and described in a report [Bargmann, Montgomery, and von Neumann, 1946] and was eventually turned into a clever and effective program in Rutishauser's contribution II/1 to the Handbook. The variable threshold strategy is presented in the unpublished report [Corneil, 1965].

The quadratic convergence rate was no secret and formal proofs were given in [Schönhage, 1961] and [Wilkinson, 1962]. A more vexing problem was the possibility that the cyclic and other more convenient strategies might not always lead to convergence to a diagonal matrix.

10

Eigenvalue Bounds

Useful information about the eigenvalues of A can be obtained from some of its submatrices. Moreover A's eigenvalues can be related to those of neighboring matrices in a way that goes far beyond standard perturbation theory. Such investigations began in the nineteenth century and continue to this day. Much of the work extends the matrix theory to cover differential and integral operators.

This chapter presents the classical matrix results and goes on to some little known refinements which have been developed in response to demands for the computation of eigenvalues of larger and ever larger matrices. All the results are inclusion theorems; that is, they describe intervals which are guaranteed to contain one or more eigenvalues of A. The later sections show how to exploit all the information which is likely to be available at the end of one step in an expensive iterative method of the sort described in Chaps. 13 and 14. By computing the appropriate optimal intervals the iteration can be stopped at the right moment for the required accuracy.

Closely related error bounds are given in the next chapter but they only make use of **norms** of certain residuals. Consequently they are less elaborate and less expensive to compute than the intervals of this chapter.

The material is based on notes and lectures by Kahan which put all the results, old and new, into a simple, unified sequence. The reader may find this chapter more difficult than the preceeding ones. The material is matrix theory rather than numerical methods and the aim is to give proofs in full detail of some hitherto unpublished results.

The convention of negative indices for the largest eigenvalues proves useful in this chapter.

10-1 CAUCHY'S INTERLACE THEOREM

How are the eigenvalues of a principal submatrix related to those of the big matrix? To pose this question precisely let

$$\mathsf{A} = \begin{bmatrix} \mathsf{H} & \mathsf{B}^* \\ \mathsf{B} & \mathsf{U} \end{bmatrix}, \qquad \begin{array}{ll} \mathsf{A} & n \text{ by } n, \\ \mathsf{H} & m \text{ by } m. \end{array} \tag{10-1-1}$$

A's eigenvalues: $\alpha_1 \leqslant \alpha_2 \leqslant \alpha_3 \cdots \leqslant \alpha_n$,

H's eigenvalues: $\theta_1 \leqslant \theta_2 \leqslant \cdots \leqslant \theta_m$,

For convenience: $\theta_i = \left\{ \begin{array}{ll} -\infty & \text{if } i \leqslant 0 \\ +\infty & \text{if } i > m \end{array} \right\}; \ \theta_{-i} = \left\{ \begin{array}{ll} -\infty & \text{if } i > m \\ +\infty & \text{if } i \leqslant 0 \end{array} \right\}.$

THEOREM For $j = 1, \ldots, m$,

$$\alpha_j \leqslant \theta_j \leqslant \alpha_{j+n-m} \quad \text{and equivalently,} \quad \alpha_{-(j+n-m)} \leqslant \theta_{-j} \leqslant \alpha_{-j}.$$

Alternatively, for $k = 1, 2, \ldots, n$,

$$\theta_{k-n+m} \leqslant \alpha_k \leqslant \theta_k \quad \text{and} \quad \theta_{-k} \leqslant \alpha_{-k} \leqslant \theta_{-(k-n+m)}. \tag{10-1-2}$$

In words, $\theta_{\pm j}$ is an **inner** bound on $\alpha_{\pm j}$ relative to A's spectrum. The proof uses Sylvester's inertia theorem (Fact 1-6).

Proof. If $\mathsf{H} - \xi$ is invertible then the block triangular decomposition (Chap. 3) of $\mathsf{A} - \xi$ is

$$\mathsf{A} - \xi = \begin{bmatrix} \mathsf{I}_m & \mathsf{O} \\ \mathsf{B}(\mathsf{H} - \xi)^{-1} & \mathsf{I}_{n-m} \end{bmatrix} \begin{bmatrix} \mathsf{H} - \xi & \mathsf{O} \\ \mathsf{O} & \mathsf{W}(\xi) \end{bmatrix} \begin{bmatrix} \mathsf{I}_m & (\mathsf{H} - \xi)^{-1}\mathsf{B}^* \\ \mathsf{O} & \mathsf{I}_{n-m} \end{bmatrix}, \tag{10-1-3}$$

where the $(n - m)$ by $(n - m)$ matrix W is given by

$$\mathsf{W}(\xi) = (\mathsf{U} - \xi) - \mathsf{B}(\mathsf{H} - \xi)^{-1}\mathsf{B}^*. \tag{10-1-4}$$

So $\mathsf{A} - \xi$ is congruent to $\text{diag}(\mathsf{H} - \xi, \mathsf{W}(\xi))$ and by Sylvester's theorem both matrices have the same inertia (π, ν, ζ) which is the sum, component by component of the inertias of $\mathsf{H} - \xi$ and $\mathsf{W}(\xi)$. This simple observation yields

$$\pi(\mathsf{H} - \xi) \leqslant \pi(\mathsf{A} - \xi) = \pi(\mathsf{H} - \xi) + \pi[\mathsf{W}(\xi)] \leqslant \pi(\mathsf{H} - \xi) + n - m, \tag{10-1-5}$$

$$\nu(\mathsf{H} - \xi) \leqslant \nu(\mathsf{A} - \xi) = \nu(\mathsf{H} - \xi) + \nu[\mathsf{W}(\xi)] \leqslant \nu(\mathsf{H} - \xi) + n - m. \tag{10-1-6}$$

The assertion $\alpha_j \leqslant \theta_j$ can be derived from (10-1-6). Suppose, to the contrary, that $\theta_j < \alpha_j$. Then it is possible to choose a ξ satisfying $\theta_j < \xi < \alpha_j$, and $\xi \neq \theta_i$, for any i. Consider the following diagram:

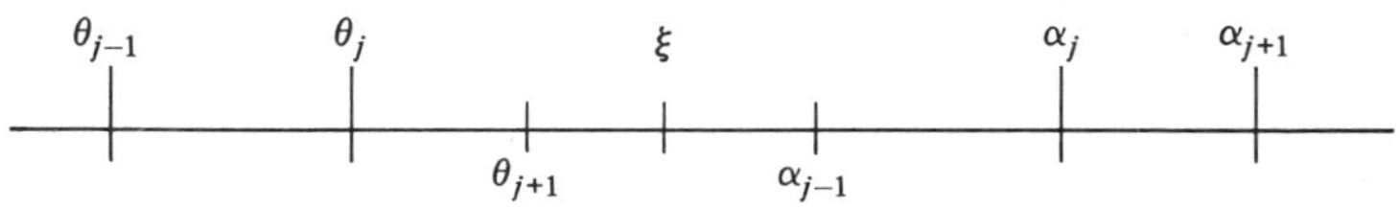

Below the line go quantities whose position relative to ξ are not determined. From the position of ξ it follows that H has at least j eigenvalues less than ξ and A has at most $j - 1$ eigenvalues less than ξ; i.e., $\nu(\mathsf{H} - \xi) \geqslant j$, $\nu(\mathsf{A} - \xi) < j$. This contradicts (10-1-6) and so $\alpha_j \leqslant \theta_j$.

The other assertions are derived similarly from either (10-1-5) or (10-1-6). □

EXAMPLE 10-1-1

$$\mathsf{H} = \begin{bmatrix} 0 & \varepsilon \\ \varepsilon & 0 \end{bmatrix}, \qquad \mathsf{A} = \begin{bmatrix} 0 & \varepsilon & 0 \\ \varepsilon & 0 & 1 \\ 0 & 1 & 0 \end{bmatrix}$$

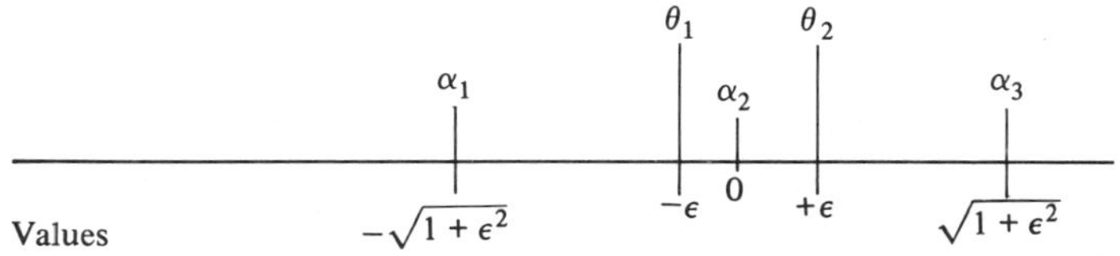

EXAMPLE 10-1-2

The eigenvalues of the famous tridiagonal matrix T_n with typical row $(\cdots -1 \quad 2 \quad -1 \cdots)$ are known to be

$$\tau_j^{(n)} = 4\sin^2 j\phi_n, \qquad j = 1, \ldots, n, \qquad \phi_n = \pi/2(n+1).$$

The interlace bounds yield the following typical results. Here $n = 100$.

$$\tau_{k-(n-m)}^{(m)} \leqslant \tau_k^{(n)} \leqslant \tau_k^{(m)} \leqslant \tau_{k+(n-m)}^{(n)}$$

Case 1. $m = 99$, $k = 45$, $1.6252 \leqslant 1.6595 \leqslant 1.6871 \leqslant 1.7210$.

Case 2. $m = 90$, $k = 45$, $1.2908 \leqslant 1.6595 \leqslant 1.9655 \leqslant 2.2790$.
As $(n - m)$ increases the bounds become very weak.

Exercises on Sec. 10-1

10-1-1. Consider the case $n - m = 1$ and show, without using the interlace theorem, that $\det(\mathsf{A} - \xi)$ changes sign between distinct eigenvalues θ_j and θ_{j+1} of H.

10-1-2. For $n - m = 1$ verify that the inequalities are best possible in the following sense. Given a set of numbers $\{\alpha_i, i = 1, \ldots, n\}$ and another set $\{\theta_j, j = 1, \ldots, n - 1\}$ which interlaces the first, i.e., $\alpha_i < \theta_i < \alpha_{i+1}$, show that there exists a matrix H with eigenvalues θ_j and a vector b and scalar π such that

$$\mathsf{A} = \begin{bmatrix} \mathsf{H} & \mathsf{b} \\ \mathsf{b}^* & \pi \end{bmatrix}$$

has eigenvalues $\{\alpha_i\}$. Note that we have written $\mathsf{B} = \mathsf{b}^*$ here.

10-1-3. Extend the result of Ex. 10-1-2 to the case in which $\alpha_i \leqslant \theta_i \leqslant \alpha_{i+1}$, i.e., suppose that $\alpha_j = \theta_j$ for some j and then construct b and π. (Harder.)

10-1-4. For $(n - m) = 1$ plot $\mathsf{W}(\xi) \equiv \omega(\xi)$ as given in (10-1-3) as a function of ξ assuming that A is tridiagonal with no zero subdiagonal elements.

10-1-5. Derive the inequalities that were not covered in the proof of Cauchy's theorem.

10-2 THE MINIMAX CHARACTERIZATIONS

In Sec. 1-4 the eigenvalues $\alpha_1, \alpha_2, \ldots, \alpha_n$ of A were declared to be the stationary values of the Rayleigh quotient function $\rho(\mathsf{x}) \equiv \mathsf{x}^*\mathsf{A}\mathsf{x}/\mathsf{x}^*\mathsf{x}$. In particular $\alpha_1 = \min \rho(\mathsf{x})$ over all $\mathsf{x} \neq \mathsf{o}$. In the same vein, for $j > 1$, α_j can be characterized as a constrained minimum, namely $\alpha_j = \min \rho(\mathsf{x})$ over all $\mathsf{x} \neq \mathsf{o}$ satisfying $\mathsf{z}_i^*\mathsf{x} = 0$ for $i < j$. From now onward we **tacitly** assume that o is excluded from the domain of ρ.

The blemish in this characterization of α_j is that it depends explicitly on the previous eigenvectors $\mathsf{z}_1, \ldots, \mathsf{z}_{j-1}$. The minimax characterization given below avoids mention of any eigenvectors and is often useful in theoretical work. For example, it provides the simplest proof that if a positive definite matrix V is added to A then **each** eigenvalue will increase, i.e., $\lambda_i[\mathsf{A} + \mathsf{V}] > \lambda_i[\mathsf{A}] i = 1, \ldots, n$.

Before confronting the general case it is helpful to look at a 3 by 3 example which permits a geometric interpretation. When A is positive definite its eigenvalues $\alpha_1, \alpha_2, \alpha_3$ are the squares of lengths of the semi-axes of the ellipsoid $\mathcal{E}$ which is the image of the unit of sphere under multiplication by $\mathsf{A}^{\frac{1}{2}}$, $\mathcal{E} = \{\mathsf{A}^{\frac{1}{2}}\mathsf{x}: \|\mathsf{x}\| = 1\}$. Fig. 10-2-1 exhibits a typical $\mathcal{E}$.

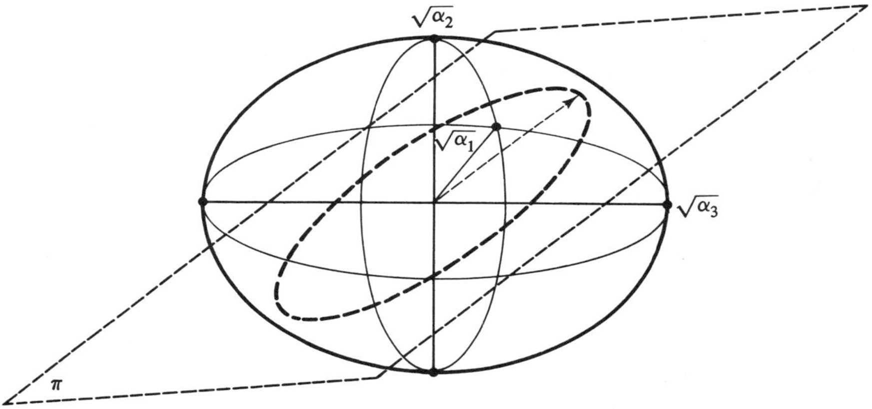

Figure 10-2-1 The minimax characterization of eigenvalues

Consider the lengths of all the rays from the origin o to points of $\mathcal{E}$. Clearly $\sqrt{\alpha_3}$ is the biggest and $\sqrt{\alpha_1}$ is the smallest of these lengths. What is a good way to describe $\sqrt{\alpha_2}$? Each plane π containing o intersects $\mathcal{E}$ in an ellipse, each ellipse has a diameter (a longest chord), and we can imagine all these diameters as π varies. The smallest of these diameters is $2\sqrt{\alpha_2}$. Moreover as x ranges over a plane π_1 the vectors $\mathsf{A}^{\frac{1}{2}}\mathsf{x}$ range over another plane π_2. Formally this characterizes $\sqrt{\alpha_2}$ as a minimum of a maximum,

$$\sqrt{\alpha_2} = \min_{\pi} \max_{\mathsf{x}\in\pi,\ \|\mathsf{x}\|=1} \|\mathsf{A}^{\frac{1}{2}}\mathsf{x}\|,$$

or, equivalently,

$$\alpha_2 = \min_{\pi} \max_{\mathsf{x}\in\pi} \rho(\mathsf{x}).$$

The theorem is stated in terms of subspaces of $\mathcal{E}^n$ (the analogues of the plane π above) and the only difficulty in the proof is keeping on top of the notation. The reader will find a full discussion of the representation of subspaces $\mathcal{S}$ in Sec. 11-1. For the proof we need three things: (1) a superscript on a subspace denotes its dimension, (2) for each $\mathcal{S}^j$ there is a n by j matrix B_j (actually many of them) whose columns form an orthonormal basis for $\mathcal{S}^j$, (3) for each B_j there is a n by $(n-j)$ matrix C_{n-j} (actually many of them) such that the n by n matrix $\mathsf{P} \equiv (\mathsf{B}_j, \mathsf{C}_{n-j})$ is orthogonal. The result itself goes back to Poincaré but is often called the Courant-Fischer theorem.

THEOREM For $j = 1, \ldots, n$,

$$\alpha_j \equiv \lambda_j[\mathsf{A}] = \min_{\mathcal{S}^j} \max_{\mathsf{u} \in \mathcal{S}^j} \rho(\mathsf{u}; \mathsf{A})$$

$$= \max_{\mathcal{R}^{j-1}} \min_{\mathsf{v} \perp \mathcal{R}^{j-1}} \rho(\mathsf{v}; \mathsf{A})$$

Here $\mathcal{S}$ and $\mathcal{R}$ are subspaces of $\mathcal{E}^n$. (10-2-1)

In words, the function ρ is continuous on the unit sphere and so must attain its maximum value on each j-dimensional unit sphere. The minimum of all these maximal values is α_j.

Proof. For each subspace $\mathcal{S}^j$ of $\mathcal{E}^n$ use (3) above to define

$$\hat{\mathsf{A}} \equiv \mathsf{P}^*\mathsf{A}\mathsf{P} = \begin{bmatrix} \mathsf{B}_j^*\mathsf{A}\mathsf{B}_j & \mathsf{B}_j^*\mathsf{A}\mathsf{C}_{n-j} \\ \mathsf{C}_{n-j}^*\mathsf{A}\mathsf{B}_j & \mathsf{C}_{n-j}^*\mathsf{A}\mathsf{C}_{n-j} \end{bmatrix}$$

The Cauchy interlace theorem applied to $\hat{\mathsf{A}}$ yields

$$\alpha_j \equiv \lambda_j[\mathsf{A}] = \lambda_j[\hat{\mathsf{A}}] \leqslant \lambda_j[\mathsf{B}_j^*\mathsf{A}\mathsf{B}_j].$$

Since $\lambda_j[\cdot]$ is the largest eigenvalue of a j by j matrix

$$\lambda_j[\mathsf{B}_j^*\mathsf{A}\mathsf{B}_j] = \max \rho(\mathsf{x}; \mathsf{B}_j^*\mathsf{A}\mathsf{B}_j), \qquad \mathsf{x} \in \mathcal{E}^j,$$

$$= \max \mathsf{x}^*\mathsf{B}_j^*\mathsf{A}\mathsf{B}_j\mathsf{x}/\mathsf{x}^*\mathsf{x},$$

$$= \max \mathsf{u}^*\mathsf{A}\mathsf{u}/\mathsf{u}^*\mathsf{u}, \qquad \mathsf{u} = \mathsf{B}_j\mathsf{x} \in \mathcal{S}^j,$$

since

$$\mathsf{u}^*\mathsf{u} = \mathsf{x}^*\mathsf{B}_j^*\mathsf{B}_j\mathsf{x} = \mathsf{x}^*\mathsf{x}.$$

This establishes that $\alpha_j \leqslant \max \rho(\mathsf{u}; \mathsf{A})$ over **any** $\mathcal{S}^j$. To prove that equality can hold consider the special subspace $\mathcal{Z}^j$, spanned by eigenvectors z_i belonging to α_i for $i = 1, 2, \ldots, j$. If α_j is a multiple eigenvalue there will be some choice for $\mathbf{z}_j$ and consequently for $\mathcal{Z}^j$ also. This lack of uniqueness is irrelevant here. The existence of $\mathcal{Z}^j$ follows from the spectral theorem (Fact 1-4) and for it alone $\lambda_i[\mathbf{B}_j^*\mathbf{A}\mathbf{B}_j] = \alpha_i$ for $i = 1, \ldots, j$.

The maximin characterization can be proved in a similar fashion by using the Cauchy result $\alpha_{-j} \geqslant \lambda_{-j}[\mathbf{B}_j^*\mathbf{A}\mathbf{B}_j] = \lambda_1[\mathbf{B}_j^*\mathbf{A}\mathbf{B}_j]$. It can also be derived from the minimax characterization by using $-\mathbf{A}$. □

One of the most important applications of this theorem is in the proof of the next one.

Exercises on Sec. 10-2

10-2-1. Without using the minimax theorem prove that $\lambda_1[\mathbf{A} + \mathbf{Y}] > \lambda_1[\mathbf{A}]$ if $\mathbf{Y}$ is positive definite. Give lower and upper bounds for the increase in terms of Y's eigenvalues.

10-2-2. Prove that $\alpha_j = \max_{\mathcal{R}^{j-1}} \min_{\mathbf{v} \perp \mathcal{R}^{j-1}} \rho(\mathbf{v}; \mathbf{A})$ by using the result $\alpha_{-k} \geqslant \lambda_{-k}[\mathbf{B}_j^*\mathbf{A}\mathbf{B}_j]$, $k \leqslant j$.

10-2-3. Prove the result in Ex. 10-2-2 by using $-\mathbf{A}$ and the minimax theorem.

10-3 THE MONOTONICITY THEOREM

A basic fact omitted from Chap. 1 is that $\lambda_i[\mathbf{W}]$ is not a linear function of $\mathbf{W}$. So what relations are there between the eigenvalues of $\mathbf{W}$, $\mathbf{Y}$, and $\mathbf{W} + \mathbf{Y}$? To simplify discussion let $\mathbf{A} = \mathbf{W} + \mathbf{Y}$ and

$$\lambda_{\pm i}[\mathbf{A}] = \alpha_{\pm i}, \quad \lambda_{\pm i}[\mathbf{W}] = \omega_{\pm i}, \quad \lambda_{\pm i}[\mathbf{Y}] = \eta_{\pm i}, \quad i = 1, \ldots, n.$$

Before stating the theorem we look at some special cases. From the minimal property of $\lambda_1[\cdot]$ it follows that

$$\omega_1 + \eta_1 \leqslant \alpha_1 \leqslant \omega_1 + \eta_{-1}.$$

It is also true, but less obviously so, that

$$\omega_j + \eta_1 \leqslant \alpha_j \leqslant \omega_j + \eta_{-1}, \quad j = 1, \ldots, n.$$

The second inequality is often put in the form

$$|\alpha_j - \omega_j| \leqslant \|\mathbf{Y}\|,$$

which is useful when $\mathbf{Y}$ is a small perturbation of $\mathbf{W}$. Less well known is the fact that Y's rank limits the extent to which α_j can stray from ω_j.

The following result was known in the nineteenth century but appears not to have been written down in full until [Weyl, 1912].

THEOREM Let $A = W + Y$. For any i, j satisfying $1 \leqslant i + j - 1 \leqslant n$, the following inequalities hold

$$\omega_i + \eta_j \leqslant \alpha_{i+j-1} \quad \text{and} \quad \alpha_{-(i+j-1)} \leqslant \omega_{-i} + \eta_{-j}. \tag{10-3-1}$$

Proof. The idea is to select appropriate subspaces in the max min characterization. Let $\mathcal{R}_W^{i-1}$ and $\mathcal{R}_Y^{j-1}$ be the subspaces of $\mathcal{E}^n$ defined implicitly by

$$\omega_i = \max_{\mathcal{R}^{i-1}} \min_{\mathbf{s} \perp \mathcal{R}^{i-1}} \rho(\mathbf{s}; W) \equiv \min_{\mathbf{s} \perp \mathcal{R}_W^{i-1}} \rho(\mathbf{s}; W),$$

$$\eta_j = \max_{\mathcal{R}^{j-1}} \min_{\mathbf{s} \perp \mathcal{R}^{j-1}} \rho(\mathbf{s}; Y) \equiv \min_{\mathbf{s} \perp \mathcal{R}_Y^{j-1}} \rho(\mathbf{s}; Y).$$

In fact $\mathcal{R}_W^{i-1}$ is the span of the eigenvectors of W belonging to the $i - 1$ smallest eigenvalues. Similarly for $\mathcal{R}_Y^{j-1}$. These subspaces may or may not have a trivial intersection. In any case we let $\mathcal{S}$ be the subspace of smallest dimension containing them both and write

$$k \equiv \dim(\mathcal{S}) + 1.$$

Dimensional considerations show that

$$k - 1 = \dim(\mathcal{S}) \leqslant (i - 1) + (j - 1) < n,$$

with equality only if the intersection, $\mathcal{R}_W^{i-1} \cap \mathcal{R}_Y^{j-1}$, is $\{\mathbf{o}\}$. So

$$\begin{aligned}
\alpha_{i+j-1} &\geqslant \alpha_k, \\
&= \max_{\mathcal{R}^{k-1}} \min_{\mathbf{x} \perp \mathcal{R}^{k-1}} \rho(\mathbf{x}; A), \quad \text{by definition of } \alpha_k, \\
&\geqslant \min_{\mathbf{x} \perp \mathcal{S}} \rho(\mathbf{x}; A), \quad \text{since } \dim \mathcal{S} = k - 1, \\
&= \min_{\mathbf{x} \perp \mathcal{S}} \left[\rho(\mathbf{x}; W) + \rho(\mathbf{x}; Y)\right], \\
&\geqslant \min_{\mathbf{u} \perp \mathcal{S}} \rho(\mathbf{u}; W) + \min_{\mathbf{v} \perp \mathcal{S}} \rho(\mathbf{v}; Y), \\
&\geqslant \min_{\mathbf{u} \perp \mathcal{R}_W^{i-1}} \rho(\mathbf{u}; W) + \min_{\mathbf{v} \perp \mathcal{R}_Y^{j-1}} \rho(\mathbf{v}; Y), \quad \text{since } \mathcal{R}_W^{i-1} \subset \mathcal{S}, \\
&\qquad\qquad \text{and } \mathcal{R}_Y^{j-1} \subset \mathcal{S}, \\
&= \omega_i + \eta_j.
\end{aligned}$$

The other inequality can be proved by using the minimax characterization. □

This is perturbation theory without the restriction that Y be small! Example 10-3-1 illustrates the monotonicity theorem applied to some small matrices.

EXAMPLE 10-3-1

$$\mathsf{W} = \begin{bmatrix} 5 & 2 & -1 \\ 2 & 3 & 1 \\ -1 & 1 & 1 \end{bmatrix}, \quad \mathsf{Y} = \begin{bmatrix} -1 & & \\ & 1 & \\ & & 2 \end{bmatrix}, \quad \mathsf{A} = \begin{bmatrix} 4 & 2 & -1 \\ 2 & 4 & 1 \\ -1 & 1 & 3 \end{bmatrix}.$$

$$\{\lambda_i[\mathsf{W}]\} = \{-0.0571, \quad 2.7992, \quad 6.2579\}$$
$$\{\lambda_1[\mathsf{Y}]\} = \{-1.0, \quad 1.0, \quad 2.0\}$$
$$\{\lambda_i[\mathsf{A}]\} = \{1.0, \quad 4.0, \quad 6.0\}$$

Table 10-3-1

i	j	$\lambda_i[\mathsf{W}] + \lambda_j[\mathsf{Y}]$	$\leqslant$	$\lambda_{i+j-1}[\mathsf{A}]$
1	1	−1.0571		1.0
2	1	1.7792		4.0
1	2	0.9429		
3	1	5.2579		6.0
2	2	3.7792		
1	3	1.9429		

Table 10-3-2

$-i$	$-j$	$\lambda_{-i}[\mathsf{W}] + \lambda_{-j}[\mathsf{Y}]$	$\geqslant$	$\lambda_{1-i-j}[\mathsf{A}]$
1	1	8.2579		6.0
2	1	4.7792		4.0
1	2	7.2579		
3	1	1.9424		1.0
2	2	3.7792		
1	3	5.2579		

The more that is known of Y the more precise the inferences we can draw. Let the inertia (defined in Sec. 1-4) of Y be (π, ν, ζ) and its rank be ρ; then every interval containing ρ ω's contains at least one α.

COROLLARY $\quad \omega_{k-\rho} \leqslant \omega_{k-\nu} \leqslant \alpha_k \leqslant \omega_{k+\pi} \leqslant \omega_{k+\rho}. \qquad$ (10-3-2)

Proof. By definition of ν, $\eta_{\nu+1} \geqslant 0$. Pick $i = k - \nu, j = \nu + 1$ in the theorem to find

$$\omega_{k-\nu} \leqslant \omega_{k-\nu} + \eta_{\nu+1} \leqslant \alpha_k.$$

Then pick $i = n + 1 - k - \pi, j = \pi + 1$ and apply the other inequality, using $\eta_{-(\pi+1)} \leqslant 0$, to get

$$\alpha_{-(n+1-k)} \leqslant \omega_{-(n+1-k-\pi)} + \eta_{-(\pi+1)} \leqslant \omega_{-(n+1-k-\pi)}.$$

On using $\alpha_{-(n+1-k)} = \alpha_k$ the assertion is verified. □

In particular, when Y is positive semi-definite, i.e., $\nu(\mathsf{Y}) = 0$, then Weyl's original **monotonicity** result is obtained

$$\omega_k \leqslant \alpha_k \leqslant \omega_{k+\rho}, \qquad k = 1, \ldots, n.$$

In words, all eigenvalues increase but the increase is limited by rank (Y) as well as by $\|\mathsf{Y}\|$.

EXAMPLE 10-3-2

Let $\mathsf{Y} = \mathsf{yy}^* \neq \mathsf{O}$ be a positive semi-definite rank one matrix; then $\rho = 1$ in the preceding result and

$$\omega_k \leqslant \alpha_k \leqslant \omega_{k+1}.$$

On the other hand we can pick $i = n + 1 - k, j = 1$ in the monoticity theorem to obtain

$$\alpha_k \leqslant \omega_k + \eta_{-1} = \omega_k + \|\mathsf{Y}\|$$

which is a special case of the result quoted at the beginning of the section.

Applications of the theorem occur in Secs. 4-5 and 11-6.

Exercise on Sec. 10-3

10-3-1. When can equality occur in the monotonicity theorem?

10-4 THE RESIDUAL INTERLACE THEOREM

This section embarks on a presentation of results which have been inspired by demands to estimate eigenvalues of larger and larger matrices. Cauchy's interlace theorem is too general to be directly applicable to the task. Recall from Sec. 10-1 that

$$\mathsf{A} = \begin{bmatrix} \mathsf{H} & \mathsf{B}^* \\ \mathsf{B} & \mathsf{Y} \end{bmatrix}, \qquad \left\{ \begin{matrix} \mathsf{H} \text{ is } m \text{ by } m, & \lambda_i[\mathsf{H}] = \theta_i \\ \mathsf{A} \text{ is } n \text{ by } n, & \lambda_i[\mathsf{A}] = \alpha_i \end{matrix} \right\}. \qquad (10\text{-}4\text{-}1)$$

Cauchy brackets each α_j using some θ_i and $\pm\infty$. These bounds are inevitably very weak when $m \leqslant n/2$ and yet interest centers on cases with $n = 1000$, $m = 100$.

What is lacking? Cauchy's theorem ignores B. Yet when $\mathsf{B} = \mathsf{O}$ each θ_i is an α_j for some j. It is plausible that when B is small, in some sense, then each θ_i must be close to one of the α_j.

In many applications B is available and the theorems which follow can be regarded as providing the best possible inferences about the $\{\alpha_i\}$ when B is known in addition to the $\{\theta_i\}$. Actually B is a **residual** matrix and Chap. 11 explains how it arises in the Rayleigh-Ritz method.

B is usually long and thin, and for the remaining results of this chapter it must be put into the more useful form $\begin{pmatrix} \mathsf{C} \\ \mathsf{O} \end{pmatrix}$, where C has only k [$\equiv$ rank (B)] rows and satisfies $\mathsf{C}^*\mathsf{C} = \mathsf{B}^*\mathsf{B}$. There are several satisfactory ways in which this can be done, as is indicated in the exercises. Error bounds utilizing only H and $\|\mathsf{B}\|$ are presented in Chap. 11.

From now on we study A's of the form

$$\mathsf{A} = \begin{bmatrix} \mathsf{H} & \mathsf{C}^* & \mathsf{O}^* \\ \mathsf{C} & \mathsf{V} & \mathsf{Z}^* \\ \mathsf{O} & \mathsf{Z} & \mathsf{W} \end{bmatrix}; \qquad \begin{array}{l} \mathsf{H} \text{ is } m \text{ by } m, \\ \mathsf{V} \text{ is } k \text{ by } k, \\ \mathsf{A} \text{ is } n \text{ by } n, \end{array} \tag{10-4-2}$$

but only H and C are available. In some applications $k = m$; in others $k \ll m \ll n$; $k = 3$, $m = 40$, $n = 1000$ is typical. The results are directly useable when C is small. However in Secs. 10-6 and 10-7 it will be shown that with additional, very crude knowledge of the missing submatrix significantly tighter bounds can be obtained. It is quite surprising that useful information about the spectrum of a 1000 by 1000 matrix can be derived from the eigenvalues of a few matrices of order 50. (See Chaps. 13, 14, and 15.)

Being ignorant of the k by k submatrix V we replace it by a k by k symmetric matrix X of our choice and in what follows X is a parameter in an auxiliary matrix of order $m + k$:

$$\mathsf{M}(\mathsf{X}) \equiv \begin{bmatrix} \mathsf{H} & \mathsf{C}^* \\ \mathsf{C} & \mathsf{X} \end{bmatrix}. \tag{10-4-3}$$

Its eigenvalues are denoted by $\mu_i = \mu_i(\mathsf{X}) \equiv \lambda_i[\mathsf{M}(\mathsf{X})]$, $i = 1, \ldots, m + k$. Appropriate choices for X will be discussed shortly. For some of the results it is necessary to assume that X satisfies the following unverifiable constraint:

$$(\mathsf{V} - \mathsf{X}) \text{ is invertible.} \tag{10-4-4}$$

With apt choices for X the μ's give much more information than the θ's alone provide. Cauchy's interlace theorem applied to the three pairs

(A, H), [A, M(V)], and [M(X), H] yields (Ex. 10-4-3),

$$\alpha_i \leqslant \theta_i \leqslant \mu_{i+k}(\mathrm{X}); \qquad \mu_{-i-k}(\mathrm{X}) \leqslant \theta_{-i} \leqslant \alpha_{-i}; \qquad \alpha_i \leqslant \mu_i(\mathrm{V}). \quad (10\text{-}4\text{-}5)$$

It is not always true that $\mu_i(\mathrm{X}) \leqslant \alpha_i$; yet the next theorem shows that the interval $[\mu_i, \mu_{i+k}]$ must contain an eigenvalue of A. To say which one is more difficult.

EXAMPLE 10-4-1

H = diag(10, 11), C = diag(0.1, 1.). The simple bounds in Sec. 4-5 dictate that [9.9, 10.1] and [10, 12] each contain an α, **possibly the same one** since the intervals overlap. Selected values of X yield the intervals $[\mu_1, \mu_3]$ and $[\mu_2, \mu_4]$ each containing its own α.

Table 10-4-1

X	μ_1	μ_2	μ_3	μ_4
diag(10, 10)	9.382	9.9	10.1	11.618
diag(11, 11)	9.9901	10.0	11.010	12.0
diag(10, 11.21)	9.9	10.1	10.1	12.11
diag(9.85, 11.1)	9.8	10.05	10.05	12.05

This tiny auxiliary calculation removes the fear that 10 and 11 are approximating a single eigenvalue α. The last two X's were chosen in the light of Theorem (10-5-3).

The following unpublished result is due to Kahan.

THEOREM Consider A as given in (10-4-2) together with eigenvalues $\mu_i(\mathrm{X})$, $i = 1, \ldots, m + k$, of the auxiliary matrix M(X) of (10-4-3) with any k by k X satisfying (10-4-4). Each interval $[\mu_j, \mu_{j+k}]$, $j = 1, \ldots, m$ contains a different eigenvalue α_J of A. In addition, there is a different eigenvalue α_I outside each open interval (μ_i, μ_{i+m}), $i = 1, \ldots, k$. (10-4-6)

The distinction between the interior and exterior intervals disappears if the real line is closed $(-\infty = +\infty)$ and if each subscript j on μ is read as $j - m - k$ if $j > m + k$.

Proof. Split A adroitly and apply the monotonicity theorem.

$$\mathsf{A} = \begin{bmatrix} \mathsf{H} & \mathsf{C}^* & \mathsf{O} \\ \mathsf{C} & \mathsf{X} & \mathsf{O} \\ \mathsf{O} & \mathsf{O} & \mathsf{W} - \mathsf{Z}(\mathsf{V}-\mathsf{X})^{-1}\mathsf{Z}^* \end{bmatrix} + \begin{bmatrix} \mathsf{O} & \mathsf{O} & \mathsf{O} \\ \mathsf{O} & \mathsf{V}-\mathsf{X} & \mathsf{Z}^* \\ \mathsf{O} & \mathsf{Z} & \mathsf{Z}(\mathsf{V}-\mathsf{X})^{-1}\mathsf{Z}^* \end{bmatrix}$$

$$= \mathsf{T}(\mathsf{X}) + \mathsf{U}(\mathsf{X}), \quad \text{defining T and U.} \tag{10-4-7}$$

The block triangular factorization of U is, by design,

$$\mathsf{U} = \begin{bmatrix} \mathsf{I}_m & \mathsf{O} & \mathsf{O} \\ \mathsf{O} & \mathsf{I}_k & \mathsf{O} \\ \mathsf{O} & \mathsf{Z}(\mathsf{V}-\mathsf{X})^{-1} & \mathsf{I}_{n-m-k} \end{bmatrix} \begin{bmatrix} \mathsf{O} & \mathsf{O} & \mathsf{O} \\ \mathsf{O} & \mathsf{V}-\mathsf{X} & \mathsf{O} \\ \mathsf{O} & \mathsf{O} & \mathsf{O} \end{bmatrix} \times \begin{bmatrix} \mathsf{I}_m & \mathsf{O} & \mathsf{O} \\ \mathsf{O} & \mathsf{I}_k & (\mathsf{V}-\mathsf{X})^{-1}\mathsf{Z}^* \\ \mathsf{O} & \mathsf{O} & \mathsf{I}_{n-m-k} \end{bmatrix}. \tag{10-4-8}$$

Denote the inertia of $\mathsf{V} - \mathsf{X}$ by $(\pi, \nu, 0)$ where, by assumption (10-4-4), $\pi + \nu = k$. Sylvester's inertia theorem applied to (10-4-8) shows that the inertia of $\mathsf{U}(\mathsf{X})$ is $(\pi, \nu, n - k)$.

Now let $\tau_i = \lambda_i[\mathsf{T}(\mathsf{X})]$, $i = 1, \ldots, n$ and apply the corollary of the monotonicity theorem to find

$$\tau_{i-\nu} \leqslant \alpha_i \leqslant \tau_{i+\pi} \leqslant \tau_{i+k}, \qquad i = 1, \ldots, n. \tag{10-4-9}$$

Since T is a direct sum of M and another matrix each μ is a τ. The theorem is simply (10-4-9) invoked for the first m values of i for which $\tau_{i-\nu}$ is a different eigenvalue of $\mathsf{M}(\mathsf{X})$. To be precise let J be the minimal index such that $\mu_j = \tau_{J-\nu}$. Then (10-4-9) yields $\mu_j \leqslant \alpha_J \leqslant \tau_{J+\pi}$. Not every τ is a μ and the smallest μ guaranteed to exceed $\tau_{J+\pi}$ is given by $\tau_{J+\pi} = \tau_{(J-\nu)+k} \leqslant \mu_{j+k}$. See the illustration below.

The final assertion of the theorem comes from the extra one-sided bounds in the monotonicity theorem, namely

$$\begin{aligned} \tau_{j-\nu} &\leqslant \alpha_j && \text{for } j = \nu + 1, \ldots, n, \\ \alpha_j &\leqslant \tau_{j+\pi} && \text{for } j = 1, 2, \ldots, n - \pi. \end{aligned} \tag{10-4-10}$$

Again apply these inequalities for just those values of j such that $\tau_{j-\nu} = \mu_{m+i}$, $i = 1, \ldots, k$. The details are left to the reader as Ex. 10-4-4. □

Illustration: $k = 2, \pi = \nu = 1$. From the illustration we may deduce bounds for α_4 and α_8:

$$j = 1; \quad J = 4 \quad \text{and} \quad \mu_1 \leqslant \alpha_4 \leqslant \mu_3,$$
$$j = 2; \quad J = 8 \quad \text{and} \quad \mu_2 \leqslant \alpha_8 \leqslant \mu_4.$$

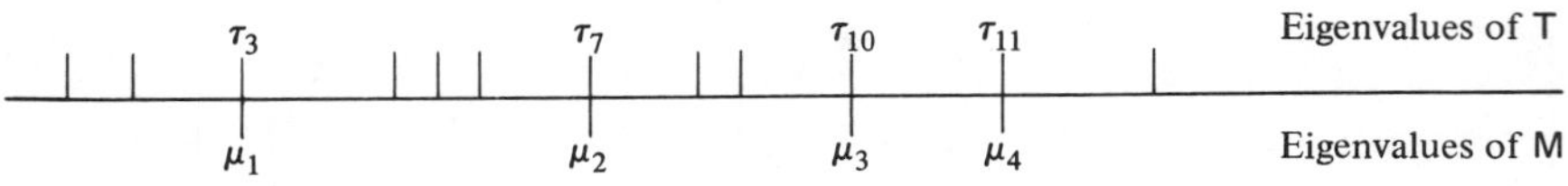

How should X be chosen? The answer depends on whether or not there is any adscititious information about A besides H and C. Various cases are considered in the following sections.

Exercises on Sec. 10-4

10-4-1. Show that if $X_1 \geqslant X_2$, in the sense that $X_1 - X_2$ is nonnegative definite, then $\mu_j(X_1) \geqslant \mu_j(X_2), j = 1, \ldots, m + k$.

10-4-2. Consider the simple choice $X = \xi$ for large scalar ξ. Use Sylvester's theorem to show that

$$\text{as } \xi \longrightarrow -\infty, \quad \mu_{i+k}(\xi) \longrightarrow \begin{cases} -\infty, & \text{for } 1 - k \leqslant i \leqslant 0, \\ \theta_i, & \text{for } 1 \leqslant i \leqslant m, \end{cases}$$

$$\text{as } \xi \longrightarrow +\infty, \quad \mu_{-i-k}(\xi) \longrightarrow \begin{cases} +\infty, & \text{for } 1 - k \leqslant i \leqslant 0, \\ \theta_{-i}, & \text{for } 1 \leqslant i \leqslant m \end{cases}$$

10-4-3. Use Cauchy's theorem to derive all the inequalities in (10-4-5).

10-4-4. Establish the right choice of index in (10-4-9) to identify the eigenvalue α outside (μ_i, μ_{i+m}).

*10-5 LEHMANN'S OPTIMAL INTERVALS

In the event of no extra outside information about A other than the submatrices H and C of (10-4-2) the best choice for the matrix X of (10-4-3) was given, implicitly, in [Lehmann, 1949] and [Lehmann, 1966]. In our development this result, namely (10-5-5), is an immediate corollary of Kahan's residual interlace theorems (10-4-6) and (10-5-3).

The word optimal has the special, but reasonable, meaning that for **any real number** ζ one can describe the X which yields the tightest bounds for A's eigenvalues α on either side of ζ. For simplicity we require that $\zeta \neq \theta_i$ ($\equiv \lambda_i[H]$), $i = 1, \ldots, m$. Recall from (10-4-6) that all the inclusion intervals derived from M(X) are of the form $[\mu_j, \mu_{j+k}]$ where k is the rank of C. Consequently, in order to have a pair of inclusion intervals abutting at ζ, say $[\mu', \zeta]$ $[\zeta, \mu'']$, it is necessary that ζ be a k-fold eigenvalue of M(X). The proper choice turns out to be the k by k matrix

$$\mathsf{X}_\zeta \equiv \zeta + \mathsf{C}(\mathsf{H} - \zeta)^{-1}\mathsf{C}^*. \tag{10-5-1}$$

> LEMMA The parameter ζ is a k-fold eigenvalue of $\mathsf{M}(\mathsf{X}_\zeta)$. Further, each eigenvalue $\mu_i(\mathsf{X}_\zeta)$ is a monotone nondecreasing function of ζ, $i = 1, 2, \ldots, m + k$. (10-5-2)

Proof. Consider the block triangular factorization of $\mathsf{M} - \zeta$

$$\mathsf{M}(\mathsf{X}_\zeta) - \zeta = \begin{bmatrix} \mathsf{H} - \zeta & \mathsf{C}^* \\ \mathsf{C} & \mathsf{C}(\mathsf{H} - \zeta)^{-1}\mathsf{C}^* \end{bmatrix}$$

$$= \begin{bmatrix} \mathsf{I} & \mathsf{O} \\ \mathsf{C}(\mathsf{H} - \zeta)^{-1} & \mathsf{I} \end{bmatrix} \begin{bmatrix} \mathsf{H} - \zeta & \mathsf{O} \\ \mathsf{O}^* & \mathsf{O} \end{bmatrix} \begin{bmatrix} \mathsf{I} & (\mathsf{H} - \zeta)^{-1}\mathsf{C}^* \\ \mathsf{O}^* & \mathsf{I} \end{bmatrix}.$$

By Sylvester's inertia theorem $\mathsf{M} - \zeta$ has the same nullity (namely, order − rank) as diag$(\mathsf{H} - \zeta, \mathsf{O})$ which is k since $\zeta \neq \lambda_i[\mathsf{H}]$. In other words ζ is a k-fold eigenvalue of $\mathsf{M}(\mathsf{X}_\zeta)$.

To show monotonicity we invoke the minimax characterization (Sec. 10-3) to find

$$\mu_i(\mathsf{X}_\zeta) = \min_{\mathcal{S}^i} \max_{\mathsf{s} \in \mathcal{S}^i} \frac{\mathsf{s}^*\mathsf{M}(\mathsf{X}_\zeta)\mathsf{s}}{\mathsf{s}^*\mathsf{s}},$$

where $\mathcal{S}^i$ is any subspace of dimension i. Partition s^* as $(\mathsf{u}^*, \mathsf{v}^*)$ where u is m by 1, v is k by 1. Then, by the definition of M_ζ in (10-4-3),

$$\mathsf{s}^*\mathsf{M}_\zeta\mathsf{s} = \mathsf{u}^*\mathsf{H}\mathsf{u} + \mathsf{u}^*\mathsf{C}^*\mathsf{v} + \mathsf{v}^*\mathsf{C}\mathsf{u} + \mathsf{v}^*(\zeta + \mathsf{C}(\mathsf{H} - \zeta)^{-1}\mathsf{C}^*)\mathsf{v},$$

and

$$\frac{d}{d\zeta}(\mathsf{s}^*\mathsf{M}_\zeta\mathsf{s}) = \mathsf{v}^*\mathsf{v} + \mathsf{v}^*\mathsf{C}(\mathsf{H} - \zeta)^{-2}\mathsf{C}^*\mathsf{v}$$

$$= \|\mathsf{v}\|^2 + \|(\mathsf{H} - \zeta)^{-1}\mathsf{C}^*\mathsf{v}\|^2 \geqslant 0. \qquad \square$$

The best interval depends on the location of ζ.

> THEOREM Let ζ be any number satisfying $\theta_j < \zeta < \theta_{j+1}$. Each interval $[\mu_j(\mathsf{X}_\zeta), \zeta]$ and $[\zeta, \mu_{j+k+1}(\mathsf{X}_\zeta)]$ contains at least one of A's eigenvalues. Moreover there is an A with eigenvalues only at the end points of these intervals. (10-5-3)

Proof. The interlace theorem (10-1-2) applied to $\mathsf{M}(\mathsf{X}_\zeta)$ yields

$$\mu_{j+\ell}(\mathsf{X}_\zeta) \leqslant \theta_{j+\ell} \leqslant \mu_{j+k+\ell}(\mathsf{X}_\zeta), \qquad \ell = 0, 1.$$

Since ζ is a k-fold eigenvalue of $\mathsf{M}(\mathsf{X}_\zeta)$ these inequalities tell us which of M's eigenvalues equal ζ;

$$\mu_j \leqslant \theta_j < \mu_{j+1} = \mu_{j+2} = \cdots = \mu_{j+k} = \zeta < \theta_{j+1} \leqslant \mu_{j+k+1}.$$

The conclusion is now a corollary of Kahan's theorem (10-4-6) and simply exploits the coincidence of all the μ's. The A which proves optimality is $\operatorname{diag}[\mathsf{M}(\mathsf{X}_\zeta), \zeta \mathsf{I}_{n-m-k}]$; all its eigenvalues, other than the μ's, are at ζ. □

It is not obvious from this result how to pick ζ to reduce the width of $[\mu_j, \zeta]$ to a minimum. Indeed no general formula is known but Fig. 10-5-1 shows how the width varies with ζ. An application was given in Example 10-4-1.

The original form in which Lehmann presented his results confirms that ζ be chosen not too far from θ_j. How that formulation is derived from the preceding theorem is indicated in Exs. 10-5-5, 10-5-6, and 10-5-7.

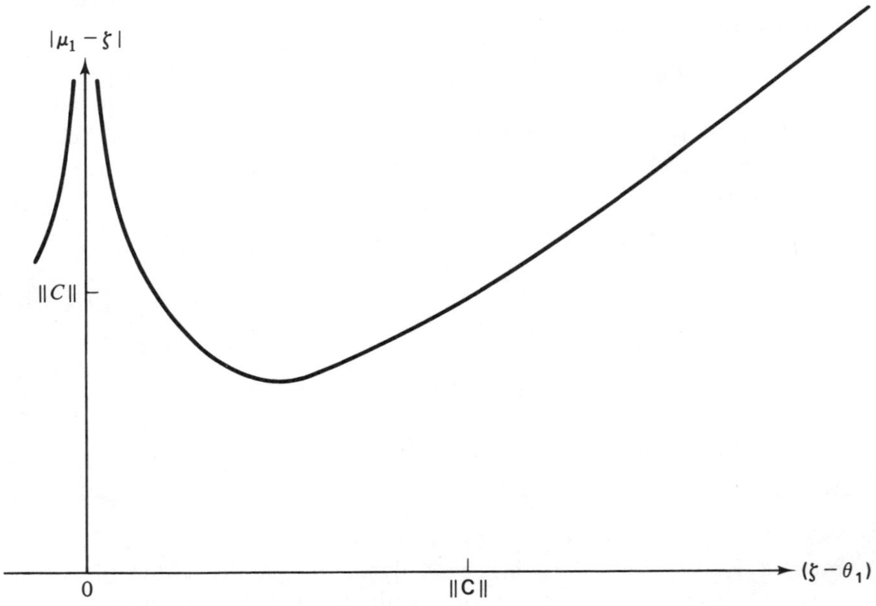

Figure 10-5-1 The width of an optimal interval as a function of $\zeta - \theta_1$

For any real number ξ Lehmann defines

$$\delta_i = \delta_i(\xi) \equiv \sqrt{\lambda_i[(\mathsf{H} - \xi)^2 + \mathsf{C}^*\mathsf{C}]}\,, \quad i = 1, \ldots, m. \qquad (10\text{-}5\text{-}4)$$

Here is the original formulation of (10-5-3) from [Lehmann, 1949].

> THEOREM For every ξ and for $i = 1, 2, \ldots, m$, A has at least i eigenvalues in $[\xi - \delta_i, \xi + \delta_i]$, has at least $m + 1 - j$ eigenvalues not in $(\xi - \delta_i, \xi + \delta_i)$, and there is an A with no eigenvalue in this open interval. (10-5-5)

In Lehmann's formulation the goal is to minimize $\delta_1(\xi)$ and by Weyl's monotonicity theorem applied to (10-5-4),

$$\delta_1(\xi)^2 \leqslant \lambda_1[(\mathsf{H} - \xi)^2] + \|\mathsf{C}\|^2.$$

In many applications $\|\mathsf{C}\|$ is small compared to $\|\mathsf{H}\|$ and, in any case, this **bound** on δ_1^2 is minimized when $\xi = \theta_j$ for some j. However the choice $\xi = \theta_j$ does not in general minimize δ_1. It turns out (Ex. 10-5-5) that ξ and $\delta_1(\xi)$ are, respectively, the midpoint and half-width of $[\mu_j, \mu_{j+k}]$ as indicated in the next figure.

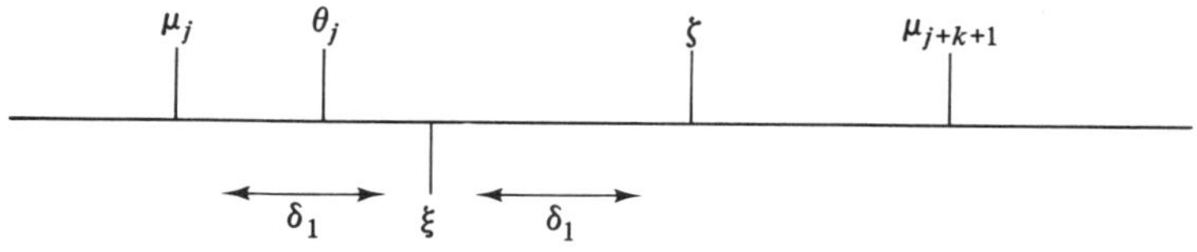

Which formulation is preferable? The matrix $(\mathsf{H} - \xi)^2 + \mathsf{C}^*\mathsf{C}$ is m by m while $\mathsf{M}(\mathsf{X}_\xi)$ is $(m + k)$ by $(m + k)$. However k of M's eigenvalues are known and the corresponding subspace is known too (Ex. 10-5-8). This information can be used to facilitate the calculation of the rest of M's eigenvalues but such efficiency considerations must take second place to those of accuracy.

Unfortunately the δ formulation forces the calculation of $\delta_1^2(\xi)$ in order to find $\delta_1(\xi)$. Now δ_1^2 is the smallest eigenvalue and so (see Sec. 2-7) will be computed with an error which is tiny compared to $\|(\mathsf{H} - \xi)^2 + \mathsf{C}^*\mathsf{C}\|$. The cases of greatest interest will have $\delta_1/|\xi|$ small, say 10^{-5}, and although δ_m^2 will be correct to working accuracy we must expect ten fewer correct decimals in δ_1^2 and hence in δ_1. This loss is enough to destroy the value of the computation just when its potential is greatest. Since $\delta_1/|\xi|$ is small there is no need for very high relative accuracy in δ_1 but there is need for some correct figures. With the M formulation the computed values μ_j and μ_{j+k} $(=\zeta)$ may be very close but that does not prevent a good program from computing them to as much accuracy as their separation warrants. To summarize, the δ formulation inevitably discards nearly twice as many figures as is necessary. Exercise 10-5-10 illustrates this phenomenon.

Exercises on Sec. 10-5

10-5-1. What happens to $\mathsf{X}_\zeta = \zeta + \mathsf{C}(\mathsf{H} - \zeta)^{-1}\mathsf{C}^*$ and $\mu_i(\mathsf{X}_\zeta)$, $i = 1, \ldots, m + k$, as $\zeta \longrightarrow \theta_j$?

10-5-2. Show that for $i \leqslant j$ the interval $[\mu_i(\mathsf{X}_\zeta), \zeta]$ contains at least as many eigenvalues α of A as $[\mu_i(\mathsf{X}_\zeta), \zeta]$ contains eigenvalues μ of $\mathsf{M}(\mathsf{X}_\zeta)$. Similarly for $i > j$ the interval $[\zeta, \mu_i(\mathsf{X}_\zeta)]$ contains at least as many eigenvalues α as $[\zeta, \mu_i(\mathsf{X}_\zeta)]$ contains μ's.

10-5-3. What is the smallest number of α's which can lie in the union of $[-\infty, \mu_i(\mathsf{X}_\zeta)]$ and (ζ, ∞)? *Hint*: Use the last part of Kahan's theorem.

10-5-4. Show that the interval in Ex. 10-5-2 is optimal in the sense that no subinterval can be proved to contain that many of A's eigenvalues.

10-5-5. (Difficult). Let $\theta_j < \zeta < \theta_{j+1}$ and let μ be any eigenvalue of $\mathsf{M}(\mathsf{X}_\zeta)$ other than ζ. Define $\xi = (\zeta + \mu)/2$, $\delta = (\zeta - \mu)/2$. Show that

$$\det\left[(\mathsf{H} - \xi)^2 + \mathsf{C}^*\mathsf{C} - \delta^2\right] = 0.$$

10-5-6. Let $\delta_i(\xi) \equiv \lambda_i[(\mathsf{H} - \xi)^2 + \mathsf{C}^*\mathsf{C}]$, $i = 1, \ldots, m$. Use Cauchy's interlace theorem applied to $(\mathsf{A} - \xi)^2$ to prove that $[\xi - \delta_j(\xi), \xi + \delta_j(\xi)]$ contains at least j of A's eigenvalues.

10-5-7. Derive Lehmann's original formulation from (10-5-3).

10-5-8. Verify that for any k-vector v the $(m + k)$-vector $[\mathsf{v}^*\mathsf{C}(\mathsf{H} - \zeta)^{-1}, -\mathsf{v}^*]^*$ is an eigenvector of $\mathsf{M}(\mathsf{X}_\zeta)$ belonging to ζ.

10-5-9. (Difficult). Deflate $\mathsf{M}(\mathsf{X}_\zeta)$ analytically to obtain a more complicated m by m matrix whose eigenvalues are $\mu_1, \ldots, \mu_j, \mu_{j+k+1}, \ldots, \mu_{m+k}$. *Hint*: Use the matrix $\begin{bmatrix} (\mathsf{H} - \zeta)^{-1}\mathsf{C}^* \\ -\mathsf{I}_k \end{bmatrix}\mathsf{R}$, where R is k by k upper triangular chosen so that the columns of the product are orthonormal, i.e., $\mathsf{R}^{-*}\mathsf{R}^{-1}$ is the Choleski decomposition (see Chap. 3) of

$$\mathsf{C}(\mathsf{H} - \zeta)^{-2}\,\mathsf{C}^* + \mathsf{I}.$$

10-5-10. Consider the case $\xi = 1$, $\mathsf{C} = (0, \gamma^2)$, and $\mathsf{H} = \begin{bmatrix} \gamma & 1 \\ 1 & \gamma \end{bmatrix}$. Suppose that the sum $1 + \gamma^4$ is computed as 1 (or try $\gamma = 10^{-3}$). Study the computation of $\delta_1(\xi)$ as given in (10-5-4). Then consider the computation of the $\mu_i(\mathsf{X}_\zeta)$ when $\zeta = 1 + \gamma^2/\sqrt{1 + \gamma^2}$.

*10-6 THE USE OF BOUNDS ON THE MISSING SUBMATRIX

Recall that we are studying matrices of the form

$$\mathsf{A} = \left[\begin{array}{c|cc} \mathsf{H} & \mathsf{C}^* & \mathsf{O}^* \\ \hline \mathsf{C} & & \\ \mathsf{O} & & \mathsf{Y} \end{array}\right], \qquad \begin{array}{l} \mathsf{A} \text{ is } n \text{ by } n, \\ \mathsf{H} \text{ is } m \text{ by } m, \end{array} \tag{10-6-1}$$

and Y is unknown. Lehmann's bounds can be improved a lot if just a little is known about Y, a bound on $\|\mathsf{Y}\|$ or $\|\mathsf{Y}^{-1}\|$ for instance. To derive these smaller intervals it is helpful to see what can be said when Y's eigenvalues $\{\eta_i, i = 1, \ldots, n-m\}$ are known. To this end set the matrix X in (10-4-3) to $\eta \mathsf{I}_k$;

$$\mathsf{M}(\eta) \equiv \begin{bmatrix} \mathsf{H} & \mathsf{C}^* \\ \mathsf{C} & \eta \mathsf{I}_k \end{bmatrix}, \quad \mu_i(\eta) \equiv \lambda_i[\mathsf{M}(\eta)], \quad i = 1, \ldots, m+k. \tag{10-6-2}$$

Convention: $\mu_i = -\infty$ when $i \leqslant 0$; $\mu_i = +\infty$ when $i > m+k$. The following blizzard of results comes from Weyl's monotonicity theorem (10-3-1). They were obtained by Kahan in 1957 but have been derived independently and published in [Weinberger 1959, and 1974].

LEMMA For $0 \leqslant i \leqslant m+1$, $0 \leqslant j \leqslant n-m$, and $\alpha_i = \lambda_i[\mathsf{A}]$,

$$\min\{\mu_i(\eta_{j+1}), \eta_{j+1}\} \leqslant \alpha_{i+j} \leqslant \max\{\mu_{i+k}(\eta_j), \eta_j\}, \tag{10-6-3}$$

$$\min\{\mu_{-i-k}(\eta_{-j}), \eta_{-j}\} \leqslant \alpha_{-i-j} \leqslant \max\{\mu_{-i}(\eta_{-j-1}), \eta_{-j-1}\}, \tag{10-6-4}$$

$$\alpha_i \leqslant \mu_i(\eta_{-1}), \qquad \mu_{-i}(\eta_1) \leqslant \alpha_{-i}. \tag{10-6-5}$$

Proof. Split A appropriately and apply the monotonicity theorem in Sec. 10-3 adroitly to get

$$\alpha_{i+j} \geqslant \lambda_i\left[\left(\begin{array}{c|cc} \mathsf{H} & \mathsf{C}^* & \mathsf{O}^* \\ \hline \mathsf{C} & & \\ \mathsf{O} & & \eta_{j+1} \end{array}\right)\right] + \lambda_{j+1}\left[\left(\begin{array}{c|cc} \mathsf{O} & \mathsf{O}^* & \mathsf{O}^* \\ \hline \mathsf{O} & & \\ \mathsf{O} & & \mathsf{Y} - \eta_{j+1} \end{array}\right)\right],$$

$$= \lambda_i\left[\begin{pmatrix} \mathsf{M}(\eta_{j+1}) & \mathsf{O}^* \\ \mathsf{O} & \eta_{j+1} \end{pmatrix}\right] + 0$$

$$\geqslant \min\{\mu_i(\eta_{j+1}), \eta_{j+1}\}.$$

Note that the second term is still 0 when $j = n - m$ and $\eta_{j+1} = +\infty$. The result is vacuously true when $i = 0$. The dual part of the monotonicity theorem leads to

$$\alpha_{-p-q} \leqslant \max\{\mu_{-p}(\eta_{-q-1}), \eta_{-q-1}\}.$$

Now translate this, using $\alpha_{-\ell} = \alpha_{n+1-\ell}$, $\mu_{-\ell} = \mu_{m+k+1-\ell}$, $\eta_{-\ell} = \eta_{n-m+1-\ell}$, with $p = m + 1 - i$, $q = n - m - j$, to get

$$\alpha_{i+j} \leqslant \max\{\mu_{i+k}(\eta_j), \eta_j\}.$$

Thus (10-6-3) is proved and (10-6-4) is equivalent to (10-6-3).
The last result, (10-6-5), is left as Ex. 10-6-1. □

The preceding results are academic; if the η_i were known then the roles of Y and H could be reversed to good effect. In practice if anything is known of Y it is likely to be a crude bound. Recall that $\eta \leqslant$ Y means that $\eta \leqslant \rho(\mathsf{x}; \mathsf{Y})$ for all $\mathsf{x} \neq \mathsf{o}$.

COROLLARY 1 For $i = 1, \ldots, m$,

if $\eta' \leqslant$ Y then

$$\min\{\mu_i(\eta'), \eta'\} \leqslant \alpha_i \text{ and } \mu_{-i}(\eta') \leqslant \alpha_{-i},$$

if Y $\leqslant \eta''$ then

$$\alpha_i \leqslant \mu_i(\eta'') \text{ and } \alpha_{-i} \leqslant \max\{\mu_{-i}(\eta''), \eta''\}. \quad (10\text{-}6\text{-}6)$$

Proof. By monotonicity of the μ_i, (10-5-2),

$$\mu_i(\eta') \leqslant \mu_i(\eta_1), \mu_i(\eta_{-1}) \leqslant \mu_i(\eta'').$$

Now apply (10-6-3) with $j = 0$ for the lower bound on α_i and (10-6-5) for the upper bound. The treatment of α_{-i} is similar. □

It is also possible to exploit a known gap in Y's spectrum to find some inclusion intervals.

COROLLARY 2 Suppose that

$$\eta_\pi \leqslant \eta' < \eta'' \leqslant \eta_{\pi+1} \text{ for some } \pi, \text{ and} \quad (10\text{-}6\text{-}7a)$$

$$\theta_j \leqslant \eta' \leqslant \theta_{j+1} \leqslant \cdots \leqslant \theta_{j+\ell} \leqslant \eta'' < \theta_{j+\ell+1}, \quad (10\text{-}6\text{-}7b)$$

then

$$\mu_i(\eta'') \leqslant \alpha_{\pi+i} \leqslant \mu_{i+k}(\eta'), \qquad j + 1 \leqslant i \leqslant j + \ell. \quad (10\text{-}6\text{-}7c)$$

The proof is left as Ex. 10-6-3. Note that it is not necessary that the index π be known for these bounds to be of use.

EXAMPLE 10-6-1

$$\mathsf{A} = \left[\begin{array}{c|cc} \theta & \gamma & 0 \\ \hline \gamma & & \\ 0 & & \mathsf{Y} \end{array}\right], \qquad \begin{array}{l} m = k = 1 \\ \mathsf{H} = \theta, \mathsf{C} = \gamma. \end{array}$$

$$\mathsf{M}(\xi) = \begin{bmatrix} \theta & \gamma \\ \gamma & \xi \end{bmatrix}, \qquad \underline{0 < \gamma \ll \theta}\,.$$

The residual norm bound of Sec. 4-5 using $\mathbf{e}_1$ as an approximate eigenvector shows that there is an α in $[\theta - \gamma, \theta + \gamma]$. Assuming nothing of Y the optimal Lehmann/Kahan interval using the value $\zeta = \theta + \gamma$ also turns out to be $[\theta - \gamma, \theta + \gamma]$ and all other ζ's produce bigger bounds. Now we try some other M's and let $\sigma \equiv \sqrt{\theta^2 + 4\gamma^2} \doteqdot \theta + 2\gamma^2/\theta$.

ξ	μ_1	μ_2
0	$\frac{1}{2}(\theta - \sigma) \doteqdot -\gamma^2/\theta$	$\frac{1}{2}(\theta + \sigma) \doteqdot \theta + \gamma^2/\theta$
θ	$\theta - \gamma$	$\theta + \gamma$
2θ	$\theta + \frac{1}{2}(\theta - \sigma) \doteqdot \theta - \gamma^2/\theta$	$\theta + \frac{1}{2}(\theta + \sigma) \doteqdot 2\theta + \gamma^2/\theta$

Assumption: $0 \leqslant \mathsf{Y} \leqslant 2\theta$.
Inference from (10-6-6):

$$-\gamma^2/\theta \leqslant \alpha_1 \leqslant \theta - \gamma^2/\theta, \qquad \theta + \gamma^2/\theta \leqslant \alpha_3 \leqslant 2\theta + \gamma^2/\theta.$$

The α's have a lot of freedom.

Assumption: $\theta \leqslant \mathsf{Y} \leqslant 2\theta$.
Inference from (10-6-6):

$$\theta - \gamma \leqslant \alpha_1 \leqslant \theta - \gamma^2/\theta, \qquad \theta + \gamma \leqslant \alpha_3 \leqslant 2\theta + \gamma^2/\theta.$$

Note that α_1 is tightly constrained.

Assumption: $\eta_1 \leqslant 0 < 2\theta \leqslant \eta_2$.
Inference from (10-6-7):

$$\alpha_2 = \theta\left[1 + \gamma^2/\theta^2 + O(\gamma^4/\theta^4)\right] \text{ !!}$$

What has not been shown is that the bounds in (10-6-6) and (10-6-7) are an improvement over Lehmann's. The difference is strongest as $\|\mathsf{C}\| \longrightarrow 0$ and $\theta_j \longrightarrow \alpha_{j+p}$ for some p. The width of Lehmann's interval is $2\delta_1(\xi)$ and from the definition of δ_j in (10-5-4),

$$\delta_1(\xi)^2 \geqslant \lambda_1[\mathsf{H} - \xi]^2. \tag{10-6-8}$$

Thus $\delta_1(\xi) = O(\|\mathsf{C}\|)$ if $|\xi - \theta_j| = O(\|\mathsf{C}\|)$ as $\|\mathsf{C}\| \longrightarrow 0$.

In contrast

$$\mu_j(\eta) \geqslant \theta_j - \frac{2\|\mathsf{C}\|^2}{\eta - \theta_j + \sqrt{(\eta - \theta_j)^2 + 4\|\mathsf{C}\|^2}} \tag{10-6-9}$$

and so, for example, the intervals $[\mu_i(\eta'), \mu_i(\eta'')]$ shrink down on α_i with $O(\|\mathsf{C}\|^2)$ provided that η', η'' are separated from θ_j and are independent of $\|\mathsf{C}\|$. A proof of (10-6-9) is given in [Wielandt, 1967, sec. 28]. Example 10-6-1 illustrates this phenomenon.

Exercises on Sec. 10-6

10-6-1. By using the Weyl monotonicity result (10-4-3) show that $\alpha_i \leqslant \mu_i(\eta_{-1})$ and $\mu_{-i}(\eta_1) \leqslant \alpha_{-i}$.

10-6-2. Show that if $\eta' \leqslant \mathsf{Y} \leqslant \eta''$ then

$$\mu_{-i}(\eta') \leqslant \alpha_{-i} \leqslant \max\{\mu_{-i}(\eta''), \eta''\}.$$

10-6-3. Prove (10-6-7) by applying the lemma with j replaced by π.

10-6-4. If the eigenvalues η_i of Y are known then they can be related to the eigenvalues θ_i of H. Prove that if $\theta_i \leqslant \eta_{j+1}$ then $\mu_i(\eta_{j+1}) \leqslant \alpha_{i+j}$ and that if $\eta_j \leqslant \theta_i$ then $\alpha_{i+j} \leqslant \mu_{i+k}(\eta_j)$.

*10-7 THE USE OF GAPS IN A'S SPECTRUM

Consider the example in Sec. 10-6 in which there is a good approximation θ to an eigenvalue α,

$$\mathsf{A} = \begin{bmatrix} \theta & \gamma & 0 \\ \gamma & & \\ 0 & & \mathsf{Y} \end{bmatrix}, \qquad \mathsf{M}(\xi) = \begin{bmatrix} \theta & \gamma \\ \gamma & \xi \end{bmatrix}.$$

With no knowledge of Y the best inference is that α lies in $[\theta - \gamma, \theta + \gamma]$. Suppose however that adscititious knowledge comes in the form that α is the only eigenvalue of A in a **larger** interval $[\theta - \beta, \theta + \beta]$ where $\gamma < \beta$. The strong conclusion is that α actually lies in $[\theta - \gamma^2/\tau, \theta + \gamma^2/\tau]$ where $\tau = (\beta + \sqrt{\beta^2 + 4\gamma^2})/2 \doteqdot \beta + \gamma^2/\beta$. If $\beta/\theta = 10^{-1}$, $\gamma/\theta = 10^{-3}$ then

θ agrees with α to five decimal figures. Observations such as this permit prompt termination of expensive iterative procedures for calculating α's.

These better bounds come from the use of two samples of the auxiliary matrix $\mathbf{M}$.

ξ	$\mu_1(\xi)$	$\mu_2(\xi)$
$\theta - \beta = \alpha'$	$\theta - \tau$	$(\theta + \gamma^2/\tau)$
$\theta + \beta = \alpha''$	$(\theta - \gamma^2/\tau)$	$\theta + \tau$

In 1929 Temple proved essentially that $\mu_1(\alpha'') \leqslant \alpha \leqslant \mu_2(\alpha')$ and the theorem below generalizes it and similar results of Kato.

Consider $\mathbf{A}$ in the form (10-6-1) with $\|\mathbf{C}\|$ small enough that an interval, $[\alpha', \alpha'']$ say, is known to contain the **same** number of α's ($\mathbf{A}$'s eigenvalues) as θ's ($\mathbf{H}$'s eigenvalues) as indicated in the figure.

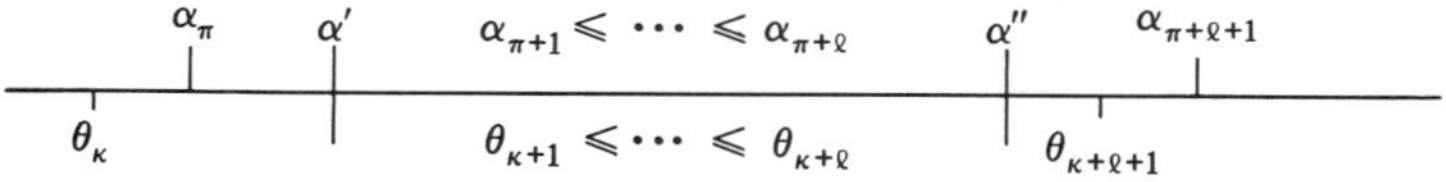

The optimal intervals are given by eigenvalues μ of the auxiliary matrix $\mathbf{M}(\mathbf{X}_\zeta)$ of (10-4-3) and (10-5-1) with appropriate choice of ζ. Theorem (10-7-1) is a corollary of (10-5-3) and was first stated in [Lehmann, 1949].

THEOREM Suppose that for some indices π, κ, and l

$$\alpha_\pi < \alpha' \leqslant \alpha_{\pi+1} \leqslant \cdots \leqslant \alpha_{\pi+l} \leqslant \alpha'' < \alpha_{\pi+l+1}, \tag{10-7-1a}$$

$$\theta_\kappa < \alpha' < \theta_{\kappa+1} \leqslant \cdots \leqslant \theta_{\kappa+l} < \alpha'' < \theta_{\kappa+l+1}, \tag{10-7-1b}$$

then

$$\kappa \leqslant \pi, \text{ and} \tag{10-7-1c}$$

$$\mu_{\kappa+j}(\mathbf{X}_{\alpha''}) \leqslant \alpha_{\pi+j} \leqslant \mu_{\kappa+j+k}(\mathbf{X}_{\alpha'}), \qquad \text{for } 1 \leqslant j \leqslant l. \tag{10-7-1d}$$

Recall that $k = \text{rank}[\mathbf{C}]$. These inequalities are best possible inferences from the data.

It is not necessary to know π in order to apply the theorem.

Proof. By Cauchy's theorem $\alpha_\kappa \leqslant \theta_\kappa$ and, by (10-7-1a) and (10-7-1b), $\theta_\kappa < \alpha' < \alpha_{\pi+1}$, which establishes (10-7-1c). Now for the harder part. By (10-7-1b) $\theta_\kappa < \alpha' < \theta_{\kappa+1}$ and so the Residual Interlace Theorem (10-5-3) may be invoked with $\zeta = \alpha'$ to deduce that $[\alpha', \mu_{\kappa+k+1}(X_{\alpha'})]$ contains at least one α and moreover $\alpha' = \mu_{\kappa+1}(X_{\alpha'}) = \cdots = \mu_{\kappa+k}(X_{\alpha'})$. By (10-7-1a) $\alpha_{\pi+1}$ must be in that interval while $\alpha_{\pi+2}$ may or may not be included. Thus

$$\alpha_{\pi+1} \leqslant \mu_{\kappa+k+1}(X_{\alpha'})$$

which establishes the second inequality in (10-7-1d) with $j = 1$.

To deal with the other values of j use either Lehmann's formulation (10-5-5) with suitable ξ or Ex. 10-5-2 to find a larger interval, $[\alpha', \mu_{\kappa+k+j}(X_{\alpha'})]$, containing at least j α's. By (10-7-1a) it must include $\alpha_{\pi+1}, \ldots, \alpha_{\pi+j}$. This establishes the second inequality in (10-7-1d) for all j up to $m - \kappa$ but only for $j \leqslant l$ will the first inequality also hold.

To establish the first inequality pick $\zeta = \alpha''$ and use Ex. 10-5-2 to conclude that $\mu_{\kappa+l}(X_{\alpha''}) = \cdots = \mu_{\kappa+l+k}(X_{\alpha''}) = \alpha''$ and that $[\mu_{\kappa+j}(X_{\alpha''}), \alpha'']$ contains at least $l + 1 - j$ α's for all j from l down to $-(\kappa - 1)$. By (10-7-1a) it must include $\alpha_{\pi+j}, \ldots, \alpha_{\pi+l}$. The only values of j for which both inequalities hold is $1 \leqslant j \leqslant l$.

Equality in (10-7-1d) holds on the left for $A = \text{diag}[M(X_{\alpha''}), \alpha'' + 1]$ and on the right for $A = \text{diag}[M(X_{\alpha'}), \alpha' - 1]$. □

Exercise on Sec. 10-7

10-7-1. Suppose that $\|C\|$ is so small that $\alpha' + \|C\| \leqslant \theta_{q+1}$ in hypothesis (b) of Theorem 10-7-1. Show that for $j = 1, \ldots, l$,

$$\theta_{q+j} \leqslant \mu_{q+j+k}(X_{\alpha'}) \leqslant \theta_{q+j} + \|C\|^2/(\theta_{q+j} - \alpha').$$

Notes and References

The interlace theorem goes back to [Cauchy, 1821] as do the simplest cases of the full monotonicity theorem [Weyl, 1912]. In fact these two theorems and the minimax characterization of eigenvalues [Fischer, 1905] form a corpus of information in which any of the results can be deduced from the others.

In the 1930s Temple began to work on error bounds for approximate eigenpairs using residuals. Others who contributed to the subject, and extended it to differential operators, were Weinstein, Kato, and Wielandt. By bringing A's spectrum into the picture interesting a priori error bounds can be derived. A recent comprehensive account of work in this field is [Weinberger, 1974]. More recent still is [Chatelin and Lemordant, 1978].

This chapter singles out one theme, the exploitation of the residual matrix itself—not just its norm— and tries to present the ideas simply and yet completely. Lehmann [1949, 1963] was the first to obtain the optimal bounds which can be derived from the given information but the approach taken in these unpublished notes of Kahan unifies all the results and has the advantage of being in English.

11

Approximations from a Subspace

11-1 SUBSPACES AND THEIR REPRESENTATION

Heavy use of the abstract notions of vector space and subspace in a discussion of numerical methods may seem to some readers unnecessarily abstruse. In fact, however, the language of subspaces simplifies such discussions by suppressing distracting details. This section reviews the way in which subspaces are handled and shows how the choosing of a basis corresponds to certain explicit matrix manipulations.

A subspace $\mathcal{S}$ of $\mathcal{E}^n$ is a subset which happens to be closed under the operation of taking linear combinations. A more useful definition is that $\mathcal{S}$ is the totality of **all** linear combinations of some small set of vectors in $\mathcal{S}$. Any small set which generates $\mathcal{S}$ is called a **spanning set** and there are (infinitely) many spanning sets for each $\mathcal{S}$ other than the trivial subspace $\{\mathbf{o}\}$. It is convenient to order the m vectors in a spanning set as columns of a matrix $\mathbf{S} = (\mathbf{s}_1, \ldots, \mathbf{s}_m)$ and to say, briefly and improperly, that $\mathbf{S}$ **spans** $\mathcal{S}$. $\mathbf{Sx}$ is a neat way to denote a linear combination of the columns of $\mathbf{S}$. There are several different ways of describing $\mathcal{S}$,

$$\mathcal{S} = \text{span } \mathbf{S} = \{\mathbf{Sx} : \mathbf{x} \in \mathcal{E}^m\} = \mathbf{S}\mathcal{E}^m. \tag{11-1-1}$$

Other names for $\mathcal{S}$ are the range, image, and column space of $\mathbf{S}$. The introduction of $\mathcal{E}^m$ in (11-1-1) is important; as $\mathbf{x}$ ranges over **all** m-vectors, i.e., over **all** of $\mathcal{E}^m$, $\mathbf{Sx}$ ranges over **part** of $\mathcal{E}^n$, namely $\mathcal{S}$ (Ex. 11-1-2).

It pays to keep spanning sets as small as possible. The minimal ones are called **bases** and all bases of $\mathcal{S}$ have the same number of vectors in them. This number is $\mathcal{S}$'s dimension. From now on the dimension of a subspace will be denoted automatically by a superscript, thus $\mathcal{S} \equiv \mathcal{S}^m$. One advantage of using an n by m basis S for $\mathcal{S}^m$ is that there is a one-one correspondence between the long n-vectors of $\mathcal{S}^m$ and the short auxiliary vectors x of $\mathcal{E}^m$ given by $\mathsf{s} = \mathsf{Sx}$.

A need arises in Sec. 11-4 to describe typical subspaces $\mathcal{G}^j$ of the subspace $\mathcal{S}^m$ of $\mathcal{E}^n$. How is this to be done?

EXAMPLE 11-1-1

Let $\mathcal{S}^3 \equiv \text{span}\left(\begin{bmatrix} 1 & 1 & 1 \\ 1 & 1 & 0 \\ 1 & 0 & 1 \\ 0 & 1 & 1 \end{bmatrix}\right) \subset \mathcal{E}^4$. With respect to this basis the subspace

$$\mathcal{G}^2 \equiv \text{span}\left(\begin{bmatrix} 1 & 1 \\ 1 & 1 \\ 1 & 0 \\ 0 & 1 \end{bmatrix}\right) = \text{span}\left(\begin{bmatrix} 1 & 1 & 1 \\ 1 & 1 & 0 \\ 1 & 0 & 1 \\ 0 & 1 & 1 \end{bmatrix}\begin{bmatrix} 1 & 0 \\ 0 & 1 \\ 0 & 0 \end{bmatrix}\right) \subset \mathcal{S}^3$$

corresponds to the subspace

$$\hat{\mathcal{G}}^2 \equiv \text{span}\left(\begin{bmatrix} 1 & 0 \\ 0 & 1 \\ 0 & 0 \end{bmatrix}\right) \subset \mathcal{E}^3.$$

A different, more interesting example shows that

$$\mathcal{G}^2 \equiv \left\{\alpha\begin{bmatrix} 1 \\ 1 \\ 1 \\ 0 \end{bmatrix} + \beta\begin{bmatrix} 1 \\ 1 \\ 0 \\ 1 \end{bmatrix} + \gamma\begin{bmatrix} 1 \\ 0 \\ 1 \\ 1 \end{bmatrix} : \alpha + \beta + \gamma = 0\right\} \subset \mathcal{S}^3$$

corresponds to the subspace

$$\hat{\mathcal{G}}^2 \equiv \left\{\begin{bmatrix} \alpha \\ \beta \\ \gamma \end{bmatrix} : \alpha + \beta + \gamma = 0\right\} \subset \mathcal{E}^3.$$

And so it is in general. Once a basis S for $\mathcal{S}^m$ is chosen there is a natural correspondence between subspaces $\mathcal{G}^j$ of the subspace $\mathcal{S}^m$ and subspaces $\hat{\mathcal{G}}^j$ of $\mathcal{E}^m$ given by $\mathcal{G}^j = \mathsf{S}\hat{\mathcal{G}}^j = \{\mathsf{Su} : \mathsf{u} \in \hat{\mathcal{G}}^j\}$, (Ex. 11-1-4). Figure 11-1-1 illustrates the matter in another way.

$$
\begin{array}{llll}
\mathcal{E}^n & & & \\
\cup & & & \\
\mathcal{S}^m & = \mathsf{S}\mathcal{E}^m, & & \mathsf{S} \text{ is } n \text{ by } m, \\
\cup & \quad\cup & & \\
\mathcal{G}^j & = \mathsf{S}\hat{\mathcal{G}}^j & = \mathsf{SG}\mathcal{E}^j, & \mathsf{G} \text{ is } m \text{ by } j.
\end{array}
$$

One–one correspondence
between subspaces $\mathcal{G}^j$ of $\mathcal{S}^m$ and $\hat{\mathcal{G}}^j$ of $\mathcal{E}^m$

Figure 11-1-1 Subspaces of a subspace

The subspace $\mathcal{S}^m$ is in $\mathcal{E}^n$, not just $\mathcal{R}^n$, and so it is both meaningful and convenient to use orthonormal bases to describe it. See Ex. 11-1-5.

Exercises on Sec. 11-1

11-1-1. Let $\mathcal{S}$ be a subset of $\mathcal{E}^n$ with the property that if u and v are in $\mathcal{S}$ then so is $\alpha\mathsf{u} + \beta\mathsf{v}$. Why must $\mathcal{S}$ also be the totality of all linear combinations of some finite subset of vectors in $\mathcal{S}$?

11-1-2. Let $\mathsf{S} = \begin{bmatrix} 2 & 4 \\ 1 & 2 \\ 0 & 0 \end{bmatrix}$ be a (wasteful) spanning set for $\mathcal{S}$. Show that every vector s in $\mathcal{S}$ can be written as Su for some u in $\mathcal{E}^2$. How many u for each s?

11-1-3. Describe the second pair of subspaces $\mathcal{G}^2$ and $\hat{\mathcal{G}}^2$ in the example as the column spaces of certain matrices G and $\hat{\mathsf{G}}$.

11-1-4. Let S be a basis for $\mathcal{S}^m$. Prove that the mapping between the subspaces of $\mathcal{E}^m$ and S^m induced by S is indeed one-one.

11-1-5. By using Gram-Schmidt, or otherwise, give orthonormal bases of the subspaces given in Sec. 11-1.

11-2 INVARIANT SUBSPACES

An eigenvector z of A may be normalized to have any convenient nonzero norm and we usually say, somewhat loosely, that z, $2\mathsf{z}$, and $-\mathsf{z}$ are the same eigenvector. It is more convenient to speak of the subspace $\mathcal{S}^1 = \text{span}(\mathsf{z})$. This subspace of $\mathcal{E}^n$ enjoys two remarkable properties:

1. $\mathcal{S}^1$ is mapped into itself by A, $\mathsf{A}\mathcal{S}^1 \subset \mathcal{S}^1$.
2. The image under A of any z in $\mathcal{S}^1$ is simply a fixed multiple of z, $\mathsf{Az} = \mathsf{z}\lambda$ and λ depends on $\mathcal{S}^1$ alone, not z.

Subspaces obeying 2. are called **eigenspaces**.

Now let $\mathsf{Z} = (\mathsf{z}_1, \ldots, \mathsf{z}_m)$ be an n by m matrix whose columns are eigenvectors of A. Then span(Z) enjoys 1. but not 2., unless $\lambda_1 = \lambda_2 = \cdots = \lambda_m$. Subspaces of $\mathcal{E}^n$ satisfying 1. are called **invariant**. Conversely, any invariant subspace has a basis of eigenvectors (Ex. 11-2-1).

For any given n by m $\mathsf{F} = (\mathsf{f}_1, \ldots, \mathsf{f}_m)$ it is desirable to have a test for span(F)'s invariance. By 1., $\mathsf{Af}_j = \Sigma \mathsf{f}_i c_{ij}$, for each $j = 1, \ldots, m$, for some unknown coefficients c_{ij}. These relations can be expressed more neatly in terms of F's **residual matrix**,

$$\mathsf{R} \equiv \boxed{\mathsf{A}}\,\boxed{\mathsf{F}} - \boxed{\mathsf{F}}\,\boxed{\mathsf{C}} = \boxed{\mathsf{O}}. \qquad (11\text{-}2\text{-}1)$$

When F has full rank m then (11-2-1) can be solved for a unique C given by $\mathsf{C} = (\mathsf{F^*F})^{-1}\mathsf{F^*AF}$. If rank($\mathsf{F}$) $< m$ then there are many solutions C of (11-2-1) but it is wasteful and unnecessary to work with such F's.

When an orthonormal basis Q of span(F) [i.e., span(Q) = span(F)] is available the test takes a more convenient form

$$\mathsf{R(Q)} \equiv \mathsf{AQ} - \mathsf{QH} = \mathsf{O}, \ \mathsf{H} \equiv \mathsf{Q^*AQ}. \qquad (11\text{-}2\text{-}2)$$

Both C and H represent the restriction of A to span(F) but H has the advantage of being symmetric. Moreover

> each eigenvector of C, or H, determines an eigenvector of A.

If $\mathsf{Cy} = \mathsf{y}\lambda$ then Fy is an eigenvector of A with eigenvalue λ. If $\mathsf{Hx} = \mathsf{x}\lambda$ then Qx is an eigenvector of A with eigenvalue λ.

EXAMPLE 11-2-1

$$\mathsf{F} = \begin{bmatrix} 3 & 0 \\ 1 & 1 \\ -1 & -1 \\ -3 & 0 \end{bmatrix}, \qquad \mathsf{Q} = \begin{bmatrix} 3 & -1 \\ 1 & 3 \\ -1 & -3 \\ -3 & 1 \end{bmatrix} \frac{1}{2\sqrt{5}},$$

$$\mathsf{A} = \begin{bmatrix} 1 & 1 & -2 & 0 \\ 1 & 1 & 0 & -2 \\ -2 & 0 & 1 & 1 \\ 0 & -2 & 1 & 1 \end{bmatrix}.$$

The computation of C and y is left as an exercise, but

$$\mathsf{H} = \tfrac{2}{5}\begin{bmatrix} 7 & 6 \\ 6 & -2 \end{bmatrix}, \qquad \mathsf{x} = \begin{bmatrix} 2 \\ 1 \end{bmatrix} \text{ or } \begin{bmatrix} 1 \\ -2 \end{bmatrix}.$$

Exercises on Sec. 11-2

11-2-1. Prove that if $\mathbb{S}$ is invariant under A then $\mathbb{S}$ has a basis of eigenvectors. Use results from Sec. 1-4.

11-2-2. Let F = QL* with Q orthonormal. Express C in terms of H and L*.

11-2-3. In the example compute C and find its eigenvectors and the corresponding ones of A.

11-3 THE RAYLEIGH-RITZ PROCEDURE

Usually the subspace $\mathbb{S}^m$ on hand turns out not to be invariant under A. If it is nearly invariant then it should contain good approximations to some eigenvectors of A. Three ingredients are required for computing the best set of approximate eigenvectors from $\mathbb{S}^m$ to eigenvectors of A:

1. S, an n by m full rank matrix whose columns are a basis for $\mathbb{S}^m$.
2. A subprogram, call it OP, which returns Ax for any given x.
3. Utility programs for orthonormalizing sets of vectors and computing eigensystems of m by m symmetric matrices.

There is no need for A to be known explicitly. In some, but not all applications, n is large ($n > 1000$) and $m \ll n$. Sometimes only a subset of p of the m approximate eigenpairs are wanted. The well-known Rayleigh-Ritz procedure (specified in Table 11-3-1) computes these approximations.

Exercises on Sec. 11-3

Apply the RR procedure in the following contexts:

11-3-1. The matrix A of Sec. 11-2, $\mathbb{S} = \text{span}[(2\ 1\ -1\ -2)^*, (2\ -1\ -1\ 2)^*]$.

11-3-2. The matrix A of Sec. 11-2, $\mathbb{S} = \text{span}[e_1, e_2, e_3]$.

11-3-3. The matrix A of Sec. 11-5, $\mathbb{S} = \text{span}$ Q.

11-3-4. Verify the operation counts giving, whenever possible, a second term in the expression. Suppose that $n = 10^3$, $m = 10^2$, $p = 10$. Calculate the op ratios for each step keeping ℓ free.

11-3-5. How do the operation counts in Steps 3 and 4 change if H happens to be tridiagonal? Does this property alter the significance of ℓ in assessing the cost of executing RR?

Table 11-3-1 Procedure RR (Rayleigh-Ritz)

	Action	*Cost* (in ops†)
1.	Orthonormalize the columns of S, if necessary, to get an orthonormal n by m Q written over S.	$m(m+1)n$
2.	Form AQ by m calls to the subprogram OP. The term ℓ is the average number of nonzero elements per row of A.	$\sim \ell mn$ (Greater if A is represented in a sophisticated way in order to conserve storage)
3.	Form m by m $\mathbf{H} = \rho(\mathbf{Q}) \equiv \mathbf{Q}^*(\mathbf{AQ})$, the (matrix) Rayleigh quotient of Q.	$\frac{1}{2}m(m+1)n$
4.	Compute the $p(\leqslant m)$ eigenpairs of H which are of interest, say $\mathbf{H}\mathbf{g}_i = \mathbf{g}_i\theta_i$, $i = 1, \ldots, p$. The θ_i are the **Ritz values.**	$\sim m^3$ (Less, if $p \ll m$)
5.	If desired compute the p **Ritz vectors** $\mathbf{y}_i = \mathbf{Q}\mathbf{g}_i$, $i = 1, \ldots, p$. The full set $\{(\theta_i, \mathbf{y}_i), i = 1, \ldots, m\}$ is the best set of approximations to eigenpairs of A which can be derived from $\mathcal{S}^m$ alone.	pmn
6.	Residual Error Bounds. Form the p residual vectors $\mathbf{r}_i = \mathbf{r}(\mathbf{y}_i) = \mathbf{A}\mathbf{y}_i - \mathbf{y}_i\theta_i = (\mathbf{AQ})\mathbf{g}_i - \mathbf{y}_i\theta_i$, using the last expression for computation. Also compute $\\|\mathbf{r}_i\\|$. Each interval $[\theta_i - \\|\mathbf{r}_i\\|, \theta_i + \\|\mathbf{r}_i\\|]$ contains an eigenvalue of A. If some of the intervals overlap then a bit more work is required to guarantee approximations to p eigenvalues of A. See Sec. 11-5.	$p(m+2)n$
7.	A nontraditional extra step is described in Sec 11-8.	$\frac{1}{2}p(p+1)n + O(p^3)$

†An op is either a multiplication or a division.

11-4 OPTIMALITY

There are three (related) ways of justifying the claim that the RR approximations $\{\theta_i, \mathsf{y}_i\}$ are optimal for the given information. The first is a natural corollary of the minimax characterization of eigenvalues. From Sec. 10-2

$$\alpha_j = \lambda_j[\mathsf{A}] = \min_{\mathcal{F}^j \subset \mathcal{E}^n} \max_{\mathsf{f} \in \mathcal{F}^j} \rho(\mathsf{f}; \mathsf{A}), \qquad (\mathsf{f} \neq 0). \qquad (11\text{-}4\text{-}1)$$

Recall that dimensions of spaces are denoted by superscripts and $\rho(\mathsf{f}; \mathsf{A}) = \mathsf{f}^*\mathsf{Af}/\mathsf{f}^*\mathsf{f}$ for $\mathsf{f} \neq \mathsf{o}$. Consequently the natural definition of the best approximation β_j to α_j from the given subspace $\mathcal{S}^m$ is to replace $\mathcal{E}^n$ by $\mathcal{S}^m$ in (11-4-1) to get

$$\beta_j \equiv \min_{\mathcal{G}^j \subset \mathcal{S}^m} \max_{\mathsf{g} \in \mathcal{G}^j} \rho(\mathsf{g}; \mathsf{A}), \qquad (\mathsf{g} \neq \mathsf{o}). \qquad (11\text{-}4\text{-}2)$$

The only difficulty in the proof of the theorem below is characterizing the subspaces $\mathcal{G}^j$ of the subspace $\mathcal{S}^m$. If Q is a fixed orthonormal basis in $\mathcal{S}^m$ then $\mathcal{S}^m = \mathsf{Q}\mathcal{E}^m \equiv \{\mathsf{Qs} : \mathsf{s} \in \mathcal{E}^m\}$. The key point is that the subspaces of $\mathcal{S}^m$ are generated by the subspaces of $\mathcal{E}^m$ from the same correspondence. This was established in Sec. 11-1, namely

$$\mathcal{G}^j \subset \mathcal{S}^m \text{ if, and only if, } \mathcal{G}^j = \mathsf{Q}\hat{\mathcal{G}}^j \text{ and } \hat{\mathcal{G}}^j \subset \mathcal{E}^m. \qquad (11\text{-}4\text{-}3)$$

THEOREM

$$\beta_j = \lambda_j[\mathsf{H}] \equiv \theta_j, \qquad j = 1, \ldots, m, \qquad \text{where } \mathsf{H} = \mathsf{Q}^*\mathsf{AQ}. \qquad (11\text{-}4\text{-}4)$$

Proof. From (11-4-2), (11-4-3), and $\mathsf{Q}^*\mathsf{Q} = \mathsf{I}_m$ it follows that

$$\begin{aligned}\beta_j &= \min_{\mathcal{G}^j \subset \mathcal{S}^m} \max_{\mathsf{g} \in \mathcal{G}^j} \rho(\mathsf{g}; \mathsf{A}), \quad \text{and } \mathsf{o} \neq \mathsf{g} = \mathsf{Qs}, \\ &= \min_{\hat{\mathcal{G}}^j \subset \mathcal{E}^m} \max_{\mathsf{s} \in \hat{\mathcal{G}}^j} \rho(\mathsf{s}; \mathsf{H}), \quad \text{since } \mathsf{G}^j = \mathsf{Q}\hat{\mathcal{G}}^j \text{ and } \mathsf{s}^*\mathsf{s} = \mathsf{g}^*\mathsf{g} \neq 0 \\ &= \lambda_j[\mathsf{H}] \equiv \theta_j. \quad \square\end{aligned}$$

This optimality result concerns only the θ's.

The second way in which the approximations are optimal concerns Q. For any m by m matrix B there is associated a residual matrix $\mathsf{R}(\mathsf{B}) \equiv \mathsf{AQ} - \mathsf{QB}$. The minimizing property of the familiar Rayleigh quotient $\rho(\mathsf{q})$ is inherited by the matrix $\mathsf{H} = \mathsf{Q}^*\mathsf{AQ} \equiv \rho(\mathsf{Q})$.

THEOREM For given orthonormal n by mQ

$$\|R(H)\| \leq \|R(B)\| \quad \text{for all } m \text{ by } m \text{ B.} \tag{11-4-5}$$

Proof. $R(B)^*R(B) = QA^2Q - B^*(Q^*AQ) - (Q^*AQ)B + B^*B$. The trick is to see that the last three terms are just $(H - B)^*(H - B) - H^2$ and so

$$R(B)^*R(B) = Q^*A^2Q - H^2 + (H - B)^*(H - B),$$
$$= R(H)^*R(H) + (H - B)^*(H - B).$$

Since $(H - B)^*(H - B)$ is positive semi-definite

$$\|R(B)\|^2 = \lambda_{-1}[R(B)^*R(B)] \geq \lambda_{-1}[R(H)^*R(H)] = \|R(H)\|^2. \quad \square$$

For uniqueness see Exs. 11-4-4 and 11-4-5.

Now let S be any orthonormal basis in $\mathcal{S}^m$ and let Δ be any diagonal matrix. Thus the pairs $\{(\delta_i, s_i), i = 1, \ldots, m\}$ are rival eigenpair approximations. However, from theorem (11-4-5), $\|AS - S\Delta\|$ is minimized over S and Δ when and only when $s_i = y_i$, $\delta_i = \theta_i$, $i = 1, \ldots, m$. The verification of this is left as an important exercise (Ex. 11-4-1). A related and useful characterization of the y_i is given in Ex. 11-4-6.

A third way in which the RR approximations are optimal is in the spirit of backward error analysis (see Chap. 2). Since $\mathcal{S}^m$ is not invariant it is meaningless to speak of the restriction of A to $\mathcal{S}^m$. The next best thing is A's **projection** onto $\mathcal{S}^m$. See Sec. 1-4 for a discussion of projections. If $P_{\mathcal{S}}$ is the orthogonal projector onto $\mathcal{S}^m$, i.e., $P_{\mathcal{S}}g$ is the closest vector in $\mathcal{S}^m$ to g, then $\mathcal{S}^m$ is invariant under $P_{\mathcal{S}}A$ and so it is meaningful to speak of its restriction to $\mathcal{S}^m$, namely $P_{\mathcal{S}}A|_{\mathcal{S}}$, and this is the desired projection. It is intimately related to the matrix $P_{\mathcal{S}}AP_{\mathcal{S}}$ which acts on the whole of $\mathcal{E}^n$ and is also called A's projection. No harm results from this ambiguity because the two projections have the same action on $\mathcal{S}^m$. Here is the third characterization of the Ritz pairs.

LEMMA The (θ_i, y_i), $i = 1, \ldots, m$ are the eigenpairs for A's projection onto $\mathcal{S}^m$. (11-4-6)

The proof constitutes Ex. 11-4-2.

EXAMPLE 11-4-1

If A rotates the plane $\mathcal{E}^2$ through 45° counterclockwise then A's projection onto any $\mathcal{S}^1$ simply shrinks $\mathcal{S}^1$ by a factor $\sqrt{2}$ as shown in the figure below.

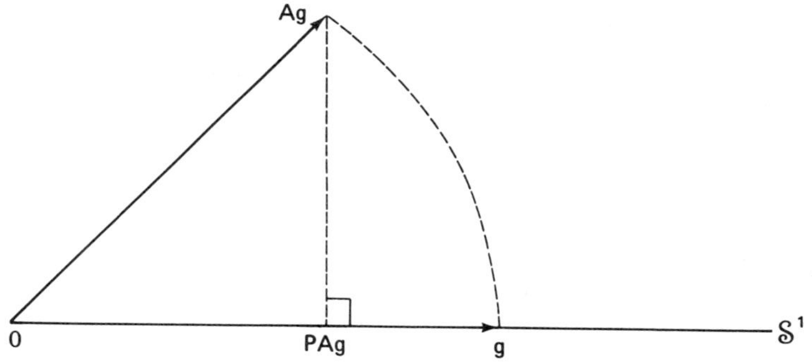

There are two ways in which the RR approximations are not optimal. In general no Ritz vector $\mathbf{y}_j$ is the closest unit vector in $\mathcal{S}^m$ to any eigenvector of A. Even more surprising perhaps is that the error bound $\|\mathbf{A}\mathbf{v} - \mathbf{v}\rho(\mathbf{v})\|/\|\mathbf{v}\|$ is not minimized over $\mathcal{S}^m$ by any of the Ritz vectors when $m > 1$. Example 11-5-1 illustrates both assertions. To summarize, by achieving **collective** optimality for m pairs the RR approximations usually relinquish optimality for any particular eigenpair.

There is little profit in approximations which are good but not known to be good. The rest of this chapter focuses on the assessment of the RR approximations.

Exercises on Sec. 11-4

11-4-1. Use theorem (11-4-5) to show that for any orthonormal basis S of $\mathcal{S}^m$ $\min\|\mathbf{AS} - \mathbf{S}\Delta\|$ over diagonal Δ and orthonormal S is achieved when $\mathbf{s}_i = \mathbf{y}_i$, $\delta_i = \theta_i$.

11-4-2. Verify that each pair $(\theta_i, \mathbf{y}_i)$ is an eigenpair $\mathbf{P}_{\mathcal{S}}\mathbf{A}|_{\mathcal{S}}$.

11-4-3. Show that $\mathcal{S}$ is also invariant under $\bar{\mathbf{A}} \equiv \mathbf{A} - \mathbf{R}(\mathbf{H})\mathbf{Q}^* - \mathbf{Q}\mathbf{R}(\mathbf{H})^*$. Verify that $\|\bar{\mathbf{A}} - \mathbf{A}\| = \|\mathbf{R}(\mathbf{H})\|$. Are the $(\theta_i, \mathbf{y}_i)$ eigenpairs of $\bar{\mathbf{A}}$? Does $\bar{\mathbf{A}} = \mathbf{P}_{\mathcal{S}}\mathbf{A}\mathbf{P}_{\mathcal{S}}$?

11-4-4. Define $\|\cdot\|_F$ by $\|\mathbf{B}\|_F^2 = \text{trace}(\mathbf{B}^*\mathbf{B})$. With respect to theorem (11-4-5) show that $\|\mathbf{R}(\mathbf{B})\|_F \geq \|\mathbf{R}(\mathbf{H})\|_F$ with equality only when $\mathbf{B} = \mathbf{H}$.

11-4-5. Find A, Q, and a 2 by 2 matrix $\mathbf{B} \neq \mathbf{H}$ such that $\|\mathbf{R}(\mathbf{B})\| = \|\mathbf{R}(\mathbf{H})\|$.

11-4-6. Show that, for x in $\mathcal{S}^m$,

$$\mathbf{A}\mathbf{x} - \mathbf{x}\rho(\mathbf{x}) \perp \mathcal{S}^m \text{ if, and only if, } \mathbf{x} = \mathbf{y}_i \text{ for some } i \leq m.$$

11-5 RESIDUAL BOUNDS ON CLUSTERED RITZ VALUES

At the completion of Step 6 in the RR procedure there are on hand θ_i, y_i, r_i, $\|\mathsf{r}_i\|$, for $i = 1, \ldots, p$. The simple error bounds in Sec. 4-5 guarantee an eigenvalue α of A in each interval $[\theta_i - \|\mathsf{r}_i\|, \theta_i + \|\mathsf{r}_i\|]$. If the intervals are disjoint then the θ_i provide approximations to p different eigenvalues of A, as desired. Better bounds require either more knowledge of A or more work as described in Sec. 11-8 and Chap. 10.

If two or more intervals overlap then it is possible that two or more different θ_i are approximating a single α as Example 11-5-1 shows.

EXAMPLE 11-5-1

$$\mathsf{A} = \begin{bmatrix} 0 & \gamma & 0 \\ \gamma & 0 & 1 \\ 0 & 1 & 0 \end{bmatrix}, \quad \mathsf{Q} = \begin{bmatrix} 1 & 0 \\ 0 & 1 \\ 0 & 0 \end{bmatrix}. \qquad \text{Take } \gamma \doteq 0.1. \quad \alpha_{\pm 1} = \mp\sqrt{1+\gamma^2}, \ \alpha_2 = 0.$$

Then compute H and

$$\theta_{\pm 1} = \mp\gamma, \qquad \mathsf{y}_i^* = (1 \quad -1 \quad 0)/\sqrt{2}, \qquad \mathsf{y}_2^* = (1 \quad 1 \quad 0)/\sqrt{2}.$$

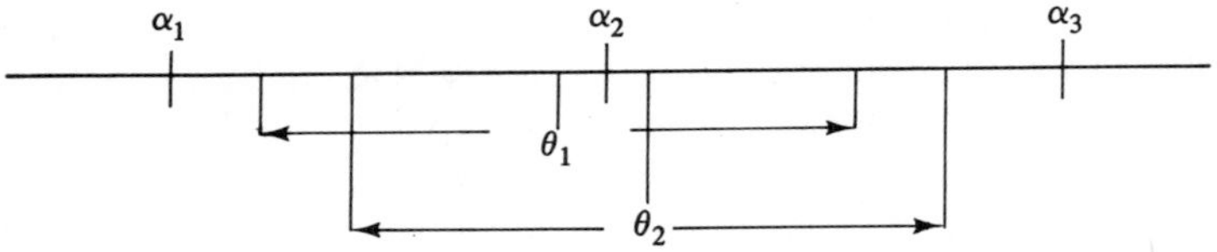

$$\mathsf{R}Y = (\mathsf{r}_1, \mathsf{r}_2) = \begin{bmatrix} 0 & 0 \\ 0 & 0 \\ 1 & 1 \end{bmatrix} \frac{1}{\sqrt{2}}, \qquad \|\mathsf{r}_1\| = \|\mathsf{r}_2\| = \frac{1}{\sqrt{2}},$$

$$\|\mathsf{R}Y\| = 1, \ \mathsf{r}_i = \mathsf{A}\mathsf{y}_i - \mathsf{y}_i\theta_i.$$

Note that θ_1 and θ_2 provide spurious evidence that there are two α's in $\left[\frac{-1}{\sqrt{2}} - \gamma, \frac{1}{\sqrt{2}} + \gamma\right]$.

The way out of the overlap difficulty is to use the norm of the associated residual matrix R. Unfortunately the following useful theorem from [Kahan, 1967] has not been published.

THEOREM Let Q be any orthonormal n by m matrix. Associated with it are H($\equiv$ Q*AQ) and R($\equiv$ AQ − QH). There are m of A's eigenvalues $\{\alpha_{j'}, j = 1, \ldots, m\}$ which can be put in one-one correspondence with the eigenvalues θ_j of H in such a way that

$$|\theta_j - \alpha_{j'}| \leqslant \|R\|, j = 1, \ldots, m. \tag{11-5-1}$$

Proof. Orthonormal columns can be appended to Q to fill out a square orthonormal matrix $P \equiv (Q, \tilde{Q})$. Then

$$P^*AP \equiv \begin{bmatrix} Q^*AQ & Q^*A\tilde{Q} \\ \tilde{Q}^*AQ & \tilde{Q}^*A\tilde{Q} \end{bmatrix} \equiv \begin{bmatrix} H & B^* \\ B & W \end{bmatrix}$$

where only H will be known explicitly. Yet

$$\begin{aligned} P^*R &= P^*AQ - P^*QH \\ &= (P^*AP)(P^*Q) - (P^*Q)H \\ &= \begin{bmatrix} H & B^* \\ B & W \end{bmatrix}\begin{bmatrix} I \\ O \end{bmatrix} - \begin{bmatrix} I \\ O \end{bmatrix}H = \begin{bmatrix} O \\ B \end{bmatrix}. \end{aligned} \tag{11-5-2}$$

By the orthogonal invariance of $\|\cdot\|$ (Fact 1-10 in Sec. 1-6)

$$\|R\| = \|P^*R\| = \left\|\begin{bmatrix} O \\ B \end{bmatrix}\right\| = \|B\|.$$

Now

$$P^*AP = \begin{bmatrix} H & O^* \\ O & W \end{bmatrix} + \begin{bmatrix} O & B^* \\ B & O \end{bmatrix} \tag{11-5-3}$$

and by the Weyl monotonicity theorem (Sec. 10-3) for $i = 1, \ldots, n$

$$\alpha_i = \lambda_i[P^*AP] \leqslant \lambda_i\left[\begin{pmatrix} H & O^* \\ O & W \end{pmatrix}\right] + \lambda_{-1}\left[\begin{pmatrix} O & B^* \\ B & O \end{pmatrix}\right]. \tag{11-5-4}$$

Now the θ_j appear somewhere in the ordered list of eigenvalues of H and W. So there exist indices j' such that

$$\lambda_{j'}\left[\begin{pmatrix} H & O^* \\ O & W \end{pmatrix}\right] = \theta_j, \qquad j = 1, \ldots, m.$$

To evaluate the second term on the right in (11-5-4) square the matrix:

$$\begin{bmatrix} O & B^* \\ B & O \end{bmatrix}^2 = \begin{bmatrix} B^*B & O^* \\ O & BB^* \end{bmatrix}. \tag{11-5-5}$$

Since B*B and BB* have the same nonzero eigenvalues (Ex. 11-5-2)

$$\lambda_{-1}\left[\begin{pmatrix} \mathsf{O} & \mathsf{B}^* \\ \mathsf{B} & \mathsf{O} \end{pmatrix}\right] = \sqrt{\lambda_{-1}[\mathsf{B}^*\mathsf{B}]} = \sqrt{\|\mathsf{B}\|^2} = \|\mathsf{R}\|. \qquad (11\text{-}5\text{-}6)$$

With $i = j'$ in (11-5-4) the theorem's inequality is obtained. □

When $m > 2$, $\|\mathsf{R}\|$ has a small but nonnegligible cost. It can be majorized via $\|\mathsf{R}\|^2 \leqslant \|\mathsf{R}\|_F^2 = \sum_{i=1}^{m} \|\mathsf{r}_i\|^2$ and the $\|\mathsf{r}_i\|^2$ are already available from Step 6 in the RR procedure. If $\|\mathsf{R}\|_F$ is to be used there is a corresponding result whose proof constitutes Ex. 11-5-3.

THEOREM	$\sum_{i=1}^{m} (\theta_j - \alpha_{j'})^2 \leqslant 2\|\mathsf{R}\|_F^2.$	(11-5-7)

In practice Theorem (11-5-1) will usually be applied not to $\mathsf{R}(\mathsf{Q})$ but to a matrix $\bar{\mathsf{R}} \equiv (\mathsf{r}_1, \ldots, \mathsf{r}_\nu)$ which corresponds to a subset of θ's whose residual intervals overlap. In the proof Q is replaced by $\bar{\mathsf{Q}} = (\mathsf{y}_1, \ldots, \mathsf{y}_\nu)$ and H by $\bar{\mathsf{H}} = \operatorname{diag}(\theta_1, \ldots, \theta_\nu)$. Then it follows that each interval $[\theta_i - \|\bar{\mathsf{R}}\|, \theta_i + \|\bar{\mathsf{R}}\|]$ contains its own α for these clustered θ_i, $i = 1, \ldots, \nu$.

Exercises on Sec. 11-5

11-5-1. Let $\mathsf{H} = \operatorname{diag}(\theta_1, \ldots, \theta_4)$, $\mathsf{C} = \operatorname{diag}(\gamma_1, \ldots, \gamma_4)$, and let $\mathsf{Q} = (\mathsf{e}_1, \ldots, \mathsf{e}_4)$. Apply theorem (11-5-1) to $\mathsf{A} = \begin{bmatrix} \mathsf{H} & \mathsf{C}^* \\ \mathsf{C} & \mathsf{H} \end{bmatrix}$. What does this example show about possible strengthening of the results?

11-5-2. There are several ways to prove that B*B and BB* have the same nonzero eigenvalues. Show that $\begin{bmatrix} \mathsf{B}^*\mathsf{B} & \mathsf{O} \\ \mathsf{B} & \mathsf{O} \end{bmatrix}$ and $\begin{bmatrix} \mathsf{O} & \mathsf{O} \\ \mathsf{B} & \mathsf{B}\mathsf{B}^* \end{bmatrix}$ are similar.

11-5-3. Apply the Wielandt-Hoffman theorem, Fact 1-11 in Sec. 1-6, to (11-5-3) to prove theorem (11-5-7).

11-5-4. Use the example in Sec. 11-2 to evaluate $\mathsf{R}(\mathsf{Q})$. How much worse is $\|\mathsf{R}(\mathsf{Q})\|$ than either of the residual vector bounds?

11-5-5. Apply theorem (11-5-1) to a tridiagonal matrix taking $\mathsf{Q} = (\mathsf{e}_1, \ldots, \mathsf{e}_m)$. Which theorem in Chap. 10 gives the same result?

11-6 NO RESIDUAL BOUNDS ON RITZ VECTORS

How well does a Ritz vector y_i approximate some eigenvector $\mathsf{z}_{i'}$? The following 2 by 2 example shows that without adscititious information no bound can be placed on the error in the Ritz vectors. Consider

$$A = \begin{bmatrix} \nu + \delta & \gamma \\ \gamma & \nu - \delta \end{bmatrix}; \qquad \begin{matrix} \alpha_1 = \nu - \sigma, \qquad \alpha_2 = \nu + \sigma, \\ \text{where } \sigma^2 = \delta^2 + \gamma^2. \end{matrix}$$

Take $\mathsf{Q} = \mathsf{e}_1$, so that $\mathsf{H} = \theta = \nu + \delta$. Then

$$|\theta - \alpha_{1'}| = \sigma - |\delta| = \gamma^2/(\sigma + |\delta|) \leqslant |\gamma| = \|\mathsf{R}\|$$

where $1' = 2$ if $\delta \geqslant 0$, $1' = 1$ if $\delta \leqslant 0$.

The "true" eigenvector and the Ritz vector are, respectively,

$$\begin{bmatrix} 1 \\ \pm\gamma/(\sigma + |\delta|) \end{bmatrix} \quad \text{and} \quad \begin{bmatrix} 1 \\ 0 \end{bmatrix}.$$

If ϕ is the acute angle between them (the error angle) then

$$\cos\phi = \left(1 + \gamma^2/(\sigma + |\delta|)^2\right)^{-1/2},$$

$$\longrightarrow \begin{Bmatrix} 1, & \text{if } \delta \neq 0 \\ 1/\sqrt{2}\,, & \text{if } \delta = 0 \end{Bmatrix} \text{ as } \gamma \longrightarrow 0.$$

So when $\delta = 0$ the "true" eigenvector is $(1, \pm 1)^*$ for **all** nonzero γ and the error angle is $\pi/4$. That is why there is no bound, in terms of $\|\mathsf{R}\|$, on the error in the Ritz vector.

What has gone wrong? Why does the method seem to break down when $|\delta|$ is small? **The answer is that the method does not break down, but the question does.** As $\delta \longrightarrow 0$ the request for **the** error $\|\mathsf{e}_1 - \mathsf{z}_1\|$ in the Ritz vector becomes sillier and sillier: a perfect example of the danger of a purely formal approach.

When $|\delta/\theta|$ is very small then θ is almost as good an approximation to α_1 as to α_2. Which of the two eigenvectors should the Ritz vector approximate? Since they are mutually orthogonal no single vector can be close to both and, with wisdom no less than Solomon's, the Ritz vector splits the difference between the two rivals: In the limit, as $\delta \longrightarrow 0$

$$\mathsf{e}_1 = \tfrac{1}{2}\left[\begin{pmatrix} 1 \\ 1 \end{pmatrix} + \begin{pmatrix} 1 \\ -1 \end{pmatrix}\right] = \lim(\mathsf{z}_1 + \mathsf{z}_2)/\sqrt{2}\,.$$

This example reflects the general situation in which the eigenvectors are not necessarily continuous functions of the matrix elements in the neighborhood of matrices with multiple eigenvalues. See Sec. 1-4. In such cases the useful objects are the invariant subspaces associated with each cluster of very close eigenvalues.

The user's difficulty has changed but it is still there. Instead of asking whether a Ritz vector is good or bad he or she may ask whether there are several of A's eigenvalues close to one of the computed θ's. The Ritz method cannot answer this question.

When extra information, beyond R, is available then it can often be used to give an error bound. In its absence we cannot do better than Ex. 11-4-3.

11-7 GAPS IN THE SPECTRUM

Consider now any unit vector y and its Rayleigh quotient $\theta = \rho(\mathsf{y})$. The simple error bound in Sec. 4-5 guarantees that there is at least one eigenvalue α of A satisfying $|\alpha - \theta| \leqslant \|\mathsf{Ay} - \mathsf{y}\theta\| = \|\mathsf{r}(\mathsf{y})\|$. However if θ is known to be well separated from all eigenvalues other than the closest one α (called the spectrum **complementary to** θ) then not only can the error bound on θ be improved but some bound can be put on the accuracy with which y approximates α's eigenvector z.

> THEOREM Let y be a unit vector with $\theta = \rho(\mathsf{y})$, let α be the eigenvalue of A closest to θ, and let z be its normalized eigenvector. The gap $\gamma \equiv \min|\lambda_i[\mathsf{A}] - \theta|$ over all $\lambda_i \neq \alpha$. Let $\psi = \angle(\mathsf{y}, \mathsf{z})$. Then
>
> $$|\sin\psi| \leqslant \|\mathsf{r}(\mathsf{y})\|/\gamma, \qquad |\theta - \alpha| \leqslant \|\mathsf{r}(\mathsf{y})\|^2/\gamma. \tag{11-7-1}$$

Proof. Decompose y in the form $\mathsf{y} = \mathsf{z}\cos\psi + \mathsf{w}\sin\psi$, where w is the unit vector in the y-z plane orthogonal to z. Hence

$$\begin{aligned} \mathsf{r}(\mathsf{y}) &= (\mathsf{A} - \theta)\mathsf{z}\cos\psi + (\mathsf{A} - \theta)\mathsf{w}\sin\psi \\ &= \mathsf{z}(\alpha - \theta)\cos\psi + (\mathsf{A} - \theta)\mathsf{w}\sin\psi \end{aligned} \tag{11-7-2}$$

since $\mathsf{Az} = \mathsf{z}\alpha$. Happily, $\mathsf{z}^*(\mathsf{A} - \theta)\mathsf{w} = 0$ (since $\mathsf{z}^*\mathsf{w} = 0$ and $\mathsf{Az} = \mathsf{z}\alpha$) and so, by Pythagoras,

$$\|\mathsf{r}(\mathsf{y})\|^2 = (\alpha - \theta)^2\cos^2\psi + \|(\mathsf{A} - \theta)\mathsf{w}\|^2\sin^2\psi. \tag{11-7-3}$$

Now (11-7-3) already yields $|\sin\psi| \leqslant \|\mathsf{r}(\mathsf{y})\|/\gamma$, Exercise 11-7-1, for any value of θ. When $\theta = \rho(\mathsf{y})$ then $\mathsf{r} = \mathsf{r}(\mathsf{y})$ is orthogonal to y, i.e.,

$$0 = \mathsf{y}^*\mathsf{r}(\mathsf{y}) = (\alpha - \theta)\cos^2\psi + \mathsf{w}^*(\mathsf{A} - \theta)\mathsf{w}\sin^2\psi. \tag{11-7-4}$$

Thus $\cos^2 \psi$ and $\sin^2 \psi$ are in the ratio $w^*(A - \theta)w : \theta - \alpha$. Using these values in (11-7-3) yields an expression for $\|r\|^2$ in terms of w,

$$\begin{aligned} \|r(y)\|^2 &= \left[(\theta - \alpha)^2 w^*(A - \theta)w + w^*(A - \theta)^2 w(\theta - \alpha)\right] \\ &\quad /w^*(A - \alpha)w, \\ &= (\theta - \alpha)w^*(A - \alpha)(A - \theta)w/w^*(A - \alpha)w. \end{aligned} \tag{11-7-5}$$

By assumption there are no eigenvalues of A separating α and θ. Thus $(A - \alpha)(A - \theta)$ is positive definite and so, if $w = \Sigma \xi_i z_i$,

$$\begin{aligned} w^*(A - \alpha)(A - \theta)w &= \Sigma |\alpha_i - \alpha|\,|\alpha_i - \theta|\xi_i^2 \\ &\geqslant \gamma \Sigma |\alpha_i - \alpha| \xi_i^2, \qquad \text{by definition of } \gamma, \\ &\geqslant \gamma |\Sigma(\alpha_i - \alpha)\xi_i^2|, \\ &= \gamma |w^*(A - \alpha)w|. \end{aligned} \tag{11-7-6}$$

Substitute (11-7-6) into (11-7-5) and the theorem's second inequality appears. □

COROLLARY When $\theta \equiv \rho(y)$ is closest to α_1 or to α_{-1} then

$$|\tan \psi| \leqslant \|r(y)\|/\gamma. \tag{11-7-7}$$

Proof. In these cases $|w^*(A - \theta)w| \geqslant \gamma$, Ex. 11-7-4, and

$$\tan^2 \psi = (\theta - \alpha)/w^*(A - \theta)w \leqslant \|r(y)\|^2/\gamma |w^*(A - \theta)w|. \quad \square$$

In many applications, but not all, γ is unknown and the bounds in the gap theorem are theoretical. In some circumstances, in the Lanczos algorithm for instance, there comes a time when it is very unlikely that there remain undetected α's hidden among the known α's. In such circumstances the gap γ_j for θ_j can be replaced by the computable quantity $\min_{\theta_i \neq \theta_j} (|\theta_i - \theta_j| - \|r_j\|)$ and the bounds become good estimates.

Theorem (11-7-1) is very satisfactory for pinning down isolated eigenvalues. It is of no use when a cluster of close α's is approximated by a cluster of θ's with overlapping interval bounds. Theorem (11-7-1) suggests, and experience corroborates, that the Ritz vectors for such close θ's are sometimes poor eigenvector approximations. To be specific, suppose that y_1, y_2, y_3 are three such Ritz vectors. It turns out that if the cluster of θ's is well separated from all the α's not in the cluster then span(y_1, y_2, y_3) is a much better approximation to the associated invariant subspace, $\mathcal{Z}^3$ say, than is any of the y's as an individual eigenvector approximation. In other words, the mismatch between the bases does not prevent the two subspaces from being close. What is needed is a measure of the closeness of two subspaces.

11-7-1 GAP THEOREMS FOR SUBSPACES

Our physical confinement to three-dimensional space gives us no intuitive feeling for the way that even a pair of planes can be related in $\mathcal{E}^4$, let alone a pair of p-dimensional subspaces nestling in $\mathcal{E}^n$. A proper treatment of this topic is beyond the scope of this book and the interested reader should consult our source [Davis and Kahan, 1970]. What follows is a summary of a part of that work.

If f and g are unit vectors in $\mathcal{E}^n$ then

$$\angle(\text{span } \mathbf{f}, \text{span } \mathbf{g}) = \cos^{-1}|\mathbf{f}^*\mathbf{g}| = \text{arc cos } |\mathbf{f}^*\mathbf{g}|.$$

Now let $\mathbf{F}$ and $\mathbf{G}$ be orthonormal n by p matrices and let $\mathcal{F}^p = \text{span}(\mathbf{F})$, $\mathcal{G}^p = \text{span}(\mathbf{G})$. It turns out that the proper measure of the closeness of $\mathcal{F}^p$ and $\mathcal{G}^p$ is a set of p numbers called the angles between $\mathcal{F}^p$ and $\mathcal{G}^p$. However we shall only need the largest of them and will take that as the angle between the spaces

$$\angle(\mathcal{F}^p, \mathcal{G}^p) \equiv \text{arc cos } \|\mathbf{F}^*\mathbf{G}\|. \tag{11-7-8}$$

Note that $0 \leq \|\mathbf{F}^*\mathbf{G}\| \leq 1$, Ex. 11-7-6, and further, Ex. 11-7-7,

$$\angle(\mathcal{F}^p, \mathcal{G}^p) = \max_{\mathbf{f}\in\mathcal{F}^p} \min_{\mathbf{g}\in\mathcal{G}^p} \angle(\mathbf{f}, \mathbf{g}). \tag{11-7-9}$$

The next task is to define the gap. By theorem (11-5-1) there are p eigenvalues $\alpha_{i'}$ of a given $\mathbf{A}$ which can be paired with the eigenvalues θ_i of $\rho(\mathbf{F}) = \mathbf{F}^*\mathbf{A}\mathbf{F}$ so that $|\alpha_{i'} - \theta_i| \leq \|\mathbf{AF} - \mathbf{F}\rho(\mathbf{F})\|$, $i = 1, \ldots, p$. (If there are more than p α's which satisfy the inequality then any selection of p of them will do.) The remaining $n - p$ α's constitute the spectrum of **A complementary to** the spectrum of $\rho(\mathbf{F})$. Let the indices of these complementary α's form the set $\mathcal{J}$. The gap between the spectrum of $\rho(\mathbf{F})$ and the complementary spectrum of $\mathbf{A}$ is defined by

$$\gamma \equiv \min\{|\theta_i - \alpha_j| : 1 \leq i \leq p, j \in \mathcal{J}\}. \tag{11-7-10}$$

The object of all this preparation is to compare $\mathcal{F}^p$ with the invariant subspace $\mathcal{Z}^p$ belonging to the p α's paired with the θ's. Davis and Kahan give the following elegant generalization of Theorem (11-7-1).

THEOREM With the notation developed above

$$\gamma \sin \angle(\mathcal{F}^p, \mathcal{Z}^p) \leqslant \|\mathsf{AF} - \mathsf{F}\rho(\mathsf{F})\|. \qquad (11\text{-}7\text{-}11)$$

One important application replaces F by the set of Ritz vectors associated with a cluster of close Ritz values θ_i and $\rho(\mathsf{F})$ by $\mathrm{diag}(\theta_1, \ldots, \theta_p)$. When there are more than p eigenvalues α bunched close together then the bound yields very little, otherwise it is very satisfactory.

Exercises on 11-7

11-7-1. Obtain a lower bound for $\|(\mathsf{A} - \theta)\mathsf{w}\|^2$, discard some information in (11-7-3), and deduce the theorem's first inequality.

11-7-2. Show that when $\theta = \rho(\mathsf{y})$, $\sin^2 \psi = (\theta - \alpha)/\mathsf{w}^*(\mathsf{A} - \alpha)\mathsf{w}$.

11-7-3. Derive the second line of (11-7-5) from the first using the fact that $\mathsf{w}^*\mathsf{w} = 1$.

11-7-4. Show that $|\mathsf{w}^*(\mathsf{A} - \theta)\mathsf{w}| \geqslant \gamma$ when θ is closest to the greatest eigenvalue $\alpha_{-1} = \|\mathsf{A}\|$.

11-7-5. Does Theorem (11-7-1) continue to hold if α is a multiple eigenvalue but γ is the distance from θ to eigenvalues other than α? If not, where does the proof break down?

11-7-6. F and G are n by p and orthonormal. Show that $\|\mathsf{F}^*\mathsf{G}\| \leqslant 1$. *Hint*: $\|\mathsf{C}\| = \max_{\mathsf{u},\mathsf{v}} |\mathsf{u}^*\mathsf{C}\mathsf{v}|/\|\mathsf{u}\| \cdot \|\mathsf{v}\|$. When can equality occur?

11-7-7. Prove (11-7-9).

11-8 CONDENSING THE RESIDUAL

Step 6 in the *RR* procedure produced the p columns of the matrix $\mathsf{R}(Y) = \mathsf{A}Y - Y\Theta$ where Θ is the diagonal matrix of Ritz values and $Y = (\mathsf{y}_1, \ldots, \mathsf{y}_p)$ contains the Ritz vectors. The previous error bounds involved $\|\mathsf{r}_i\|$, $i = 1, \ldots, p$, or possibly $\|\mathsf{R}(Y)\|$ if the θ's were tightly bunched. Improved bounds can be had by using the matrix $\mathsf{R}(Y)$ itself.

If the Gram-Schmidt process is applied to $\mathsf{R}(Y)$ it produces n by p orthonormal S and p by p upper triangular C such that $\mathsf{R}(Y) = \mathsf{SC}$. We hasten to add that there is no need to do this in practice because $\mathsf{R}(Y)^*\mathsf{R}(Y) = \mathsf{C}^*\mathsf{S}^*\mathsf{SC} = \mathsf{C}^*\mathsf{C}$. Thus C is the Choleski factor of the p by p matrix $\mathsf{R}(Y)^*\mathsf{R}(Y)$. It is C which provides the extra information needed to improve the previous bounds. To this end the RR procedure should be supplemented with an extra step.

In Step 7 the p by p matrix $W = R(Y)^*R(Y)$ is formed. The cost is $\frac{1}{2}p(p+1)n$ ops. Next the triangular Choleski factor C of W is computed. The cost is $p^3/6$ ops. The optimal bounds developed in Secs. 10-6, 10-7, and 10-8 can now be applied with Θ replacing H. The cost is $0(p^3)$ ops.

Note that if $R(Y)$ does not have full rank then C will have to be in upper echelon form, the error bounds of Chap. 10 require a little less computation, and the derivation of C is a nuisance.

Justification for Step 7 comes from

LEMMA A is orthogonally similar to

$$\bar{A} = \left[\begin{array}{c|cc} \Theta & C^* & O^* \\ \hline C & & \\ O & U & \end{array}\right]$$

where U is unknown. (11-8-1)

The proof constitutes Ex. 11-8-2.

Exercises on Sec. 11-8

11-8-1. Use the fact that Θ is the (matrix) Rayleigh quotient of Y to show that $Y^*S = O$ when $R(Y) = SC$ has full rank.

11-8-2. Let $P = (Y, S, J)$ be an n by n orthonormal matrix. Use it to prove Lemma (11-8-1).

11-8-3. Show that $\|R(Y)\| = \|R(Q)\|$ when $p = m$.

11-8-4. Compare the operation counts for computing C by (a) forming W and doing a Choleski factorization and (b) using modified Gram-Schmidt and discarding the columns of S. What about storage requirements?

*11-9 A PRIORI BOUNDS FOR INTERIOR RITZ APPROXIMATIONS

The subspace $\mathcal{S}^m$ yields Rayleigh-Ritz approximations (θ_i, y_i) to eigenpairs $(\alpha_{i'}, z_{i'})$ for $i = 1, \ldots, m$. The results of this section are of interest when $\mathcal{S}^m$ is sufficiently well placed in $\mathcal{E}^n$ that α_i itself is the closest eigenvalue to θ_i. The bounds are a priori and not computable. Their value lies in assessing the **potential** accuracy of the RR approximations from $\mathcal{S}^m$ (see Chap. 12). The idea is to use the error in the extremal Ritz vector y_1 to obtain a simple error bound for θ_2, and then to use the errors in y_1 and θ_2 to bound the error in y_2, and so on, moving toward the interior of the spectrum.

By the minimax characterization of eigenvalues, $\alpha_1 \leqslant \theta_1 \leqslant \rho(\mathbf{s})$ for any $\mathbf{s}$ in $\mathcal{S}^m$. A clever choice of $\mathbf{s}$ (near the eigenvector $\mathbf{z}_1$) will lead to as good a bound as the circumstances warrant. However the corresponding bounds for θ_2, $\alpha_2 \leqslant \theta_2 \leqslant \rho(\mathbf{s})$, hold only if $\mathbf{s}^*\mathbf{y}_1 = 0$ and $\mathbf{s} \in \mathcal{S}^m$ (Ex. 11-9-1). And there lies the difficulty because the condition $\mathbf{s}^*\mathbf{y}_1 = 0$ demands exact knowledge of $\mathbf{y}_1$ and that is not allowed in a strict a priori analysis. The remedy is to take two useable conditions (1) $\mathbf{s}^*\mathbf{z}_1 = 0$ and (2) a bound on $\angle(\mathbf{y}_1, \mathbf{z}_1)$ and then derive modified bounds of the form $\alpha_2 \leqslant \theta_2 \leqslant \rho(\mathbf{s})$ + "a little something."

Let $\phi_i = \angle(\mathbf{y}_i, \mathbf{z}_i)$, $i = 1, \ldots, m$.

LEMMA For each $j \leqslant m$ and for any $\mathbf{s} \in \mathcal{S}^m$ which satisfies

$$\mathbf{s}^*\mathbf{z}_i = 0, \qquad i = 1, \ldots, j-1,$$

$$\alpha_j \leqslant \theta_j \leqslant \rho(\mathbf{s}) + \sum_{i=1}^{j-1} (\alpha_{-1} - \theta_i) \sin^2 \phi_i$$

$$\leqslant \rho(\mathbf{s}) + \sum_{i=1}^{j-1} (\alpha_{-1} - \alpha_i) \sin^2 \phi_i. \qquad (11\text{-}9\text{-}1)$$

Proof. The first and third inequalities follow from the Cauchy interlace theorem. To establish the middle one take $\mathbf{s}$ to be a unit vector and decompose it as

$$\mathbf{s} = \mathbf{t} + \sum_{i=1}^{j-1} \mathbf{y}_i\gamma_i, \text{ where } \mathbf{t}^*\mathbf{y}_i = 0, i = 1, \ldots, j-1. \qquad (11\text{-}9\text{-}2)$$

The hypothesis $\mathbf{s}^*\mathbf{z}_i = 0$ leads straight to a bound on γ_i,

$$|\gamma_i| = |\mathbf{s}^*\mathbf{y}_i| = |\mathbf{s}^*(\mathbf{y}_i - \mathbf{z}_i \cos \phi_i)| \leqslant \|\mathbf{s}\| \cdot |\sin \phi_i| = |\sin \phi_i|. \qquad (11\text{-}9\text{-}3)$$

Since $\mathbf{t}^*\mathbf{y}_i = 0$ and $\mathbf{y}_i^*\mathbf{y}_k = 0$,

$$\rho(\mathbf{s}) = \mathbf{t}^*\mathbf{A}\mathbf{t} + \Sigma(\mathbf{y}_i^*\mathbf{A}\mathbf{y}_i)\gamma_i^2. \qquad (11\text{-}9\text{-}4)$$

In order to turn (11-9-4) into the desired inequality all the terms must be negative; so we shift $\mathbf{A}$ by α_{-1};

$$\begin{aligned} \rho(\mathbf{s}) - \alpha_{-1} &= \mathbf{t}^*(\mathbf{A} - \alpha_{-1})\mathbf{t} + \Sigma(\theta_i - \alpha_{-1})\gamma_i^2, \\ &\geqslant \mathbf{t}^*(\mathbf{A} - \alpha_{-1})\mathbf{t}/\mathbf{t}^*\mathbf{t} + \Sigma(\theta_i - \alpha_{-1})\gamma_i^2, \quad \text{since } \|\mathbf{t}\| \leqslant 1, \\ &\geqslant \rho(\mathbf{t}) - \alpha_{-1} + \sum_{i=1}^{j-1} (\theta_i - \alpha_{-1}) \sin^2 \phi_i, \quad \text{by (11-9-3).} \end{aligned}$$

Finally note that $\rho(\mathrm{t}) \geqslant \theta_j = \min \rho(\mathrm{u})$ over $\mathrm{u} \in \mathcal{S}^m$, $\mathrm{u} \perp \mathrm{y}_i$, $i < j$. Add α_{-1} to each side and the middle inequality is established. □

The next task is to see how each angle ϕ_j can be estimated in terms of the previous ones. To do this we introduce, temporarily, $\phi_{ij} = \angle(\mathrm{z}_i, \mathrm{y}_j)$, $i = 1, \ldots, n; j = 1, \ldots, m$. Then $\phi_i = \phi_{ii}$. The following trigonometric facts are needed; for each j,

$$\mathrm{y}_j = \sum_{i=1}^{n} \mathrm{z}_i \cos \phi_{ij}, \tag{11-9-5}$$

$$|\cos \phi_{ij}| \leqslant |\sin \phi_i|, \qquad \text{(Ex. 11-9-3)}, \tag{11-9-6}$$

$$\sum_{i=j+1}^{n} \cos^2 \phi_{ij} = \sin^2 \phi_j - \sum_{i=1}^{j-1} \cos^2 \phi_{ij}. \tag{11-9-7}$$

LEMMA For each $j = 1, \ldots, m$,

$$\sin^2 \phi_j \leqslant \left[(\theta_j - \alpha_j) + \sum_{i=1}^{j-1} (\alpha_{j+1} - \alpha_i) \sin^2 \phi_i\right] / (\alpha_{j+1} - \alpha_j). \tag{11-9-8}$$

Proof. By (11-9-5) and Ex. 11-9-2

$$\rho(\mathrm{y}_j; \mathrm{A} - \alpha_j) = \theta_j - \alpha_j = \sum_{i=1}^{n} (\alpha_i - \alpha_j) \cos^2 \phi_{ij}.$$

Take terms involving the previous ϕ_{ij} to the other side and use the ordering $\alpha_1 \leqslant \alpha_2 \leqslant \cdots$ to find

$$\begin{aligned}
(\theta_j - \alpha_j) &+ \sum_{i=1}^{j-1} (\alpha_j - \alpha_i) \cos^2 \phi_{ij} \\
&= \sum_{i=j+1}^{n} (\alpha_i - \alpha_j) \cos^2 \phi_{ij}, \\
&\geqslant (\alpha_{j+1} - \alpha_j) \Sigma \cos^2 \phi_{ij}, \\
&= (\alpha_{j+1} - \alpha_j)\left(\sin^2 \phi_j - \sum_{i=1}^{j-1} \cos^2 \phi_{ij}\right), \text{ by (11-9-7).}
\end{aligned}$$

Solve for $\sin^2 \phi_j$ and use (11-9-6) to obtain desired inequality. □

The case $j = 1$ yields $\sin^2 \phi_1 \leqslant (\theta_1 - \alpha_1)/(\alpha_2 - \alpha_1)$. This is a disguised form of the error estimate for Rayleigh quotients. True a priori bounds are obtained by choosing suitable vectors s and then using (11-9-1) and (11-9-8) alternatively to majorize $\sin^2 \varphi_j$ and $\theta_j - \alpha_j$. The price paid for spurning explicit use of θ_j is rather high as Sec. 12-4 reveals.

Exercises on Sec. 11-9

11-9-1. Prove that $\alpha_2 \leqslant \theta_2 \leqslant \rho(\mathsf{s})$ if and only if $\mathsf{s} \in \mathcal{S}^m$ and $\mathsf{s}^*\mathsf{y}_1 = 0$ by using the minimax characterization of eigenvalues.

11-9-2. Show that $\rho(\mathsf{y}_i) = \theta_i$ and $\mathsf{y}_i^*\mathsf{y}_j = 0$ by using the definition of y_i.

11-9-3. Establish (11-9-6) by using the results of Ex. 11-9-2.

11-9-4. Prove that $\alpha_{-j} \geqslant \theta_{-j} \geqslant \rho(\mathsf{s}) - \sum_{i=1}^{j-1} (\alpha_i - \alpha_1) \sin^2 \phi_{-i}$ where

$$\phi_{-i} = \angle(\mathsf{y}_{-i}, \mathsf{z}_{-i}) \quad \text{and} \quad \mathsf{s}^*\mathsf{z}_{-i} = 0, i = 1, \ldots, j-1.$$

11-9-5. What is the analogue of Lemma (11-9-8) for $\sin^2 \phi_{-j}$?

11-9-6. Show that Lemma (11-9-1) continues to hold when α_{-1} is replaced by θ_{-1}.

*11-10 NONORTHOGONAL BASES

In principle it is always possible to choose an orthonormal basis for the subspace $\mathcal{S}^m$ but in practice it is not always convenient to do so. Some techniques, such as inverse iteration, produce approximate eigenvectors which fail to be mutually orthogonal when the eigenvalues are huddled close together. When n is large it is tempting to skip the precaution of reorthogonalizing computed eigenvectors, particularly when vectors are nearly orthogonal. What is there to lose?

In order to answer the question quantitatively let S be an n by p matrix with normalized columns, probably a matrix of "Ritz" vectors, and let T be any p by p symmetric matrix, probably $\mathsf{T} = \operatorname{diag}(\theta_1, \ldots, \theta_p)$. In any case $\theta_i = \lambda_i[\mathsf{T}]$. The nearly orthogonal condition is that

$$0 \leqslant \mathsf{u}^*(\mathsf{I}_p - \mathsf{S}^*\mathsf{S})\mathsf{u} < 1 \text{ for all } \mathsf{u} \in \mathcal{E}^p. \tag{11-10-1}$$

This guarantees that S has full rank. The lower bound results from normalizing the columns of S. The proper measure of the degree of linear independence among S's columns is $\sigma_1(\mathsf{S})$, the smallest singular value of S. By definition

$$\sigma_1^2(\mathsf{S}) \equiv \lambda_1[\mathsf{S}^*\mathsf{S}]. \tag{11-10-2}$$

Thus $\sigma_1 = 0$ signals linear dependence, $\sigma_1 = 1$ guarantees orthonormality.

It turns out that the computable error bounds presented earlier fail gracefully as $\sigma_1(\mathsf{S})$ declines from the value 1.

The simple error bounds of Chap. 4 made no use of orthogonality and remain intact. Of course, the approximations $(\theta_i, \mathbf{s}_i)$ are no longer the optimal Ritz approximations but it is still possible to compute the residual matrix $\mathsf{R}(\mathsf{S}, \mathsf{T}) = \mathsf{AS} - \mathsf{ST}$ and use it for clustered θ_i as indicated in the following unpublished result in [Kahan, 1967].

THEOREM For any p by p T and any n by p S satisfying (11-10-1) there are (at least) p eigenvalues $\alpha_{i'}$ of A which can be paired with the θ_i so that

$$|\alpha_{i'} - \theta_i| \leqslant \sqrt{2}\, \|\mathsf{R}(\mathsf{S}, \mathsf{T})\| / \sigma_1(\mathsf{S}), \qquad i = 1, \ldots, p. \qquad (11\text{-}10\text{-}3)$$

Proof. Since the statement of the theorem is coordinate-free there is no loss of generality in choosing an orthonormal basis in which $\mathsf{S} = \begin{bmatrix} \mathsf{J} \\ \mathsf{O} \end{bmatrix}$ where J is diagonal. Now partition A conformably and split it adroitly just as in the proof of Theorem (11-5-1).

$$\mathsf{A} = \begin{bmatrix} \mathsf{H} & \mathsf{B}^* \\ \mathsf{B} & \mathsf{U} \end{bmatrix} = \begin{bmatrix} \mathsf{T} & \mathsf{O}^* \\ \mathsf{O} & \mathsf{U} - \mathsf{X} \end{bmatrix} + \begin{bmatrix} \mathsf{H} - \mathsf{T} & \mathsf{B}^* \\ \mathsf{B} & \mathsf{X} \end{bmatrix} \qquad (11\text{-}10\text{-}4)$$

where X is to be chosen later. By the monotonicity theorem (Sec. 10-3), there are p eigenvalues $\alpha_{i'}$ of A such that, for $i = 1, \ldots, p$,

$$|\alpha_{i'} - \theta_i| \leqslant \left\| \begin{bmatrix} \mathsf{H} - \mathsf{T} & \mathsf{B}^* \\ \mathsf{B} & \mathsf{X} \end{bmatrix} \right\|. \qquad (11\text{-}10\text{-}5)$$

At this point the extension theorem of Sec. 11-11 can be invoked to say that there exists X such that

$$\left\| \begin{bmatrix} \mathsf{H} - \mathsf{T} & \mathsf{B}^* \\ \mathsf{B} & \mathsf{X} \end{bmatrix} \right\| = \|\hat{\mathsf{R}}\|, \qquad \hat{\mathsf{R}} = \begin{bmatrix} \mathsf{H} - \mathsf{T} \\ \mathsf{B} \end{bmatrix}. \qquad (11\text{-}10\text{-}6)$$

However both H and B are unknown. What is known is $\|\mathsf{R}\|$ where

$$\mathsf{R} \equiv \mathsf{A}\begin{bmatrix} \mathsf{J} \\ \mathsf{O} \end{bmatrix} - \begin{bmatrix} \mathsf{J} \\ \mathsf{O} \end{bmatrix}\mathsf{T} = \begin{bmatrix} \mathsf{HJ} - \mathsf{JT} \\ \mathsf{BJ} \end{bmatrix}. \qquad (11\text{-}10\text{-}7)$$

It remains to relate $\|\hat{\mathsf{R}}\|$ to $\|\mathsf{R}\|$ in the nontrivial case when $\mathsf{JT} \neq \mathsf{TJ}$.

First note that, by (11-10-2)

$$\|\mathsf{J}^{-1}\|^2 = \|(\mathsf{J}^*\mathsf{J})^{-1}\| = \|(\mathsf{S}^*\mathsf{S})^{-1}\| = \sigma_1^{-2}, \qquad (11\text{-}10\text{-}8)$$

and so

$$\|\mathsf{B}\| = \|(\mathsf{BJ})\mathsf{J}^{-1}\| \leqslant \|\mathsf{BJ}\|\sigma_1^{-1}. \qquad (11\text{-}10\text{-}9)$$

To cope with the top component of R rewrite it as

$$\mathsf{HJ} - \mathsf{JT} = (\mathsf{H} - \mathsf{T})\mathsf{J} + (\mathsf{TJ} - \mathsf{JT}) \qquad (11\text{-}10\text{-}10)$$

whose second term is skew. Thus for any eigenvector u of H − T with eigenvalue λ,

$$\mathsf{u}^*(\mathsf{HJ} - \mathsf{JT})\mathsf{u} = \mathsf{u}^*(\mathsf{H} - \mathsf{T})\mathsf{Ju} + 0,$$
$$= \lambda \mathsf{u}^*\mathsf{Ju}\,. \qquad (11\text{-}10\text{-}11)$$

Take u to be the dominant normalized eigenvector so that

$$|\lambda| = \|\mathsf{H} - \mathsf{T}\|.$$

Then

$$\|\mathsf{HJ} - \mathsf{JT}\| = \max |\mathsf{v}^*(\mathsf{HJ} - \mathsf{JT})\mathsf{w}|, \quad \text{over all unit vectors v, w in } \mathcal{E}^p,$$
$$\geqslant \max |\mathsf{v}^*(\mathsf{HJ} - \mathsf{JT})\mathsf{v}|, \quad \text{over all unit vectors v in } \mathcal{E}^p,$$
$$\geqslant |\mathsf{u}^*(\mathsf{HJ} - \mathsf{JT})\mathsf{u}|,$$
$$= \|\mathsf{H} - \mathsf{T}\| \cdot |\mathsf{u}^*\mathsf{Ju}|, \quad \text{by (11-10-11)},$$
$$\geqslant \|\mathsf{H} - \mathsf{T}\|\sigma_1\,, \qquad (11\text{-}10\text{-}12)$$

since $\mathsf{J} = \mathrm{diag}(\sigma_p, \ldots, \sigma_1)$. Each component of $\hat{\mathsf{R}}$ has been majorized and so

$$\|\hat{\mathsf{R}}^*\hat{\mathsf{R}}\| = \|(\mathsf{H} - \mathsf{T})^2 + \mathsf{B}^*\mathsf{B}\|,$$
$$\leqslant \|\mathsf{H} - \mathsf{T}\|^2 + \|\mathsf{B}\|^2,$$
$$\leqslant (\|\mathsf{HJ} - \mathsf{JT}\|^2 + \|\mathsf{BJ}\|^2)/\sigma_1^2, \text{ by (11-10-12) and (11-10-9)},$$
$$\leqslant 2\|\mathsf{R}\|^2/\sigma_1^2. \qquad (11\text{-}10\text{-}13)$$

This inequality when used in (11-10-6) and (11-10-5) yields the conclusion of the theorem. □

It seems plausible that the factor 2 which appears in the last step of the proof is superfluous. In any case $\sigma_1(\mathsf{S})$ can drop down to $1/\sqrt{2}$ and the error bounds merely have to be doubled, a minor modification when orders of magnitude, base 10, are all that is required in the bounds.

In point of fact the computed "Ritz" vectors $\mathsf{y}_i(= \mathsf{Sg}_i)$, $i = 1, \ldots, p$ are more nearly orthogonal than the original basis S from which they are derived when the Lanczos process is used with selective orthogonalization, as described in Chap. 13.

* 11-11 AN EXTENSION THEOREM

The theorem given below is needed to complete the proof of the residual bound theorem of the previous section. However it is also of interest in its own right as part of the study, initiated by Krein in 1946, of the extensions of an operator from a subspace to the whole space in such a way as to preserve certain properties. The result here has never been published and furnishes an explicit formula for the unique extension. It appeared in [Kahan, 1967].

THEOREM Let $R = \begin{bmatrix} H \\ B \end{bmatrix}$, where H is square. There exists a W such that the "extended" matrix $A = \begin{bmatrix} H & B^* \\ B & W \end{bmatrix}$ satisfies $\|A\| = \|R\|$.

(11-11-1)

Proof. By the Weyl monotonicity theorem (Sec. 10-4) the norm of a submatrix cannot exceed the norm of the matrix. Let $\rho = \|R\|$. Thus for any choice of W, $\rho^2 \leqslant \|A^2\|$. The theorem requires that for some W the matrix $\rho^2 - A^2$ be positive semi-definite.

The proof begins by taking any $\sigma > \rho$ and showing that $\sigma^2 - A^2$ is positive definite for some W depending on σ. Then a limiting argument shows that, as $\sigma \longrightarrow \rho +$, lim $W(\sigma)$ exists. For any W define

$$\tilde{R} \equiv \begin{bmatrix} B^* \\ W \end{bmatrix}, \qquad A = (R, \tilde{R});$$

then 2 by 2 block triangular factorization yields

$$\sigma^2 - A^2 = \begin{bmatrix} I & O \\ L & I \end{bmatrix} \begin{bmatrix} \sigma^2 - R^*R & O \\ O & U(\sigma) \end{bmatrix} \begin{bmatrix} I & L^* \\ O & I \end{bmatrix}$$

and

$$\sigma^2 - RR^* = \begin{bmatrix} I & 0 \\ K & I \end{bmatrix} \begin{bmatrix} \sigma^2 - H^2 & O \\ O & V(\sigma) \end{bmatrix} \begin{bmatrix} I & K^* \\ O & I \end{bmatrix}$$

where

$$U(\sigma) = \sigma^2 - \tilde{R}^*\left[I + R(\sigma^2 - R^*R)^{-1}R^*\right]\tilde{R},$$

$$V(\sigma) = \sigma^2\left[I - B(\sigma^2 - H^2)^{-1}B^*\right].$$

Recall that

$$\sigma^2 > \rho^2 = \|R\|^2 = \|R^*R\| = \|RR^*\|$$

and so the three matrices $\sigma^2 - R^*R$, $\sigma^2 - RR^*$, and $\sigma^2 - H^2$ are all positive definite. By Sylvester's inertia theorem $V(\sigma)$ must be positive definite too. The matrix $U(\sigma)$ depends on W and its signature is in doubt. The trick of the proof is to find a W such that $U(\sigma) = V(\sigma)$

and then, by Sylvester again, $\sigma^2 - A^2$ is positive definite. It is left as an exercise to verify that the magic W is given by

$$W(\sigma) = -BH(\sigma^2 - H^2)^{-1}B^* = -B(\sigma^2 - H^2)^{-1}HB^*.$$

It appears plausible that the construction of W and the ensuing equality of $U(\sigma)$ and $V(\sigma)$ breaks down when $\sigma \longrightarrow \rho +$ because, for one thing, $\rho^2 - R^*R$ is not invertible. However it is only the formulas which may break down. The matrix $W(\sigma)$ is a rational, and therefore meromorphic, function of the complex variable σ. As such its only singularities are poles in any neighborhood of which $\|W\|$ must be unbounded. However $\|W\| \leqslant \|A\| < \sigma$ for all real $\sigma > \rho$ and thus $W(\sigma)$ must be regular at $\sigma = \rho$ and so $W(\rho) = \lim W(\sigma)$ as $\sigma \longrightarrow \rho$. Moreover, by continuity of the norm,

$$\|A(\rho)\| = \lim_{\sigma \to \rho +} \|A(\sigma)\| = \rho. \qquad \square$$

Exercises on Sec. 11-11

11-11-1. Obtain the minimal norm extension when

$$H = \theta \quad \text{and} \quad B^* = (\beta \cos \phi, \beta \sin \phi).$$

11-11-2. Show that $\|R\| \leqslant \|A\|$ for all W. Why is it is not sufficient to prove that $\rho - A$ is positive semi-definite?

11-11-3. Verify the triangular factorizations of $\sigma^2 - A^2$ and $\sigma^2 - RR^*$ given in the proof. What are L and K? Observe that

$$I + H^2(\sigma^2 - H^2)^{-1} = \sigma^2(\sigma^2 - H^2)^{-1}.$$

11-11-4. Prove that whenever FG and GF are both defined they have the same nonzero eigenvalues. Consider the matrices

$$\begin{bmatrix} FG & O \\ G & O \end{bmatrix} \quad \text{and} \quad \begin{bmatrix} O & O \\ G & GF \end{bmatrix}.$$

Deduce that $\|R^*R\| = \|RR^*\|$.

11-11-5. Verify that the formula given for $W(\sigma)$ does ensure that $U(\sigma) = V(\sigma)$. This is a tricky calculation. Note that $H(\sigma^2 - H^2)^{-1} = (\sigma^2 - H^2)^{-1}H$.

11-11-6. Show that $(\rho^2 - H^2)C^* = B^*$ can always be solved for C^* even when $\rho^2 - H^2$ is singular. Deduce a formula for $W(\rho)$ and thus show how to avoid the limiting argument in the proof of the extension theorem.

Notes and References

The Rayleigh-Ritz approximation procedure has become a standard tool in many branches of mathematics and engineering. The original references are [Rayleigh, 1899] and [Ritz, 1909]. It is easy to get confused over the senses in which the approximations are optimal and the senses in which they are not. We have not found a text which sets the matter out clearly.

The residual bound on clustered eigenvalues appears in the unpublished report [Kahan, 1967]. The eigenvalue bounds which are based on gaps in the spectrum have their origins in [Temple, 1933], [Weinstein, 1935], and [Kato, 1949] but the material for Sec. 11-7 came from [Davis and Kahan, 1970].

In a different vein, complementing the a posteriori results discussed so far, come the inequalities which show how the Ritz approximations to inner eigenvalues are affected by the errors in the approximations to outer eigenvalues. The source is [Kaniel, 1966] and some unpublished work of Paige. The application comes in the next chapter.

The extension of all the previous results to situations in which the bases fail to be orthonormal is of considerable practical importance. The results are taken from the unpublished report [Kahan, 1967]. The Hahn-Banach-like extension theorem (11-11-1) is of interest in its own right. It has been generalized to Hilbert space. See [Parrott, 1978] which also references earlier work on this problem.

12

Krylov Subspaces

12-1 INTRODUCTION

Of considerable importance in the theory of various methods for computing eigenpairs of A is a simple type of subspace which is determined by a single nonzero vector, f say. Krylov matrices $K^m(f)$ and Krylov subspaces $\mathcal{K}^m(f)$ are defined by

$$K^m(f) = (f, Af, \ldots, A^{m-1}f),$$
$$\mathcal{K}^m(f) = \text{span } K^m(f).$$

The dimension of $\mathcal{K}^m$ will usually be m unless either f is specially related to A or $m > n$.

When started from f the power method, described in Chap. 4, will compute the columns of $K^m(f)$ one by one. However each column is written over its predecessor and so only the latest column is retained, thereby economizing on storage. In principle the whole of $K^m(f)$ could be saved and the RR (Rayleigh-Ritz, see Chap. 11) approximations from $\mathcal{K}^m(f)$ could be computed. For $m \geqslant 2$ the RR approximations will be better than the one from the power method, but they will be more expensive. Are they cost effective? That is a nice technical question involving f, the $\alpha_i (\equiv \lambda_i[A])$, storage capacity, and the ease with which A may be manipulated, but the answer briefly is a resounding yes.

Estimates of the comparative accuracy of the two methods will be taken up in this chapter and then will come a description of how approximations from $\mathcal{K}^m(\mathbf{f})$ can be computed far more economically than appears possible at first sight. The following examples may provide incentive for tackling the details of the analysis. We hope that the whole theory will seem to emerge as a **natural** consequence of using Krylov subspaces.

EXAMPLE 12-1-1

$m = 2, n = 3$. "A plane is better than its axes." $\mathbf{A} = \mathrm{diag}(3, 2, 1)$, $\mathbf{f} = (1, 1, \eta)^*$, η small. Let $(\theta_i^{(m)}, \mathbf{y}_i^{(m)})$, $i = 1, \ldots, m$ be the RR approximations from $\mathcal{K}^m(\mathbf{f})$. Then, for $\eta < 0.01$,

$m = 2$	$\angle(\mathbf{y}_2^{(2)}, \mathbf{e}_1) \leqslant \sqrt{2}\,\eta$	$\angle(\mathbf{A}^2\mathbf{f}, \mathbf{e}_1) \doteq 4/9$
$m = 3$	$\angle(\mathbf{y}_3^{(3)}, \mathbf{e}_1) = 0$	$\angle(\mathbf{A}^3\mathbf{f}, \mathbf{e}_1) \doteq 8/27$

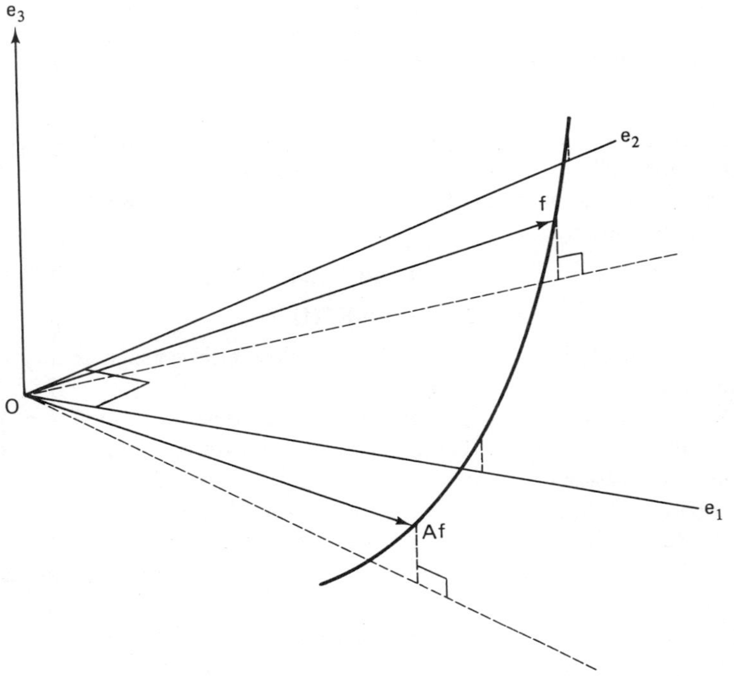

Figure 12-1-1 A plane is better than its axes

This simple example brings out the fact that the power method requires, in principle, an infinite number of iterations to capture an eigenvector whereas the RR approximations are exact for $\mathcal{K}^n(\mathbf{f})$. The case $m = 2$ shows how much better the plane $\mathcal{K}^2(\mathbf{f})$ can be than either of its given axes, $\mathbf{f}$ and $\mathbf{Af}$, as Fig. 12-1-1 suggests.

EXAMPLE 12-1-2

$$\alpha_i = \lambda_i[\mathbf{A}] = i, i = 0, \ldots, n + 1.$$

The power method will converge to $\mathbf{z}_{-1}$, the dominant eigenvector. This problem is moderately difficult when n is large. The following table gives the number of steps m required to guarantee that, for any $\mathbf{f}$, (final error angle) < (initial error angle)/100. The eigenvalue error will then be reduced by a factor of 10^4 approximately. The expressions come from Sec. 12-5.

n	PM: $(\mathbf{A} - \alpha_1)^m\mathbf{f}$	$\mathcal{K}^m(\mathbf{f})$
10^2	693 (463)	38 (32)
10^3	8061 (4607)	139 (110)
10^4	92105 (46054)	500 (380)
n	$n \ln[100\sqrt{n}]$ $(n \ln 100)$	$\frac{1}{2}\sqrt{n} \ln[200\sqrt{n}]$ $(\frac{1}{2}\sqrt{n} \ln 200)$

The first number allows for perversely chosen $\mathbf{f}$'s such as $\mathbf{z}_1 + \mathbf{z}_2 + 10^{-6}\mathbf{z}_{-1}$. The second number, in parentheses, omits the common factor in the error bounds in Sec. 12-5 gives a good estimate of m when $\mathbf{f}$ is a **reasonable** starting vector.

12-2 BASIC PROPERTIES

12-2-1 A THEORETICAL LIMITATION

To any $\mathbf{f}$ there corresponds a distinguished set of eigenvectors, namely the projections of $\mathbf{f}$ onto $\mathbf{A}$'s eigenspaces. It can happen that some of these projections are the zero vector, i.e., $\mathbf{f}$ is orthogonal to the eigenspace. Such eigenspaces are also orthogonal to $\mathcal{K}^m(\mathbf{f})$ for all m (Ex. 12-2-1) and so the vectors in them can never be detected by calculations which are based solely on the $\mathcal{K}^m(\mathbf{f})$. This is an inevitable theoretical limitation on both the power method and Krylov subspaces.

In order to describe the situation formally it helps to introduce (a) $\Sigma(\mathsf{f})$, the spectrum of A associated with f and defined by $\Sigma(\mathsf{f}) = \{\alpha : \alpha = \lambda[\mathsf{A}] \text{ and } \mathsf{H}_\alpha \mathsf{f} \neq \mathsf{o}\}$ and (b) $\mathcal{G}(\mathsf{f})$, the smallest invariant subspace of $\mathcal{E}^n$ which contains f. It can be shown (Ex. 12-2-2) that $\mathcal{G}(\mathsf{f}) = \text{span}\{\mathsf{H}_\alpha \mathsf{f} : \alpha \in \Sigma(\mathsf{f})\}$ and that A_f, the restriction of A to $\mathcal{G}(\mathsf{f})$, has simple eigenvalues (Ex. 12-2-3). Moreover the Krylov subspaces eventually fill up $\mathcal{G}(\mathsf{f})$, thus for some $\ell \leqslant n$,

$$\text{span } \mathsf{f} \subset \mathcal{K}^2(\mathsf{f}) \subset \cdots \subset \mathcal{K}^\ell(\mathsf{f}) = \mathcal{K}^{\ell+1}(\mathsf{f}) = \mathcal{G}(\mathsf{f}).$$

There are two ways of adjusting the approximation theory to this limitation. Either assume, in fine academic style, that $\mathcal{G}(\mathsf{f}) = \mathcal{E}^n$ and $\mathsf{A}_f = \mathsf{A}$ or complicate the statement of all results by referring to A_f instead of A. The fact remains that numerical methods based on powering may, in exact arithmetic, fail to detect some eigenvectors of some A's and they must, again in exact arithmetic, fail to detect the multiplicity of **any** eigenvalue they compute.

Fortunately roundoff errors make the assumption $\mathsf{A}_f = \mathsf{A}$ a realistic one in practice. This comment will be amplified during the discussion of the Lanczos algorithm in Chap. 13.

12-2-2 INVARIANCE PROPERTIES

The subspace $\mathcal{K}^m(\mathsf{f})$ depends on A and, when necessary, it is also denoted by $\mathcal{K}^m(\mathsf{f}; \mathsf{A})$. The following invariance properties are valuable.

I.	Scaling: $\mathcal{K}^m(\sigma\mathsf{f}; \tau\mathsf{A}) = \mathcal{K}^m(\mathsf{f}; \mathsf{A})$,	$\sigma \neq 0$, $\tau \neq 0$
II.	Translation: $\mathcal{K}^m(\mathsf{f}; \mathsf{A} - \sigma) = \mathcal{K}^m(\mathsf{f}; \mathsf{A})$	
III.	Change of Basis: $\mathcal{K}^m(\mathsf{Pf}; \mathsf{PAP}^*) = \mathsf{P}\mathcal{K}^m(\mathsf{f}; \mathsf{A})$,	$\mathsf{P}^* = \mathsf{P}^{-1}$

Verification is left to the reader but some comments are in order. The error bounds associated with approximations from $\mathcal{K}^m(\mathsf{f}; \mathsf{A})$ should also be invariant in the same way.

Property II is puzzling at first because it fails completely for the associated Krylov matrices $\mathsf{K}^m(\mathsf{f}; \mathsf{A})$. Thus the power method depends strongly on any shift σ, but each σ merely induces a different basis $\mathsf{K}^m(\mathsf{f}; \mathsf{A} - \sigma)$ for the same subspace $\mathcal{K}^m(\mathsf{f}; \mathsf{A})$. One consequence of properties I and II is that there is no loss of generality in supposing that $\lambda_1[\mathsf{A}] = -1$ and $\lambda_{-1}[\mathsf{A}] = +1$. This normalization reveals why, as m increases, the space $\mathcal{K}^m(\mathsf{f}; \mathsf{A})$ moves close to the eigenvectors belonging to extreme eigenvalues. Not surprisingly the error bounds involve ratios of differences of A's eigenvalues; such quantities are properly invariant.

Property III reflects the fact that an orthogonal basis change in $\mathcal{E}^n$ induces an orthogonal similarity transformation of $\mathbf{A}$. There is no loss of generality in studying diagonal matrices $\mathbf{A}$.

Exercises on Sec. 12-2

12-2-1. Show that if $\mathbf{A}\mathbf{z} = \mathbf{z}\alpha$ and $\mathbf{z} \perp \mathbf{f}$ then $\mathbf{z} \perp \mathcal{K}^m(\mathbf{f})$ for all $m > 1$.

12-2-2. Show that $\mathcal{G}(\mathbf{f}) = \text{span}\{\mathbf{H}_\alpha \mathbf{f} : \alpha \in \Sigma(\mathbf{f})\}$ by using $\mathbf{A}\mathbf{H}_\alpha = \mathbf{H}_\alpha \mathbf{A} = \alpha \mathbf{H}_\alpha$.

12-2-3. Show that $\mathbf{A}_f \equiv \mathbf{A}|_{\mathcal{G}(\mathbf{f})}$ has simple eigenvalues.

12-2-4. Consider $\mathbf{A} = \text{diag}(-1, -0.5, 0, 0.5, 1)$ and $\mathbf{f} = (1, 1, \ldots, 1)^*$. Compute the angle between $\mathbf{A}^k\mathbf{f}$ and the $(\mathbf{e}_1, \mathbf{e}_5)$ plane for $k = 0, 1, 2, 3, 4$.

12-2-5. Verify the invariance properties I, II, and III.

12-3 REPRESENTATION BY POLYNOMIALS

Each element $\mathbf{s}$ in $\mathcal{K}^m(\mathbf{f})$ has the special form

$$\mathbf{s} = \sum_{i=0}^{m-1} (\mathbf{A}^i\mathbf{f})\gamma_i = \sum_{i=0}^{m-1} (\gamma_i \mathbf{A}^i)\mathbf{f} = \pi(\mathbf{A})\mathbf{f} \qquad (12\text{-}3\text{-}1)$$

where $\pi(\xi) \equiv \Sigma \gamma_i \xi^i$ is a polynomial of degree $< m$. This chapter will make heavy use of the vector space $\mathcal{P}^k$ of all real polynomials of degree not exceeding k. For example (12-3-1) has the nice interpretation

$$\mathcal{K}^m(\mathbf{f}) = \{\pi(\mathbf{A})\mathbf{f} : \pi \epsilon \mathcal{P}^{m-1}\} \qquad (12\text{-}3\text{-}2)$$

and provides a one-one correspondence between $\mathcal{K}^m$ and $\mathcal{P}^{m-1}$ thanks to our tacit assumption that dim $\mathcal{K}^m = m$.

Not surprisingly polynomials play a large role in describing approximations from $\mathcal{K}^m$. For future reference we describe the Ritz vectors. Recall from Chap. 11 that for any subspace $\mathcal{S}^m$ the Ritz vectors $\mathbf{y}_i$, $i = 1, \ldots, m$ are mutually orthogonal and are characterized by the property

$$\mathbf{A}\mathbf{y}_i - \mathbf{y}_i\theta_i \perp \mathcal{S}^m, \qquad i = 1, \ldots, m. \qquad (12\text{-}3\text{-}3)$$

In Krylov subspaces it is easy to characterize vectors orthogonal to a particular $\mathbf{y}_k$.

LEMMA Let $(\theta_i, \mathbf{y}_i)$, $i = 1, \ldots, m$ be the Rayleigh-Ritz approximations from $\mathcal{K}^m(\mathbf{f})$. If $\omega \epsilon \mathcal{P}^{m-1}$ then

$$\omega(\mathbf{A})\mathbf{f} \perp \mathbf{y}_k \text{ if, and only if, } \omega(\theta_k) = 0. \qquad (12\text{-}3\text{-}4)$$

Proof. Suppose first that ω is in $\mathcal{P}^m$, not just $\mathcal{P}^{m-1}$, and that $\omega(\xi) = (\xi - \theta_k)\pi(\xi)$ where π is in $\mathcal{P}^{m-1}$. Thus

$$\pi(\mathsf{A})\mathsf{f} \in \mathcal{K}^m(\mathsf{f}) \tag{12-3-5}$$

and

$$\begin{aligned} \mathsf{y}_k^*\omega(\mathsf{A})\mathsf{f} &= \mathsf{y}_k^*(\mathsf{A} - \theta_k)\pi(\mathsf{A})\mathsf{f} \\ &= [(\mathsf{A} - \theta_k)\mathsf{y}_k]^*\pi(\mathsf{A})\mathsf{f}, \text{ since } \mathsf{A} = \mathsf{A}^*, \\ &= 0, \quad \text{by (12-3-3) and (12-3-5).} \end{aligned}$$

This establishes sufficiency with a little extra in hand. Next observe that $\mathsf{y}_k \perp \mathcal{S}_k$ where

$$\mathcal{S}_k \equiv \{\tau(\mathsf{A})\mathsf{f} : \tau \in \mathcal{P}^{m-1}, \tau(\theta_k) = 0\} = (\mathsf{A} - \theta_k)\mathcal{K}^{m-1}(\mathsf{f})$$

is a subspace of $\mathcal{K}^m$ of dimension $m - 1$. Since the set of all vectors in $\mathcal{K}^m$ which are orthogonal to y_k is a subspace of dimension $m - 1$ it must coincide with $\mathcal{S}_k$. □

This result yields a description of y_k. It is natural to define

$$\mu(\xi) \equiv \prod_{i=1}^{m} (\xi - \theta_i) \text{ and } \pi_k(\xi) \equiv \mu(\xi)/(\xi - \theta_k). \tag{12-3-6}$$

From the result established in the proof of Lemma (12-3-4) follows:

COROLLARY

$$\begin{aligned} \mathsf{y}_k &= \pi_k(\mathsf{A})\mathsf{f}/\|\pi_k(\mathsf{A})\mathsf{f}\|, \\ \beta_m &\equiv \|\mu(\mathsf{A})\mathsf{f}\| = \min \|\omega(\mathsf{A})\| \end{aligned} \tag{12-3-7}$$

where the minimum is over all monic polynomials ω of degree m.

The proof is left as an exercise. Naturally μ is called the *minimal polynomial* of f of degree m. Note also that $\|\mu(\mathsf{A})\mathsf{f}\|$ is the distance of $\mathsf{A}^m\mathsf{f}$ from $\mathcal{K}^m$.

The next step is to establish a basic lemma which will be used to derive bounds on $(\theta_j - \alpha_j)$ for each $j = 1, \ldots, m$. The role of f in these bounds can be captured adequately by two numbers: $\angle(\mathsf{f}, \mathsf{z}_j)$ is one and the other is $\angle(\mathsf{f}, \mathcal{Z}^j)$ where $\mathcal{Z}^j = \text{span}(\mathsf{z}_1, \ldots, \mathsf{z}_j)$. When $j > 1$ the latter angle may be much smaller than the former.

LEMMA Let h be the normalized projection of f orthogonal to $\mathcal{Z}^j$. For each $\pi \in \mathcal{P}^{m-1}$ and each $j \leq m$ the Rayleigh quotient ρ satisfies

$$\rho(\pi(\mathsf{A})\mathsf{f}; \mathsf{A} - \alpha_j) \leq (\alpha_n - \alpha_j)\left[\frac{\sin \angle(\mathsf{f}, \mathcal{Z}^j)}{\cos \angle(\mathsf{f}, \mathsf{z}_j)} \frac{\|\pi(\mathsf{A})\mathsf{h}\|}{\pi(\alpha_j)}\right]^2. \quad (12\text{-}3\text{-}8)$$

Proof. Let $\psi = \angle(\mathsf{f}, \mathcal{Z}^j)$ and let g be the normalized projection of f onto $\mathcal{Z}^j$ so that

$$\mathsf{f} = \mathsf{g} \cos \psi + \mathsf{h} \sin \psi$$

is an orthogonal decomposition of f. Since $\mathcal{Z}^j$ is invariant under A,

$$\mathsf{s} \equiv \pi(\mathsf{A})\mathsf{f} = \pi(\mathsf{A})\mathsf{g} \cos \psi + \pi(\mathsf{A})\mathsf{h} \sin \psi$$

is an orthogonal decomposition of s. A little calculation yields

$$\rho(\mathsf{s}; \mathsf{A} - \alpha_j) = \big[\mathsf{g}^*(\mathsf{A} - \alpha_j)\pi^2(\mathsf{A})\mathsf{g} \cos^2 \psi$$
$$+ \mathsf{h}^*(\mathsf{A} - \alpha_j)\pi^2(\mathsf{A})\mathsf{h} \sin^2 \psi\big]/\|\pi(\mathsf{A})\mathsf{f}\|^2. \quad (12\text{-}3\text{-}9)$$

The eigenvalues of A are labeled so that $\alpha_1 \leq \alpha_2 \leq \ldots \leq \alpha_n$ and

(a) $\mathsf{v}^*(\mathsf{A} - \alpha_j)\mathsf{v} \leq 0$ for all $\mathsf{v} \in \mathcal{Z}^j$, in particular $\mathsf{v} = \pi(\mathsf{A})\mathsf{g}$,
(b) $\mathsf{w}^*(\mathsf{A} - \alpha_j)\mathsf{w} \leq (\alpha_n - \alpha_j)\|\mathsf{w}\|^2$ for all $\mathsf{w} \in \mathcal{Z}^j$, in particular $\mathsf{w} = \pi(\mathsf{A})\mathsf{h}$.

When (a) and (b) are used to simplify (12-3-9) it becomes:

$$\rho(\mathsf{s}; \mathsf{A} - \alpha_j) \leq (\alpha_n - \alpha_j)\big[\|\pi(\mathsf{A})\mathsf{h}\| \sin \psi / \|\pi(\mathsf{A})\mathsf{f}\|\big]^2. \quad (12\text{-}3\text{-}10)$$

The proof is completed by using an eigenvector expansion of f

$$\|\mathsf{s}\|^2 = \|\pi(\mathsf{A})\mathsf{f}\|^2$$
$$= \sum_{i=1}^{n} \pi^2(\alpha_i) \cos^2(\mathsf{f}, \mathsf{z}_i) \geq \pi^2(\alpha_j) \cos^2 \angle(\mathsf{f}, \mathsf{z}_j). \qquad \square$$

The inequalities (a) and (b) used to obtain (12-3-10) are crude but without dragging in more information they cannot be improved. Since $\angle(\mathsf{f}, \mathcal{Z}^j) \leq \angle(\mathsf{f}, \mathsf{z}_j)$ it is valid to replace the sine and cosine in the lemma by $\tan \angle(\mathsf{f}, \mathsf{z}_j)$ but when $j > 1$ this increases the bound unnecessarily. There is no loss of generality in assuming that all our angles are acute.

Exercises on Sec. 12-3

12-3-1. Prove both parts of (12-3-7). *Hint*: Consider the direction of $\mu(\mathsf{A})\mathsf{f}$.

12-3-2. Show that for each $\pi \in \mathcal{P}^{m-1}$ and $j \leqslant m$,

$$|\tan \angle(\mathsf{z}_j, \mathcal{K}^m)| \leqslant |\tan \angle(\mathsf{z}_j, \mathsf{f})| \cdot \|\pi(\mathsf{A})\mathsf{f}_j\|/|\pi(\alpha_j)|$$

where f_j is the normalized projection of f orthogonal to z_j.

*12-4 THE ERROR BOUNDS OF KANIEL AND SAAD

This section develops a priori error bounds on the Rayleigh-Ritz approximations (θ_i, y_i), $i = 1, \ldots, m$, from $\mathcal{K}^m(\mathsf{f}; \mathsf{A})$ to the *corresponding* eigenpairs (α_i, z_i) of A. These results should be contrasted with the computable residual error bounds (Chap. 11) which relate each θ_i to some close eigenvalue $\alpha_{i'}$ of unknown index. Naturally the a priori bounds are of most interest when f is so nicely chosen that α_i is indeed the closest eigenvalue to θ_i. There are analogous bounds relating $(\theta_{-i}, \mathsf{y}_{-i})$ to $(\alpha_{-i}, \mathsf{z}_{-i})$ for $i = 1, \ldots, m$, and consequently each θ_i is compared tacitly with α_i and with α_{i-m-1} for $i = 1, \ldots, m$. At most one of the bounds will be small.

The error bound on $\theta_j - \alpha_j$ depends on several quantities: the starting vector f, the $\angle(\mathsf{y}_i, \mathsf{z}_i)$ for $i < j$, and the **spread** $(\alpha_n - \alpha_1)$ of A's spectrum. However, as will be seen shortly, the leading role is played by the Chebyshev polynomial T_{m-j} whose steep climb outside the interval $[-1, 1]$ helps to explain the excellent approximations obtained from Krylov subspaces.

The error bounds come from choosing a polynomial π in Lemma (12-3-8) such that, among other things

I. $|\pi(\alpha_j)|$ is large while $\|\pi(\mathsf{A})\mathsf{h}\|$ is small, and

II. $\rho(\mathsf{s}; \mathsf{A} - \alpha_j) \geqslant 0$ where $\mathsf{s} = \pi(\mathsf{A})\mathsf{f}$.

It is I that brings in the Chebyshev polynomials. Note that

$$\begin{aligned}\|\pi(\mathsf{A})\mathsf{h}\|^2 &= \sum_{i=j+1}^{n} \pi^2(\alpha_i)\cos^2\angle(\mathsf{f}, \mathsf{z}_i) \Big/ \sum_{i=j+1}^{n} \cos^2\angle(\mathsf{f}, \mathsf{z}_i), \\ &\leqslant \max_{i>j} \pi^2(\alpha_i).\end{aligned}$$

There are $n - j$ α's exceeding α_j but π has, it turns out, only $m - j$ disposable zeros and usually $m \ll n$. The best π will depend strongly on the actual distribution of A's eigenvalues. In the interests of simplicity it is customary to majorize $\|\pi(\mathsf{A})\mathsf{h}\|$ still further by

$$\|\pi(\mathsf{A})\mathsf{h}\|^2 \leq \max_{i>j} \pi^2(\alpha_i) \leq \max \pi^2(\tau) \text{ over all } \tau \text{ in } [\alpha_{j+1}, \alpha_n].$$

It is a scaled Chebyshev polynomial which minimizes the term on the right when $\pi(\alpha_j)$ is given.

Requirement II concerns the left side of the inequality in Lemma (12-3-8), namely $\rho(\mathsf{s}; \mathsf{A} - \alpha_j)$. The following facts are known:

(A) $0 \leq \theta_j - \alpha_j$ (from Sec. 10-1),

(B) $\theta_j - \alpha_j \leq \rho(\mathsf{s}; \mathsf{A} - \alpha_j)$, if $\mathsf{s} \perp \mathsf{y}_i$ for all $i < j$ from (11-4-6),

(C) $\theta_j - \alpha_j \leq \rho(\mathsf{s}; \mathsf{A} - \alpha_j) + \sum_{i=1}^{j-1} (\alpha_n - \alpha_i) \sin^2 \angle(\mathsf{y}_i, \mathsf{z}_i)$,
if $\mathsf{s} \perp \mathsf{z}_i$ for all $i < j$ (from Sec. 11-9).

It is clear from (A) that if $\rho(\mathsf{s}; \mathsf{A} - \alpha_j) < 0$ then, a fortiori, $\rho(\mathsf{s}; \mathsf{A} - \alpha_j) < \theta_j - \alpha_j$ and Lemma (12-3-8) cannot be used to bound $\theta_j - \alpha_j$. Hence II turns our attention to either (B), which yields Saad's bounds, or (C) which yields Kaniel's bound. We take (B) first.

THEOREM Let $\theta_1 \leq \ldots \leq \theta_m$ be the Ritz values derived from $\mathcal{K}^m(\mathsf{f})$ and let (α_i, z_i) be the eigenpairs of A. For each $j = 1, \ldots, m$

$$0 \leq \theta_j \leq (\alpha_n - \alpha_j) \left[\frac{\sin \angle(\mathsf{f}, \mathcal{Z}^j)}{\cos \angle(\mathsf{f}, \mathsf{z}_j)} \cdot \frac{\prod_{\nu=1}^{j-1} \left(\frac{\theta_\nu - \alpha_n}{\theta_\nu - \alpha_j} \right)}{T_{m-j}(1 + 2\gamma)} \right]^2$$

and

$$\tan \angle(\mathsf{z}_j, \mathcal{K}^m) \leq \frac{\sin \angle(\mathsf{f}, \mathcal{Z}^j)}{\cos \angle(\mathsf{f}, \mathsf{z}_j)} \cdot \frac{\prod_{v=1}^{j-1} \left(\frac{\alpha_\nu - \alpha_n}{\alpha_\nu - \alpha_j} \right)}{T_{m-j}(1 + 2\gamma)},$$

where $\gamma \equiv (\alpha_j - \alpha_{j+1})/(\alpha_{j+1} - \alpha_n)$. (12-4-1)

Proof. Apply Lemma (12-3-8). To ensure (B) the trial vector $\mathsf{s} = \pi(\mathsf{A})\mathsf{f}$ must be orthogonal to $\mathsf{y}_1, \ldots, \mathsf{y}_{j-1}$. By Lemma (12-3-4) it suffices to consider polynomials π of the form

$$\pi(\xi) = (\xi - \theta_1) \ldots (\xi - \theta_{j-1}) \tilde{\pi}(\xi), \qquad \tilde{\pi} \in \mathcal{P}^{m-j}.$$

Note that for such π

$$\frac{\|\pi(\mathsf{A})\mathsf{h}\|}{|\pi(\alpha_j)|} \leq \frac{\|(\mathsf{A}-\theta_1)\ldots(\mathsf{A}-\theta_{j-1})\|\cdot\|\tilde{\pi}(\mathsf{A})\mathsf{h}\|}{|(\alpha_j-\theta_1)\ldots(\alpha_j-\theta_{j-1})|\cdot|\tilde{\pi}(\alpha_j)|},$$

$$\leq \prod_{i=1}^{j-1}\left[\frac{\alpha_n-\theta_i}{\alpha_j-\theta_i}\right]\max_{\tau}\frac{|\tilde{\pi}(\tau)|}{|\tilde{\pi}(\alpha_j)|} \text{ over } \tau \text{ in } [\alpha_{j+1},\alpha_n], \tag{12-4-2}$$

since $\mathsf{h}\perp\mathcal{Z}^j$. The problem has been reduced to finding $\tilde{\pi}\in\mathcal{P}^{m-j}$ that minimizes the ratio on the right side of (12-4-2). The well-known solution (see Appendix B) is the Chebyshev polynomial adapted to $[\alpha_{j+1},\alpha_n]$. In fact

$$\min_{\tilde{\pi}}\max_{\tau}\frac{|\tilde{\pi}(\tau)|}{|\tilde{\pi}(\alpha_j)|} = \frac{\max_{\tau} T_{m-j}(\tau;[\alpha_{j+1},\alpha_n])}{T_{m-j}(\alpha_j;[\alpha_{j+1},\alpha_n])},$$

$$= \frac{1}{T_{m-j}(1+2\gamma)}, \tag{12-4-3}$$

where the **gap ratio** γ is defined in the statement of the theorem. On combining (B), Lemma (12-3-8), (12-4-2), and (12-4-3) the first of the results is obtained.

The second result comes from decomposing f as

$$\mathsf{f} = \bar{\mathsf{g}}\cos\angle(\mathsf{f},\mathcal{Z}^{j-1}) + \mathsf{z}_j\cos\angle(\mathsf{f},\mathsf{z}_j) + \mathsf{h}\sin\angle(\mathsf{f},\mathcal{Z}^j).$$

This time π is chosen to satisfy $\pi(\alpha_i)=0$ for $i=1,\ldots,j-1$, so that

$$\mathsf{s} = \pi(\mathsf{A})\mathsf{f} = \mathsf{o} + \mathsf{z}_j\pi(\alpha_j)\cos\angle(\mathsf{f},\mathsf{z}_j) + \pi(\mathsf{A})\mathsf{h}\sin\angle(\mathsf{f},\mathcal{Z}^j).$$

This is an orthogonal decomposition of s and therefore

$$\tan\angle(\mathsf{s},\mathsf{z}_j) = \frac{\sin\angle(\mathsf{f},\mathcal{Z}^j)\|\pi(\mathsf{A})\mathsf{h}\|}{\cos\angle(\mathsf{f},\mathsf{z}_j)\,|\pi(\alpha_j)|}.$$

The proof is completed by taking the same $\tilde{\pi}$ as above. □

Kaniel's bounds make explicit reference to the Ritz vectors y_i, $i=1,\ldots,j-1$. He substitutes the same s as above, namely

$$\mathsf{s} = \pi(\mathsf{A})\mathsf{f} = (\mathsf{A}-\alpha_1)\ldots(\mathsf{A}-\alpha_{j-1})\tilde{\pi}(\mathsf{A})$$

into Lemma (11-9-1) which compensates, as shown in (C), for the fact that this s does not satisfy $\mathsf{s}^*\mathsf{y}_i = O$, $i<j$. The same $\tilde{\pi}$ is used as in the proof of Theorem (12-4-1) to obtain

THEOREM The Rayleigh-Ritz approximations (θ_j, y_j) from $\mathcal{K}^m(\mathsf{f})$ to (α_j, z_j) satisfy

$$0 \leqslant \theta_j - \alpha_j \leqslant \alpha_n - \alpha_j \leqslant \left[\frac{\sin \angle(\mathsf{f}, \mathcal{Z}^j)}{\cos \angle(\mathsf{f}, \mathsf{z}_j)} \cdot \frac{\sum_{\nu=1}^{j-1} \left(\frac{\alpha_\nu - \alpha_n}{\alpha_\nu - \alpha_j} \right)}{T_{m-j}(1 + 2\gamma)} \right]^2 + \sum_{\nu=1}^{j-1} (\alpha_n - \alpha_\nu) \sin^2 \angle(\mathsf{y}_\nu, \mathsf{z}_\nu),$$

where

$$\gamma = (\alpha_j - \alpha_{j+1})/(\alpha_{j+1} - \alpha_n), \quad \text{and, from (11-9-8),}$$

$$\sin^2 \angle(\mathsf{y}_\nu, \mathsf{z}_\nu) \leqslant \left[(\theta_\nu - \alpha_\nu) + \sum_{\mu=1}^{\nu-1} (\alpha_{\nu+1} - \alpha_\mu) \sin^2 \angle(\mathsf{y}_\mu, \mathsf{z}_\mu) \right] / (\alpha_{\nu+1} - \alpha_\nu). \qquad (12\text{-}4\text{-}4)$$

Theorem (12-4-1) does not give an explicit bound on $\angle(\mathsf{y}_\nu, \mathsf{z}_\nu)$ and to repair that omission Saad relates this angle to the smaller one $\angle(\mathsf{z}_\nu, \mathcal{K}^m)$ which was majorized in Theorem (12-4-1). Reference to Fig. 12-4-1 shows that

$$\sin^2 \angle(\mathsf{y}_j, \mathsf{z}_j) = PQ^2 + QR^2 = \sin^2 \phi + \sin^2 \omega \cos^2 \phi. \qquad (12\text{-}4\text{-}5)$$

As indicated the scene is three dimensional; $\{\mathsf{y}_j, \mathsf{u}_j, \mathsf{w}_j\}$ is an orthonormal set of vectors of which y_j and u_j are in $\mathcal{K}^m$ while $\mathsf{w}_j \perp \mathcal{K}^m$. The normalized projection of z_j onto $\mathcal{K}^m$ is v_j and the following relations are needed for the proof of the next theorem.

$$\mathsf{z}_j = \mathsf{v}_j \cos \phi + \mathsf{w}_j \sin \phi, \qquad (12\text{-}4\text{-}6)$$

$$\mathsf{v}_j = \mathsf{y}_j \cos \omega + \mathsf{u}_j \sin \omega. \qquad (12\text{-}4\text{-}7)$$

In order to discuss Ritz vectors such as y_j we need the (orthogonal) projector onto $\mathcal{K}^m$. This would be the matrix $\mathsf{Q}_m\mathsf{Q}_m^*$ in Sec. 11-4 but we will call it H here. Recall that (θ_j, y_j) is an eigenpair of the projection of A on $\mathcal{K}^m$, namely the operator HA restricted to $\mathcal{K}^m$. The analysis which follows gives a nice example of the importance of the domain for we shall also be interested in HA restricted to $(\mathcal{K}^m)^\perp$. The distinction between the two operators can be made by writing HAH for the former and HA(1 − H) for the latter. Theorem (12-4-9) below makes use of an interesting quantity called the **variation of $\mathcal{K}^m$ by A**,

$$\beta_m \equiv \|\mathsf{HA}(1 - \mathsf{H})\| = \|\mathsf{HA} \text{ restricted to } (\mathcal{K}^m)^\perp\| \qquad (12\text{-}4\text{-}8)$$

It is left as Ex. 12-4-2 to show that this is the same β_m as appeared in (12-3-7).

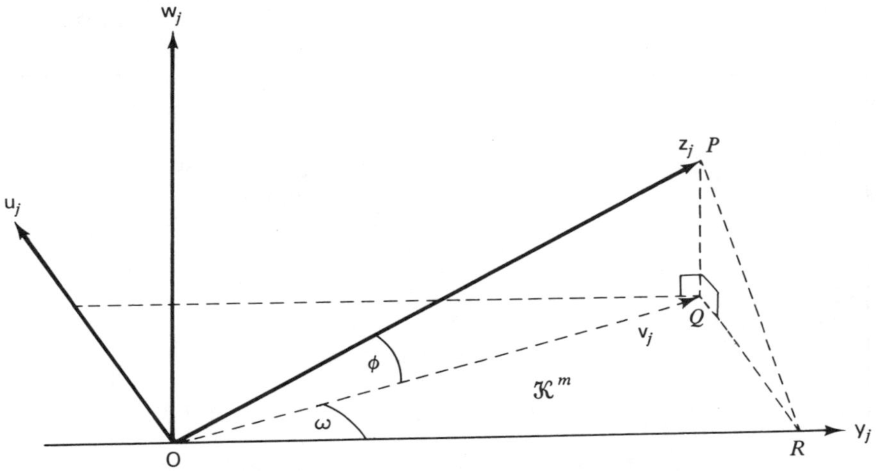

Figure 12-4-1 Projection of z_j onto $\mathcal{K}^m$ and span (y_j)

$\phi = \angle(z_j, v_j)$, $\omega = \angle(v_j, y_j)$

THEOREM Let z_j be the normalized projection of f onto the eigenspace of α_j for some $j \leq m$ and let (θ_i, y_i) $i = 1, \ldots, m$ be the Ritz approximations from $\mathcal{K}^m$. Then

$$\sin^2 \angle(z_j, y_j) \leq (1 + \beta_m^2/\gamma_j^{(m)2}) \sin^2 \angle(z_j, \mathcal{K}^m),$$

where $\gamma_j^{(m)} \equiv \min |\alpha_j - \theta_i|$ over $i \neq j$. (12-4-9)

Proof. Using (12-4-6) rewrite the defining property of z_j, namely $(A - \alpha_j)z_j = o$ in the form

$$(A - \alpha_j)v_j \cos \phi = -(A - \alpha_j)w_j \sin \phi.$$

This vector is not in $\mathcal{K}^m$ and the trick is to consider its projection onto $\mathcal{K}^m$. Hence

$$\|H(A - \alpha_j)v_j \cos \phi\| = \|H(A - \alpha_j)w_j \sin \phi\|. \quad (12\text{-}4\text{-}10)$$

The right side of (12-4-10) is easily majorized by using (12-4-8),

$$\|H(A - \alpha_j)w_j \sin \phi\| \leq \|H(A - \alpha_j)(1 - H)\| \cdot \|w_j\| \sin \phi$$
$$= \beta_m \sin \phi. \quad (12\text{-}4\text{-}11)$$

Next observe that y_j is an eigenvector of $H(A - \alpha_j)$ and so (Ex. 12-4-3) pre-multiplication of (12-4-7) by $H(A - \alpha_j)$ yields an orthogonal decomposition of the vector on the left of (12-4-10), namely

$$H(A - \alpha_j)v_j = (\theta_j - \alpha_j)y_j \cos \omega + H(A - \alpha_j)u_j \sin \omega. \quad (12\text{-}4\text{-}12)$$

Hence

$$\|H(A - \alpha_j)v_j\| \geqslant \|H(A - \alpha_j)u_j\| \sin \omega. \qquad (12\text{-}4\text{-}13)$$

Finally note that $u_j \epsilon (y_j^{\perp} \cap \mathcal{K}^m)$ which is invariant under $H(A - \alpha_j)$. The restriction of $H(A - \alpha_j)H$ to y_j has eigenvalues $\{\theta_i - \alpha_j, i = 1, \ldots, m, i \neq j\}$ and so

$$\|H(A - \alpha_j)u_j\|^2 = \rho\left(u_j; \left[H(A - \alpha_j)H\right]^2\right) \geqslant \min_{i \neq j} (\theta_i - \alpha_j)^2 = \gamma_j^{(m)2}. \qquad (12\text{-}4\text{-}14)$$

The inequalities (12-4-11), (12-4-13), and (12-4-14) applied to (12-4-10) yield

$$\gamma_j^{(m)} \sin \omega \cos \phi \leqslant \beta_m \sin \phi \qquad (12\text{-}4\text{-}15)$$

and the result follows from (12-4-5). □

The simplest choice for $\tilde{\pi}$, as made in Theorems (12-4-1) and (12-4-4), will not always give the best bound as the following figure indicates. Take $j = 3$, take α_4 close to α_3, and α_{n-1} well separated from α_n.

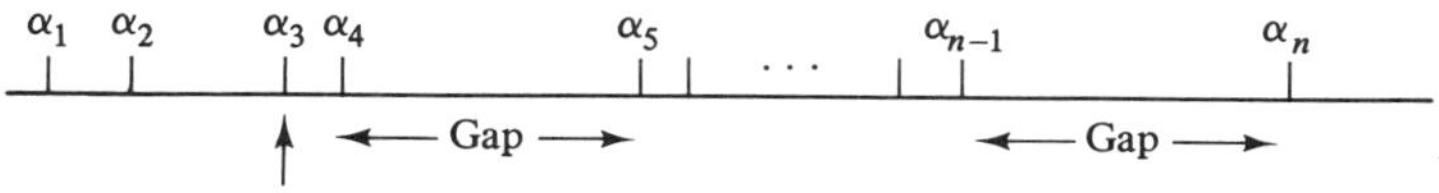

In this situation it is better to force $\tilde{\pi}(\alpha_4) = 0$, $\tilde{\pi}(\alpha_n) = 0$, and then fit T_{m-j-2} to $[\alpha_5, \alpha_{n-1}]$. The more that is known about A the better can $\tilde{\pi}$ be chosen. In general let $\mathcal{J} = \{j + 1, \cdots, k - 1, \ell + 1, \ldots, n\}$ be the index set of those α's to be taken as zeros of $\tilde{\pi}$ and define the **gap ratio** $\gamma_{jk\ell}$ by

$$\gamma_{jk\ell} \equiv (\alpha_j - \alpha_k)/(\alpha_k - \alpha_\ell), j < k < \ell. \qquad (12\text{-}4\text{-}16)$$

> In both theorems in this section it is legitimate to replace $T_{m-j}(1 + 2\gamma_{j,j+1,n})$ by
>
> $$\prod_{\mu=j+1}^{k-1} \left(\frac{\alpha_j - \alpha_\mu}{\alpha_\mu - \alpha_\ell}\right) \prod_{\nu=\ell+1}^{n} \left(\frac{\alpha_j - \alpha_\nu}{\alpha_k - \alpha_\nu}\right) T_{m-j-|\mathcal{J}|}(1 + 2\gamma_{jk\ell}).$$
>
> where $|\mathcal{J}|$ denotes the number of elements in $\mathcal{J}$. (12-4-17)

All the bounds become worthless as j increases too much because there are only m θ's to n α's and quite soon α_j will not be the closest eigenvalue to θ_j. By the invariance properties of $\mathcal{K}^m(f)$ there is no preferred end to the spectrum. The results dual to Theorems (12-4-1) and (12-4-4) involve little more than negating indices. The precise formulation of them is a useful exercise.

EXAMPLE 12-4-1

The bounds of Saad [Theorem (12-4-1)] and Kaniel [Theorem (12-4-4)] will be compared on an example close to one used by Kaniel in his original paper. For simplicity we replace $\sin \angle(\mathbf{f}, \mathcal{Z}^j)/\cos \angle(\mathbf{f}, \mathbf{z}_j)$ by $\tan \phi_j$, where $\phi_j \equiv \angle(\mathbf{f}, \mathbf{z}_j)$ and $\mathbf{f}$ is the starting vector for $\mathcal{K}^m$ with $m = 53$. The data on $\mathbf{A}$'s eigenvalues are $\alpha_1 = 0$, $\alpha_2 = 0.01$, $\alpha_3 = 0.04$, $\alpha_4 = 0.1$, $\alpha_n = 1.0$.

Saad's bounds are of the form $\theta_j - \alpha_j \leq (\alpha_n - \alpha_j)[(\tan \phi_j)\kappa_j^{(m)}/T_{m-j}(1 + 2\gamma)]^2$, where $\kappa_j^{(m)}$ is a function of θ's and α's. In this example $\kappa_j^{(m)}$ equals the corresponding factor κ_j in Kaniel's bounds to the given accuracy. Consequently Saad's bounds are simpler and tighter.

There is no loss in taking all angles to be acute.

$\boxed{j = 1}$ $\quad \gamma_{12n} = \frac{1}{99}, \quad T_{52}(1 + 2\gamma) = 1.73 \times 10^4, \quad \kappa_1^{(m)} = \kappa_1 = 1.0$

$$\theta_1 - \alpha_1 \leq (\tan \phi_1 \times 5.77 \times 10^{-5})^2,$$

$$\sin^2 \angle(\mathbf{y}_1, \mathbf{z}_1) \leq (\theta_1 - \alpha_1)/(\alpha_2 - \alpha_1) \leq (\tan \phi_1 \times 5.77 \times 10^{-4})^2.$$

$\boxed{j = 2}$ $\quad \gamma_{23n} = \frac{1}{32}, \quad T_{51}(1 + 2\gamma) = 3.39 \times 10^7,$

$$\kappa_2 \equiv (\alpha_n - \alpha_1)/(\alpha_2 - \alpha_1) = 100 = \kappa_2^{(m)}.$$

Saad: $\quad \theta_2 - \alpha_2 \leq 0.99(\tan \phi_2 \times 2.95 \times 10^{-6})^2$

Kaniel: $\quad \theta_2 - \alpha_2 \leq 0.99(\tan \phi_2 \times 2.95 \times 10^{-6})^2 + 1.0(\tan \phi_1 \times 5.77 \times 10^{-4})^2$

$$\sin^2 \angle(\mathbf{y}_2, \mathbf{z}_2) \leq 33.3(\theta_2 - \alpha_2) + 1.33 \sin^2 \angle(\mathbf{y}_1, \mathbf{z}_1),$$
$$\leq (\tan \phi_2 \times 1.70 \times 10^{-5})^2 + (\tan \phi_1 \times 3.40 \times 10^{-3})^2.$$

$\boxed{j = 3}$ $\quad \gamma_{34n} = \frac{1}{15}, \quad T_{50}(1 + 2\gamma) = 8.17 \times 10^{10}$

$$\kappa_3 = (\alpha_n - \alpha_1)(\alpha_n - \alpha_2)/(\alpha_3 - \alpha_1)(\alpha_3 - \alpha_2)$$
$$= 825 = \kappa_3^{(m)}.$$

Saad: $\quad \theta_3 - \alpha_3 \leq 0.96(\tan \phi_3 \times 1.00 \times 10^{-8})^2.$

Kaniel: $\quad \theta_3 - \alpha_3 \leq 0.96(\;''\;)^2 + 0.99 \sin^2 \angle(\mathbf{y}_2, \mathbf{z}_2) + 1.0 \sin^2 \angle(\mathbf{y}_1, \mathbf{z}_1)$

$$\leq (\tan \phi_3 \times 0.98 \times 10^{-8})^2 + (\tan \phi_2 \times 1.70 \times 10^{-5})^2 + (\tan \phi_1 \times 3.45 \times 10^{-3})^2$$

$$\sin^2 \angle(\mathbf{y}_3, \mathbf{z}_3) \leq 16.7(\theta_3 - \alpha_3) + 1.5 \sin^2 \angle(\mathbf{y}_2, \mathbf{z}_2) + 1.67 \sin^2 \angle(\mathbf{y}_1, \mathbf{z}_1).$$

Saad does not provide an a priori bound on $(\mathbf{y}_j, \mathbf{z}_j)$ but his a posteriori bound [Theorem (12-4-9)] is

$$\sin^2 \angle(\mathbf{y}_j, \mathbf{z}_j) \leqslant \left(1 + \beta_m^2/\text{gap}^2\right)\left(\kappa_j^{(m)}/T_{m-j}\right)^2 \tan^2 \phi_j$$

and in this example $(1 + \beta^2/\text{gap}^2)$ is very likely to be around 1 in magnitude. Chapter 13 will show how β_m comes to be computed in the course of the Lanczos algorithm.

The next example shows that all the bounds so far are likely to be extreme overestimates.

EXAMPLE 12-4-2

Refined error bounds. This example continues the previous one but employs the more complicated choice of π described in (12-4-17) which replaces T_{m-j} by $\sigma T_{m-j-\nu}$. Extra datum: $\theta_{n-1} = 0.9$ (to illustrate the effect of forcing π to vanish at isolated zeros at **both** ends of the spectrum).

$\boxed{j = 1}$ $\quad \gamma_{1,4,n-1} = 0.125, \qquad T_{49}(1 + 2\gamma) = 5.58 \times 10^{14},$

$$\sigma \equiv \frac{(\alpha_1 - \alpha_2)(\alpha_1 - \alpha_3)[\alpha_1 - \alpha_n]}{(\alpha_{n-1} - \alpha_2)(\alpha_{n-1} - \alpha_3)[\alpha_4 - \alpha_n]}$$

$$= \frac{1 \times 2 \times 10}{89 \times 43 \times 9} = 5.81 \times 10^{-4},$$

$$\sigma \times T_{49} = 3.24 \times 10^{11},$$

$$\theta_1 - \alpha_1 \leqslant (\tan \phi_1 \times 3.09 \times 10^{-12})^2$$

$$\sin^2 \angle(\mathbf{y}_1, \mathbf{z}_1) \leqslant (\tan \phi_1 \times 3.09 \times 10^{-11})^2.$$

$\boxed{j = 2}$ $\quad \gamma_{2,4,n-1} = \frac{9}{80}, \qquad T_{49}(1 + 2\gamma) \doteqdot 9.38 \times 10^{13}$

$$\sigma = \frac{(\alpha_3 - \alpha_2)(\alpha_n - \alpha_2)}{(\alpha_{n-1} - \alpha_3)[\alpha_n - \alpha_4]} = \frac{33}{860}, \qquad \sigma \times T_{49} \doteqdot 3.6 \times 10^{12},$$

$$\kappa_2 = \left(\frac{\alpha_n - \alpha_1}{\alpha_2 - \alpha_1}\right) = 100.$$

Saad: $\quad \theta_2 - \alpha_2 \leqslant 0.99(\tan \phi_2 \times 2.76 \times 10^{-11})^2$

Kaniel: $\quad \theta_2 - \alpha_2 \leqslant 0.99(\tan \phi_2 \times 2.76 \times 10^{-11})^2 + (\tan \phi_1 \times 3.09 \times 10^{-11})^2$

$$\sin^2 \angle(\mathbf{y}_2, \mathbf{z}_2) \leqslant \tfrac{1}{3}\left[100(\theta_2 - \phi_2) + 4 \sin^2 \angle(\mathbf{y}_1, \mathbf{z}_1)\right].$$

$$\boxed{j = 3} \quad \gamma_{3,4,n-1} = \tfrac{6}{80}, \qquad T_{49}(1 + 2\gamma) = 2.26 \times 10^{11}$$

$$\sigma = \left(\frac{\alpha_3 - \alpha_n}{\alpha_4 - \alpha_n}\right) = \frac{16}{15}, \qquad \sigma \times T_{49} \doteq 2.41 \times 10^{11},$$

$$\kappa_3 = \frac{(\alpha_n - \alpha_1)(\alpha_n - \alpha_2)}{(\alpha_3 - \alpha_1)(\alpha_3 - \alpha_2)} = 825.$$

Saad: $\theta_3 - \alpha_3 \leqslant 0.96[\tan \phi_3 \times 3.41 \times 10^{-9}]^2$

Kaniel: $\theta_3 - \alpha_3 \leqslant 0.96[\tan \phi_3 \times 3.41 \times 10^{-9}]^2$

$$+ [\tan \phi_2 \times 2.76 \times 10^{-10}]^2 + [\tan \phi_1 \times 6.18 \times 10^{-11}]^2$$

$$\sin^2 \angle(\mathbf{y}_3, \mathbf{z}_3) \leqslant 16.7(\theta_3 - \alpha_3) + 1.5 \sin^2 \angle(\mathbf{y}_2, \mathbf{z}_2) + 1.67 \sin^2 \angle(\mathbf{y}_1, \mathbf{z}_1).$$

Exercises on 12-4

12-4-1. State carefully and prove the results dual to Theorems (12-4-1) and (12-4-4). Does γ need to be redefined?

12-4-2. Show that $\beta_m = \| \mu(\mathsf{A})\mathsf{f}\| = \|\mathsf{HA}(\mathsf{I} - \mathsf{H})\|$ where μ was defined in Sec. 12-3.

12-4-3. Show that $\mathsf{y}_j \perp \mathsf{H}(\mathsf{A} - \alpha_j)\mathsf{u}_j$ in (12-4-12).

12-4-4. Justify (12-4-17).

12-4-5. Assume that $\alpha_1 = 0, \alpha_3 = 1, \alpha_n = 1001$. For $m = 10$ and $m = 30$ determine the values of α_2 such that the bound of $\theta_1 - \alpha_1$ in Theorem (12-4-1) [or Theorem (12-4-4)] is as good as the bound from (12-4-17) with $\mathcal{J} = \{2\}$.

12-5-6. Let $\alpha_i = \beta^i, i = 1, \ldots, n, \beta > 0$. Take $n = 100$ and $n = 1000$ and determine m such that

$$\sin(\mathsf{y}_i, \mathsf{z}_i) \leqslant \sin(\mathsf{z}_i, \mathsf{f})/100 \text{ for (a) } i = 1, 2 \text{ and (b) } i = -1, -2.$$

12-5 COMPARISON WITH THE POWER METHOD

After $(m - 1)$ steps of the power method the best approximation to an eigenvalue is $\rho(\mathsf{A}^{m-1}\mathsf{f})$. Lemma (12-3-8) can be applied to give an error bound provided that $\pi(\xi) = \xi^{m-1}$. The power method is not translation invariant and $\mathsf{A}^{m-1}\mathsf{f}/\|\mathsf{A}^{m-1}\mathsf{f}\|$ converges to either z_1 or z_n as $m \longrightarrow \infty$. When $\|\mathsf{A}\| = -\alpha_1$ the limit is z_1, and the bounds analogous to Theorem (12-4-4) are

$$\rho(\mathsf{A}^{m-1}\mathsf{f}) - \alpha_1 \leqslant (\alpha_n - \alpha_1)[\tan \angle(\mathsf{f}, \mathsf{z}_1)/(\alpha_1/\alpha_2)^{m-1}]^2,$$

$$\sin^2 \angle(\mathsf{A}^{m-1}\mathsf{f}, \mathsf{z}_1) \leqslant (\rho(\mathsf{A}^{m-1}\mathsf{f}) - \alpha_1)/(\alpha_2 - \alpha_1). \qquad (12\text{-}5\text{-}1)$$

A cleaner comparison with Krylov subspaces emerges if we consider a matrix in which $\|\mathsf{A}\| = \alpha_n$ and the smallest eigenvalue α_1 is approximated by using the power method with $\mathsf{A} - \alpha_n$. The bounds become

$$\rho((\mathsf{A} - \alpha_n)^{m-1}\mathsf{f}) - \alpha_1 \leqslant (\alpha_n - \alpha_1)[\tan \angle(\mathsf{f}, \mathsf{z}_1)/(1 + \gamma_{12n})^{m-1}]^2,$$

where γ_{12n} is defined in (12-4-16), and

$$\sin^2 \angle((\mathsf{A} - \alpha_n)^{m-1}\mathsf{f}, \mathsf{z}_1)$$
$$\leqslant \{\rho[(\mathsf{A} - \alpha_n)^{m-1}\mathsf{f}] - \alpha_1\}/(\alpha_2 - \alpha_1). \qquad (12\text{-}5\text{-}2)$$

Comparison of Theorem (12-4-1) with (12-5-2) shows that both angle bounds contain the factor $(\alpha_n - \alpha_1)/(\alpha_2 - \alpha_1)$ which is needed to cope with perversely chosen or unfortunate f's. It is posssible to replace the simple expression $(\alpha_n - \alpha_1)/(\alpha_2 - \alpha_1)$ with a more complicated one involving all the $\angle(\mathsf{f}, \mathsf{z}_i)$, $i = 1, \ldots, n$, and when f is reasonably well chosen the simple expression is a severe overestimate. The following table summarizes the bounds on the ratio (final error angle)/ $\angle(\mathsf{f}, \mathsf{z}_1)$ in terms of the relative gap ratio γ. Appendix B gives the estimates for the Chebyshev polynomials T_m.

Power method: $(\mathsf{A} - \alpha_n)^m\mathsf{f}$	$\mathcal{K}^m(\mathsf{f})$
$(1+\gamma)^{-m} \sim \begin{cases} e^{-m\gamma} & (\gamma \to 0) \\ \gamma^{-m} & (\gamma \to \infty) \end{cases}$	$\dfrac{1}{T_m(1+2\gamma)} \sim \begin{cases} 2e^{-2m\sqrt{\gamma}} & (\gamma \to 0) \\ 2(4\gamma)^{-m} & (\gamma \to \infty) \end{cases}$

It is the $2\sqrt{\gamma}$ which makes the difference in difficult problems when γ is small (like 10^{-4}). When γ is large (like 10^4) then m will be small (like 1 or 2) and the methods coalesce.

The basis has been laid for a comparison of approximations from $\mathcal{K}^m(\mathsf{f})$ with various other methods. The block power method starts with an n by p matrix F and, for at least p times the cost, computes approximations from span$((\mathsf{A} - \alpha_n)\mathsf{F})$. The effect on the bounds in (12-5-2) is to replace

γ_{12n} by $\gamma_{1,p+1,n}$ and $\angle(\mathsf{f}, \mathsf{z}_1)$ by $\angle[\mathsf{f}, \mathcal{Z}^p]$. If the change in angles is ignored it is clear that the block power method produces results as good as does $\mathcal{K}^m(\mathsf{f})$ when $\gamma_{1,p+1,n}/\gamma_{1,2,n} = 2/\sqrt{\gamma_{1,2,n}}$. For equispaced α_i equality is achieved only when $p = 2n/(\sqrt{n} + 2) \doteqdot 2\sqrt{n}$, an impracticably large block.

It is also instructive to compare inverse iteration with $\mathcal{K}^m(\mathsf{f})$. [Such a comparison is somewhat unnatural because if it is possible to compute $(\mathsf{A} - \sigma)^{-m}\mathsf{f}$ then it is also possible to use $\mathcal{K}^m(\mathsf{f}; (\mathsf{A} - \sigma)^{-1})$.] If σ is close to an eigenvalue, $\sigma < \alpha_1$ say, then the error bound (12-5-2) still applies provided that γ_{12n} is replaced by $\tilde{\gamma} = (\alpha_2 - \alpha_1)/(\alpha_1 - \sigma)$. (Why?) So inverse iteration will produce approximations as good as those from $\mathcal{K}^m(\mathsf{f}; \mathsf{A})$, for equispaced α_i, whenever $\ln \tilde{\gamma} > 2/\sqrt{n}$. If subspace iteration, or block inverse iteration, is used in the same equispaced problem then its approximation, from span$[(\mathsf{A} - \sigma)^{-m+1}\mathsf{F}]$ will better $\mathsf{y}_1^{(m)}$ whenever $\ln p\hat{\gamma} > 2/\sqrt{n}$. This indicates how powerful inverse iteration can be.

So far the cost of the various techniques has been ignored, as has the fact that $\mathcal{K}^m(\mathsf{f})$ produces approximations to several eigenpairs of A. There may be cases in which A is so large and intractable that vector multiplication is the only thing that can be done with A; there certainly are cases in which A, though large, does permit the occasional solution of $(\mathsf{A} - \sigma)\mathsf{u} = \mathsf{v}$ for u but at a cost significantly greater than one product Av; there are also cases in which the solution of $(\mathsf{A} - \sigma)\mathsf{u} = \mathsf{v}$ takes less time than say three products Av. Also important are the storage requirements and it is time to take up the problem of computing the RR approximations from $\mathcal{K}^m(\mathsf{f})$. The table at the beginning of Sec. 12-4 indicates that for fairly difficult cases m will have to exceed 25 and may rise into the 100's.

Perhaps we have been unfair to the power method by a factor of 2. If the optimal shift $(\alpha_2 + \alpha_{-1})/2$ were used instead of α_{-1} then the convergence factor would change from $(1 + \gamma)^{-1}$ to $(1 + 2\gamma)^{-1}$. In practice this optimal shift is more difficult to estimate than is α_{-1}.

12-6 PARTIAL REDUCTION TO TRIDIAGONAL FORM

There is an intimate connection between Krylov subspaces and tridiagonal matrices. Let us start with $\mathcal{K}^m(\mathsf{f}, \mathsf{A})$. For theoretical work, such as the Saad-Kaniel theory, the natural basis for $\mathcal{K}^m$ is the Krylov basis $\mathsf{K}^m(\mathsf{f}, \mathsf{A})$ of Sec. 12-1. For practical work there is a *distinguished orthonormal basis*, $\mathsf{Q}_m \equiv (\mathsf{q}_1, \ldots, \mathsf{q}_m)$, which is the result of applying the Gram-Schmidt orthonormalizing process to the columns of $\mathsf{K}^m(\mathsf{f})$ in the natural order f, Af, ... Here the dimension of $\mathcal{K}^m$ must be m; i.e., $\mathsf{K}^m(\mathsf{q}_1)$ must have full rank. We will not consider any m with $\mathcal{K}^{m+1} = \mathcal{K}^m$. For reasons which become clear in Chap. 13 we call Q_m the **Lanczos basis** of $\mathcal{K}^m(f; A)$. In matrix terminology we can write

$$K_m(f) = Q_m C_m^{-1} \tag{12-6-1}$$

as the QR factorization of K_m. The columns of the upper triangular matrix C_m contain the coefficients of the Lanczos polynomials introduced in Chap. 7, but this fact will not be exploited here.

For general vector sequences the Gram-Schmidt process is burdensome and becomes more so as the number of vectors increases. In contrast, for Krylov sequences $K^m(f)$ the process simplifies dramatically to yield a three-term recurrence connecting the columns of $Q_m \equiv (q_1, \ldots, q_m)$.

Before tackling the general case we consider an example. Observe first that since $q_3 \in \mathcal{K}^3(q_1)$,

$$\begin{aligned} \text{span}(q_1, q_2, q_3, Aq_3) &= \text{span}(q_1, q_2, q_3, A(\gamma_2 A^2 + \gamma_1 A + \gamma_0 I)q_1), \\ &= \text{span}(q_1, q_2, q_3, A^3 q_1), \\ &= \mathcal{K}^4(q_1). \end{aligned}$$

Thus to complete the basis for $\mathcal{K}^4$ it suffices to orthogonalize Aq_3 against q_3, q_2, and q_1. It turns out that Aq_3 is already orthogonal to q_1 (Ex. 12-6-1).

THEOREM When $K^m(f; A)$ has full rank and Q_m is defined by (12-6-1) then $Q_m^* A Q_m$ is an unreduced tridiagonal matrix. (12-6-2)

Proof. The characteristic property of $\mathcal{K}^m(f)$ is that for each $j < m$, $A\mathcal{K}^j \subset \mathcal{K}^{j+1}$. In particular $q_i \perp \mathcal{K}^{i-1}$ and $Aq_j \in \mathcal{K}^{j+1}$. Consequently

$$q_i^*(Aq_j) = 0 \text{ for each } i > j + 1. \tag{12-6-3}$$

By the symmetry of A, $q_j^*(Aq_i) = q_i^*(Aq_j) = 0$ for all $j < i - 1$. This establishes the tridiagonal nature of $Q_m^* A Q_m$. Note that

$$\begin{aligned} \mathcal{K}^{j+1} &= \text{span}(K^j, A^j f) \\ &= \text{span}(K^j, Aq_j) \\ &= \text{span}(K^j, q_{j+1}). \end{aligned}$$

If K^m has full rank and $j < m$ then Aq_j and q_{j+1} cannot be orthogonal, i.e., $q_{j+1}^*(Aq_j) \neq 0$. □

If we write $\alpha_j = \mathbf{q}_j^* \mathbf{A} \mathbf{q}_j$, $\beta_j = \mathbf{q}_{j+1}^* \mathbf{A} \mathbf{q}_j$, then the tridiagonal matrix $\mathbf{Q}_m^* \mathbf{A} \mathbf{Q}_m$ will be denoted by

$$\mathbf{T}_m = \mathbf{T}_{1,m} = \begin{bmatrix} \alpha_1 & \beta_1 & & & \\ \beta_1 & \alpha_2 & \beta_2 & & \\ & \beta_2 & \cdot & \cdot & \\ & & \cdot & \cdot & \beta_{m-1} \\ & & & \beta_{m-1} & \alpha_m \end{bmatrix} \tag{12-6-4}$$

With this extra notation we can give a useful consequence of the theorem

COROLLARY For each $j < m$,

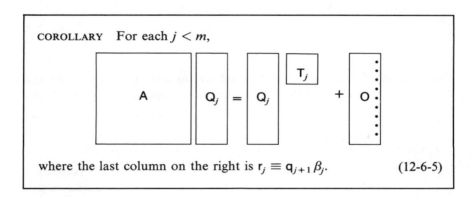

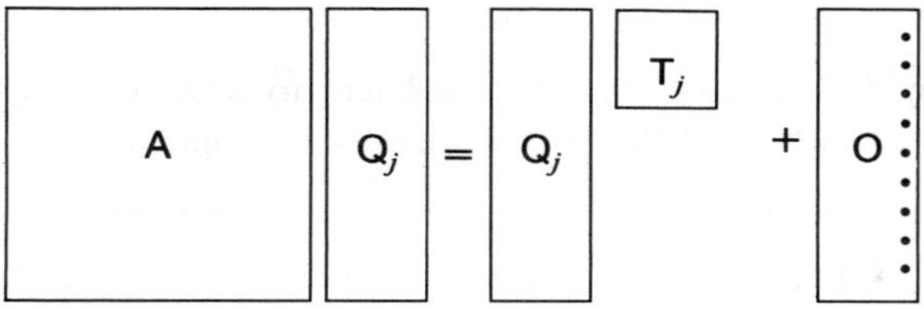

where the last column on the right is $\mathbf{r}_j \equiv \mathbf{q}_{j+1}\beta_j$. (12-6-5)

Proof. The verification that for $i \leqslant j$,

$$\mathbf{A}\mathbf{q}_i = \mathbf{q}_{i-1}\beta_{i-1} + \mathbf{q}_i\alpha_i + \mathbf{q}_{i+1}\beta_i, \tag{12-6-6}$$

is left as Ex. 12-6-2. The corollary merely expresses this in compact form. It is only necessary to note that $\beta_j = \mathbf{q}_{j+1}^* \mathbf{A} \mathbf{q}_j = \|\mathbf{r}_j\|$. □

The second matrix on the right can be written as $\mathbf{r}_j \mathbf{e}_j^*$ where $\mathbf{e}_j^* = (0, \ldots, 0, 1)$ has only j elements while all the other vectors have n elements.

Observe that the fundamental Krylov matrix $\mathbf{K}^m(\mathbf{f})$ has faded from the picture and $\mathbf{Q}_j$ is directly related to $\mathbf{A}$ in (12-6-5). The corollary suggests a way in which all the $\mathbf{Q}_j$ and $\mathbf{T}_j$, $j = 1, \ldots, m$ can be built up from $\mathbf{A}$ and $\mathbf{Q}_1 = \mathbf{q}_1 = \mathbf{f}/\|\mathbf{f}\|$. At the beginning of the j-th step $\mathbf{T}_{j-1}$, β_{j-1}, and $\mathbf{Q}_j$ are on hand.

From (12-6-5),

$$\mathsf{r}_j = \mathsf{r}_j\mathsf{e}_j^*\mathsf{e}_j = (\mathsf{AQ}_j - \mathsf{Q}_j\mathsf{T}_j)\mathsf{e}_j$$
$$= \mathsf{Aq}_j - \mathsf{q}_{j-1}\beta_{j-1} - \mathsf{q}_j\alpha_j, \tag{12-6-7}$$

where

$$\alpha_j = \mathsf{q}_j^*(\mathsf{Aq}_j), \text{ since the } \mathsf{q}\text{'s are orthogonal.} \tag{12-6-8}$$

Next

$$\beta_j = \|\mathsf{q}_{j+1}\beta_j\| = \|\mathsf{r}_j\|, \text{ from (12-6-5).} \tag{12-6-9}$$

If $\beta_j > 0$ then $\mathsf{q}_{j+1} = \mathsf{r}_j/\beta_j$ and the step is complete. If $\beta_j = 0$ then $\mathsf{AQ}_j = \mathsf{Q}_j\mathsf{T}_j$ and the algorithm halts with

$$\text{span } \mathsf{Q}_j = \mathcal{K}^j(\mathsf{f}) = \mathcal{K}^{j+1}(\mathsf{f}) = \mathcal{I}(\mathsf{f}), \tag{12-6-10}$$

the smallest invariant subspace containing f. In the 1950s, when the goal was to compute T_n, the possibility that $\beta_m = 0$ for $m < n$ was regarded as a mild nuisance. Today, seeking a few eigenvectors, early termination is an outcome devoutly to be wished because then each eigenvalue of T_m is an eigenvalue of A. In the extreme case, when f is an eigenvector the algorithm will halt after one step.

Formula (12-6-6) is the three-term recurrence relating the columns of Q_m.

The cost of the step is dominated by the formation of Aq_j, the only appearance of A in the algorithm.

Exercises on 12-6

12-6-1. Show directly, using $\mathsf{A}^* = \mathsf{A}$ and $\mathsf{q}_i^*\mathsf{q}_k = 0$, $i \neq k$, that Aq_3 is orthogonal to q_1.

12-6-2. Show that (12-6-5) and (12-6-6) are equivalent and also express theorem (12-6-2) with the notation of (12-6-4). Show that $\beta_j = \mathsf{q}_{j+1}^*\mathsf{Aq}_j = \|\mathsf{r}_j\|$.

12-6-3. Compute T_3 when $\mathsf{A} = \begin{bmatrix} 7 & 5 & 3 \\ 5 & 3 & 1 \\ 3 & 1 & 0 \end{bmatrix}$, $\mathsf{f} = \begin{bmatrix} 1 \\ -2 \\ 1 \end{bmatrix}$. Does $\beta_3 = \|\mathsf{r}_3\| = 0$?

Notes and References

The idea of the power method is a very natural one. The civil engineers call it **Stodola's iteration**. In [Krylov, 1931] the sequence $\{\mathsf{x}, \mathsf{Ax}, \mathsf{A}^2\mathsf{x}, \ldots\}$ is actually used as a means to find the coefficients of the characteristic polynomial and, despite that unfortunate goal, Krylov's name has become securely attached to the sequence.

The chapter presents the bounds in [Kaniel, 1966] (with corrections from [Paige, 1971]) which show what very good eigenvalue approximations can be obtained from Krylov subspaces of modest dimension. Our aim was to make the results intelligible as well as quotable. The recent and better bounds to be published in [Saad, 1980] were shown to me as this book was undergoing final revision. The original proofs have been simplified in order to blend with the neighboring material.

13

Lanczos Algorithms

13-1 KRYLOV + RAYLEIGH-RITZ = LANCZOS

The Lanczos algorithm has had a checkered history since its debut in 1950. Although Lanczos pointed out that his method could be used to find a few eigenvectors of a symmetric matrix it was heralded at that time as a way to reduce the whole matrix to tridiagonal form. In this capacity it flopped unless very expensive modifications were incorporated. Twenty years later Paige showed that despite its sensitivity to roundoff the simple Lanczos algorithm is nevertheless an effective tool for computing some outer eigenvalues and their eigenvectors.

The exact algorithm can be presented in various ways. In fact it has already made its appearance in Sec. 7-2 where the constructive proof that q_1 completely determines the reduction of A to tridiagonal form T ($= Q^*AQ$) is simply a description of the Lanczos algorithm.

A different approach is given in Sec. 12-6 which shows that the columns of $Q_m = (q_1, q_2, \ldots, q_m)$ form a distinguished basis for the Krylov subspace $\mathcal{K}^m(q_1)$ whose virtues are extolled at length in Chap. 12. In this basis A's projection onto $\mathcal{K}^m(q_1)$ is represented by a tridiagonal matrix T_m. The algorithm is summarized by two equations, at step j, namely

$$AQ_j - Q_jT_j = r_je_j^*, \qquad \text{where } r_j = q_{j+1}\beta_j, \tag{13-1-1}$$

and

$$I - Q_j^* Q_j = O. \tag{13-1-2}$$

A third approach is to see the Lanczos algorithm as the natural way to implement the Rayleigh-Ritz procedure (RR in Sec. 11-3) on the **sequence** of Krylov subspaces $\mathcal{K}^j(f)$, $j = 1, 2, \ldots$. At each step the subspace dimension grows by one and the best approximate eigenvectors in the subspace are computable in a straightforward manner. The general and rather costly RR procedure is dramatically simplified when not one but a sequence of Krylov subspaces is used. Let us see how this economy comes about.

The first move in RR is to determine an orthonormal basis for the subspace but, in our case, at step j the basis $\{q_1, \ldots, q_{j-1}\}$ of $\mathcal{K}^{j-1}$ is already computed and only one vector need be added. This vector q_j has to be the component of Aq_{j-1} orthogonal to $\mathcal{K}^{j-1}$ (see Sec. 12-6) and, as we shall see below, q_j is already on hand but not in normalized form. Thus the first step is a simple normalization.

The next two steps in RR compute the Rayleigh quotient matrix $\rho(Q_j) \equiv Q_j^*(AQ_j)$. In our case, by theorem 12-6-2, $\rho(Q_j)$ (i.e., T_j) is tridiagonal and is formed from T_{j-1} by adding the elements β_{j-1} and α_j in the appropriate positions. Again it turns out that β_{j-1} is on hand and so the first real computation is the formation of $\alpha_j = q_j^* u_j$ and $u_j = Aq_j$.

Next in RR comes the computation of as many eigenvalues and eigenvectors $(\theta_i^{(j)}, s_i^{(j)})$ of T_j as are required. The tridiagonal form facilitates this step considerably (see Chap. 8). In Sec. 11-4 it is shown that the collectively best approximate eigenpairs from $\mathcal{K}^j$ are the Ritz pairs $(\theta_i^{(j)}, y_i^{(j)})$ where $y_i^{(j)} = Q_j s_i^{(j)}$, $i = 1, \ldots, j$. We shall drop the superscript (j) whenever possible.

That is the end of the procedure except for assessing the accuracy of the Ritz pairs, the topic of Sec. 13-2. For the assessment it is necessary to orthogonalize u_j $(= Aq_j)$ against q_j and q_{j-1} and so obtain a useful residual vector r_j. It turns out that β_j $(\equiv \|r_j\|)$ is needed both in Sec. 13-2 and at the next step of the Lanczos algorithm. As a bonus r_j is a multiple of q_{j+1}. Now everything is ready for the RR approximations from $\mathcal{K}^{j+1}$.

For completeness we lay out the computations in one step of the process. It is left as Ex. 13-1-1 to explain why the algorithm differs slightly from the description given above.

13-1-1 SIMPLE LANCZOS

r_0 is given, $\beta_0 = \|r_0\| \neq 0$. For $j = 1, 2, \ldots$ repeat:

1. $q_j \leftarrow r_{j-1}/\beta_{j-1}$
2. $u_j \leftarrow Aq_j$

3. $r_j \leftarrow u_j - q_{j-1}\beta_{j-1} \qquad (q_0 = o)$
4. $\alpha_j \leftarrow q_j^* r_j$
5. $r_j \leftarrow r_j - q_j \alpha_j$
6. $\beta_j \leftarrow \|r_j\|$
7. Compute θ_i, s_i, y_i as desired
8. If satisfied then stop

The important observations for large matrix calculations are given in boxes (13-1-3) and (13-1-4).

> After substep 3 above q_{j-1} may be put into a secondary storage device. It is not needed again until some Ritz vector $y_i^{(m)}$ is to be formed, from $y_i^{(m)} \equiv Q_m s_i$, at say the m-th Lanczos step when $y_i^{(m)}$ has converged sufficiently. (13-1-3)

Only three n-vectors are needed in the fast store, an attractive feature when $n > 10^3$. Paige pointed out that it is often possible to get away with two n-vectors. See Ex. 13-1-2.

> The matrix A can be represented by a subprogram which computes Ax for given x, and therein exploits any known special properties of A. (13-1-4)

The Kaniel-Saad theory (Sec. 12-5) suggests that good RR approximations to the outer eigenvalues and eigenvectors will emerge for j as small as $2\sqrt{n}$ and it is for such calculations that the Lanczos algorithm is ideally suited. It is not necessary to fix the last step j $(= m)$ in advance. The process goes on until the wanted Ritz pairs $(\theta_i, y_i^{(j)})$, $i = 1, \ldots, p$ are deemed satisfactory. In exact arithmetic this must occur by $j = n$ but usually it will be much sooner. Typical values to bear in mind are $p = 10$, $m = 300$, $n = 10^4$.

The initial vector f (or q_1) is best selected by the user to embody any available knowledge concerning A's wanted eigenvectors. In the absence of such knowledge either a random vector or $(1, 1, \ldots, 1)^*$ is used for f.

In order to specify the Lanczos process completely we must say when the algorithm should bother to compute some eigenvalues of T_j. The rather surprising answer is that it should do so at every step. The next section, which continues to assume that exact arithmetic is being used, explains why.

Exercises on Sec. 13-1

13-1-1. Steps 3 to 5 do not correspond to the verbal description of the Lanczos process. Show that the two versions are equivalent in exact arithmetic. Explain the differences. *Hint*: Modified Gram-Schmidt (Sec. 6-6).

13-1-2. Assume that the operation $u \longleftarrow Av$ is implemented in the form $u \longleftarrow u + Av$. Rearrange the simple algorithm to use only two n-vectors (q and r say). *Hint*: A swap will be needed.

13-2 ASSESSING ACCURACY

The Kaniel-Saad theorem (Chap. 12) tells us to expect increasingly good approximations to the outer eigenvalues from $\mathcal{K}^j(q_1)$ as j increases. In practice we need a posteriori bounds to apply in each specific case. The gap theorems in Sec. 11-6 show that the residual norm $\|Ay - y\theta\|$ is a good measure of the accuracy of the RR pair (θ, y).

In principle it is possible to compute θ_i and y_i from T_j at each step in the Lanczos algorithm. Fortunately it is possible to compute $\|Ay_i - y_i\theta_i\|$ without computing y_i! To see how let us drop the subscript i and observe that

$$\begin{aligned} \|Ay - y\theta\| &= \|AQs - Qs\theta\|, && \text{since } y = Qs, \\ &= \|(AQ - QT)s\|, && \text{since } s\theta = Ts, \\ &= \|(\beta_j q_{j+1} e_j^*)s\|, && \text{using (13-1-1)} \\ &= \beta_j |e_j^* s|, && \text{since } \|q_{j+1}\| = 1. \end{aligned} \qquad (13\text{-}2\text{-}1)$$

So the bottom elements of the normalized eigenvectors of T_j signal convergence and there is no need to form y until its accuracy is satisfactory. This result explains why some Ritz values can be very accurate even when β_j is not small.

EXAMPLE 13-2-1 Dwindling Eigenvectors of T

$S = (s_{ij})$ is the matrix of normalized eigenvectors of a random symmetric tridiagonal matrix.

$\tilde{S} = (\tilde{s}_{ij}); \ \tilde{s}_{ij} = -\log_{10}|s_{ij}|.$

If each entry ν in $\tilde{S}$, shown below, is replaced by $10^{-\nu}$ then the new table gives the absolute values of the elements of S. The bottom row of this S has some elements as small as 10^{-13} and these are associated with the isolated eigenvalues λ_1 and λ_{25}. This phenomenon is typical.

Table 13-2-1 The matrix $\tilde{S}$

j	1	2	3	4	5	$\cdots$	11	12	13	14	15	$\cdots$	21	22	23	24	25
	6	3	1	4	1	·	2	2	1	1	1	·	2	7	3	4	6
	⋮	⋮	⋮	⋮	⋮	⋮	⋮	⋮	⋮	⋮	⋮	⋮	⋮	⋮	⋮	⋮	⋮
	4	3	4	2	2	·	2	1	1	1	1	·	3	2	4	4	5
	5	3	4	2	3	·	2	1	1	1	1	·	3	2	4	4	5
$\tilde{s}_j$	6	4	5	1	4	·	2	1	1	2	1	·	4	1	5	5	7
	7	4	5	1	4	·	2	1	1	2	1	·	4	1	5	5	7
	7	5	5	1	3	·	2	1	1	2	1	·	4	1	6	6	8
	8	5	5	1	3	·	2	1	1	3	1	·	5	2	7	7	9
	10	7	6	2	5	·	3	2	1	1	1	·	6	3	8	9	11
	11	7	7	3	5	·	3	2	1	1	1	·	6	4	9	9	12
	11	7	7	3	5	·	3	2	1	1	1	·	6	4	10	10	12
	13	9	9	4	7	·	4	2	1	1	1	·	8	6	11	12	14

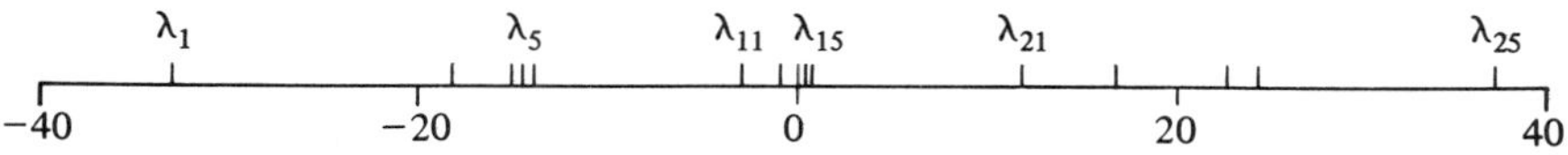

Let $s_{ji}^{(j)} \equiv \mathbf{e}_j^* \mathbf{s}_i$ and define the useful numbers

$$\beta_{ji} \equiv \beta_j |s_{ji}^{(j)}|, \qquad i = 1, \ldots, j. \tag{13-2-2}$$

Without loss of generality we may take $s_{ji}^{(j)} > 0$, for all $i \leqslant j$. At the j-th step of Lanczos, by Theorem 4-5-1

$$|\theta_i - \lambda[\mathsf{A}]| \leqslant \beta_{ji} \tag{13-2-3}$$

for some eigenvalue $\lambda[\mathsf{A}]$ which depends on i. Moreover, by using the θ_k as approximate eigenvalues in order to compute a gap $\gamma_i \equiv \min|\theta_i - \theta_k|$ over $k \neq i$, Theorem (11-7-6) can be invoked to give ultimately realistic estimates (not bounds)

$$|\sin \angle(\mathbf{y}_i^{(j)}, \mathbf{z})| \doteq \beta_{ji}/\gamma_i \tag{13-2-4}$$

for the Ritz vectors, and

$$|\theta_i - \lambda| \doteq \beta_{ji}^2/\gamma_i \tag{13-2-5}$$

for the Ritz values, where $(\lambda, \mathbf{z})$ is an eigenpair of A.

When several intervals $[\theta_i - \beta_{ji}, \theta_i + \beta_{ji}]$, $i = \ell, \ell + 1, \ell + 2$ say, happen to overlap then Theorem (11-5-1) can and should be used; namely let $\sigma \equiv \sum_{i=\ell}^{\ell+2} \beta_{ji}$; then each interval $[\theta_i - \sigma, \theta_i + \sigma]$, $i = \ell, \ell + 1, \ell + 2$ contains its own eigenvalue (Ex. 13-2-1).

These estimates and bounds are attractive but they suggest that some eigenvectors $\mathbf{s}_i^{(j)}$ of T_j must be computed in order to form β_{ji}. Although such a computation requires only $O(j)$ ops it is worth mentioning that there are at least two ways of computing $s_{ji}^{(j)}$ without computing all of $\mathbf{s}_i^{(j)}$. Of course these alternatives also require $O(j)$ ops and all three techniques are satisfactory. See Exs. 13-2-2 and 13-2-3.

A key issue is the cost of computing the Ritz values θ_i. By using the root-free QL algorithm with judicious shifts **all** the θ's can be computed in approximately $9j^2$ multiplications. On the other hand by combining the QL algorithm with spectrum slicing the p extreme eigenvalues at each end can be computed in a total of approximately $36pj$ multiplications. Here we allow 1.8 QL iterations per eigenvalue and $10j$ multiplications per QL iteration. The count of $36pj$ is so low that it is a negligible fraction of the cost of one Lanczos step when $j \ll n$.

These a posteriori error estimates help keep down the number of Lanczos steps by terminating the algorithm as soon as it attains the desired accuracy.

Exercises on Sec. 13-2

13-2-1. Show why $\sigma = \Sigma\, \beta_{ji}$ is a bound on the norm of the residual matrix $\mathsf{A}Y - Y\Theta$ where $Y = (\mathbf{y}_\ell, \mathbf{y}_{\ell+1}, \mathbf{y}_{\ell+2})$, $\Theta = \mathrm{diag}(\theta_\ell, \theta_{\ell+1}, \theta_{\ell+2})$.

13-2-2. Use the results in Sec. 7-9 to compute s_{ji}^2 assuming that the old θ's, $\theta_i^{(j-1)}$, $i = 1, \ldots, j-1$ are saved along with the $\theta_i^{(j)}$.

13-2-3. It is possible to modify the inner loop of the standard QL algorithm so that the vector $\mathbf{e}_j^*\mathsf{S}$ is updated by each plane rotation. How many ops will be added to the inner loop?

13-3 THE EFFECTS OF FINITE PRECISION ARITHMETIC

The preceding sections paint a very rosy picture of the Lanczos algorithm. However it was known to Lanczos, when he presented the algorithm in 1950, that the computed quantities could deviate greatly from their theoretical counterparts. The second basic equation, $\mathsf{Q}_j^*\mathsf{Q}_j = \mathsf{I}_j$, is utterly destroyed by roundoff and the algorithm has been described, somewhat unfairly, as unstable. What makes the practical algorithm interesting is that despite this gross deviation from the exact model, it nevertheless delivers fully accurate RR approximations. Its fault is in not quitting while it is ahead because it continues to compute many redundant copies of each Ritz pair.

In order to analyze the process without drowning in irrelevant details we make an important, and standard, change of notation. From now on Q_j and T_j denote the quantities stored in the computer under these names. No further attention will be paid to their Platonic images. Moreover the vector $y_i^{(j)}$ ($\equiv Q_j s_i^{(j)}$) will be called a "Ritz vector" even when Q_j is far from orthonormal. The quotes remind us that it is not a true Ritz vector from span Q_j.

We will now describe the curious way the Lanczos algorithm behaves in practice, and also give an example, before embarking on the analysis.

For the first few steps, maybe 3, maybe 30, the results are indistinguishable from the exact process. Then a new Lanczos vector q fails to be orthogonal, to working precision, to its predecessors. A few steps later Q_j does not even have full rank, i.e., the Lanczos vectors are linearly dependent. This looks like disaster because there is then no guarantee that T_j will bear any useful relation to A. Nevertheless an odd thing happens; at the same time as orthogonality among the $\{q_i, i = 1, \dots, j\}$ disappears a Ritz pair $(\theta_i^{(j)}, y_i^{(j)})$, for some i, converges to an eigenpair of A. As the algorithm proceeds further it "forgets" that it has found that eigenpair and starts to compute it again. Soon there will be two Ritz pairs accurately approximating that single eigenpair of A and hence the two Ritz vectors must be multiples of each other—almost—and this can only happen if Q_j has linearly dependent columns—almost.

The exact Lanczos algorithm must terminate ($\beta_j = 0$) for some $j \leqslant n$, but in practice the process grinds on forever, computing more copies of outer eigenvectors for each new inner pair it discovers.

The loss of orthogonality makes itself felt in all parts of the algorithm. For example, the relation $\|Ay_i - y_i\theta_i\| = \beta_{ji}$ no longer holds. Instead we have

$$|\theta_i - \lambda[A]| \leqslant \|Ay_i - y_i\theta_i\|/\|y_i\| \leqslant (\beta_{ji} + \|F_j\|)/\|y_i\| \qquad (13\text{-}3\text{-}1)$$

where F_j accounts for roundoff and is harmless. See (13-4-1) for more details. Recall that we do not want to compute y_i at each step and are faced with the real possibility that y_i might be small since the only bound we have is

$$\|y_i\|^2 \geqslant \lambda_1[Q_j^* Q_j]. \qquad (13\text{-}3\text{-}2)$$

We shall return to this matter later. Of course it is still necessary that β_{ji} be small for (θ_i, y_i) to be a good approximation. Except in rare cases no β_j will ever be tiny.

This odd behavior of the algorithm is certainly not what we wanted but neither is it a disaster.

The following example [Scott, 1978] was rigged to have an isolated dominant eigenvalue in order to induce rapid convergence and the consequent arrival of a duplicate Ritz pair. The linear independence of the $\mathbf{q}_i$ is best measured by $\sigma_1(\mathsf{Q}_j)$, the smallest singular value of Q_j, and $\sigma_1(\mathsf{Q}_j)^2 \equiv \lambda_1[\mathsf{Q}_j^*\mathsf{Q}_j]$.

EXAMPLE 13-3-1

$$\mathsf{A} = \text{diag}(0, 1 \times 10^{-3}, 2 \times 10^{-3}, 3 \times 10^{-3}, 4 \times 10^{-3}, 1.0),$$

$$\mathbf{q}_1 = (1, 1, 1, 1, 1, 1)^*/\sqrt{6}\,,$$

$$\text{roundoff unit} = 10^{-7}.$$

Table 13-3-1 History of the Lanczos run

j	α_j	β_j	$\sigma_1(\mathsf{Q}_j)$
1	0.1668333	0.3726035	1.0
2	0.8333665	0.0003464	0.9999999
3	0.0002004	0.0003094	0.9997097
4	0.1464297	0.3532944	0.0760186
5	0.9998344	0.0001098	*0.0000004* ←

Table 13-3-2 Selected Ritz values and residual norms

j	*Ritz Values* $\theta_i^{(j)}$	$\|\mathsf{A}\mathbf{y}_i - \mathbf{y}_i\theta_i\|$
3	$\theta_3^{(3)} = 1.000000$	0.48×10^{-7}
5	$\theta_4^{(5)} = 0.9999996$	*0.41* $\times$ *10*$^{-4}$ ←
	$\theta_5^{(5)} = 1.0000001$	*0.68* $\times$ *10*$^{-5}$ ←

Exercise on Sec. 13-3

13-3-1. Show that $\mathbf{y}_i$ ($\equiv \mathsf{Q}_j\mathbf{s}_i$) satisfies (13-3-2).

13-4 PAIGE'S THEOREM

Examination of the elements of $\mathsf{Q}_j^*\mathsf{Q}_j$ in a variety of cases suggests that orthogonality loss among Q_j's columns is both widespread and featureless. Such observations prompted the crushingly expensive remedy of explicitly orthogonalizing each new $\mathbf{q}_{j+1}$ against **all** the previous $\mathbf{q}$'s. However when the "Ritz vectors," $\mathbf{y}_i$ ($\equiv \mathbf{Q}_j\mathbf{s}_i^{(j)}$), $i = 1, \ldots, j$ are examined the hidden pattern becomes clearer.

The theorem below, which embodies the insights (13-4-6) and (13-4-7), is due to Paige but this part of his thesis has never been published. A little more preparation is needed before presenting his results.

By the end of the j-th step the Lanczos algorithm has produced $\mathbf{Q}_j$—the matrix of Lanczos vectors, $\mathbf{T}_j$—the tridiagonal matrix embodying the three-term recurrence, and the residual vector $\mathbf{r}_j \equiv (\mathbf{A}\mathbf{Q}_j - \mathbf{Q}_j\mathbf{T}_j)\mathbf{e}_j$. Information about the algorithm can be condensed into two fundamental relations which govern the computed quantities. They are

$$\overset{n \text{ by } n}{\mathbf{A}}\ \overset{n \text{ by } j}{\mathbf{Q}_j} - \overset{n \text{ by } j}{\mathbf{Q}_j}\ \overset{j \text{ by } j}{\mathbf{T}_j} = \overset{n \text{ by } 1}{\mathbf{r}_j}\ \overset{1 \text{ by } j}{\mathbf{e}_j^*} + \overset{n \text{ by } j}{\mathbf{F}_j}, \tag{13-4-1}$$

and

$$\mathbf{I}_j - \mathbf{Q}_j^*\mathbf{Q}_j = \mathbf{C}_j^* + \Delta_j + \mathbf{C}_j \tag{13-4-2}$$

where j by j $\mathbf{C}_j$ is strictly upper triangular and Δ_j is diagonal. $\mathbf{F}$, $\mathbf{C}$, and Δ account for the effects of roundoff error but at this stage there is no need to see precisely what they are like. It turns out that $\|\mathbf{F}_j\|$ remains tiny (like ε) relative to $\|\mathbf{A}\|$, for all j, but $\|\mathbf{C}_j\|$ rises to 1 as soon as a Ritz pair is duplicated.

The "Ritz vectors" $\mathbf{y}_i^{(j)} \equiv \mathbf{Q}_j\mathbf{s}_i^{(j)}$ come from the normalized eigenvectors $\mathbf{s}_i^{(j)}$ of $\mathbf{T}_j$; $\mathbf{T}_j\mathbf{s}_i^{(j)} = \mathbf{s}_i^{(j)}\theta_i^{(j)}$, $i = 1, \ldots, j$.

The effects of most of the roundoff errors are negligible and the way to keep the analysis clean is to ignore those which are inconsequential. We will give a rigorous analysis of a model of the computation, a model which captures all the important features. The first assumption is that $\mathbf{S}_j$ and Θ_j are exact, namely

$$\mathbf{T}_j = \mathbf{S}_j\Theta_j\mathbf{S}_j^*, \quad \mathbf{S}_j^* = \mathbf{S}_j^{-1}, \quad \Theta_j = \operatorname{diag}(\theta_1, \ldots, \theta_j). \tag{13-4-3}$$

The second assumption is that **local** orthogonality is maintained, i.e.,

$$\mathbf{q}_{i+1}^*\mathbf{q}_i = 0, \qquad i = 1, \ldots, j-1, \qquad \text{and } \mathbf{r}_j^*\mathbf{q}_j = 0, \tag{13-4-4}$$

or, equivalently, in terms of (13-4-2), $(\mathbf{C}_j)_{i,i+1} = 0$. The justification for (13-4-4) is that α_i is chosen to force the condition to working accuracy. Were it not for a later application we would also assume that $\Delta_j = \mathbf{O}$ in (13-4-2), i.e., that $\|\mathbf{q}_i\| = 1$ for all i.

It is important to remember that j, the number of Lanczos steps, can increase without bound. Only in exact arithmetic is $j \leqslant n$. We are now ready to establish Paige's interesting results for this model.

THEOREM Assume that the simple Lanczos algorithm satisfies (13-4-1) through (13-4-4) above. Let K_j and N_j be respectively the strictly upper triangular parts of the skew symmetric matrices

$$\mathsf{F}_j^*\mathsf{Q}_j - \mathsf{Q}_j^*\mathsf{F}_j \qquad \text{and} \quad \Delta_j\mathsf{T}_j - \mathsf{T}_j\Delta_j, \tag{13-4-5}$$

and let $\mathsf{G}_j = \mathsf{S}_j^*(\mathsf{K}_j + \mathsf{N}_j)\mathsf{S}_j$. Then the "Ritz vectors" y_i $(\equiv \mathsf{Q}_j\mathsf{s}_i)$, $i = 1, \ldots, j$ satisfy

$$\mathsf{y}_i^*\mathsf{q}_{j+1} = \gamma_{ii}^{(j)}/\beta_{ji} \tag{13-4-6}$$

and for $i \neq k$,

$$(\theta_i - \theta_k)\mathsf{y}_i^*\mathsf{y}_k = \gamma_{ii}^{(j)}\left(\frac{s_{jk}}{s_{ji}}\right) - \gamma_{kk}^{(j)}\left(\frac{s_{ji}}{s_{jk}}\right) - (\gamma_{ik}^{(j)} - \gamma_{ki}^{(j)}), \tag{13-4-7}$$

where $\mathsf{G}_j \equiv (\gamma_{ik}^{(j)})$ and $\beta_{ji} \equiv \beta_j s_{ji}$.

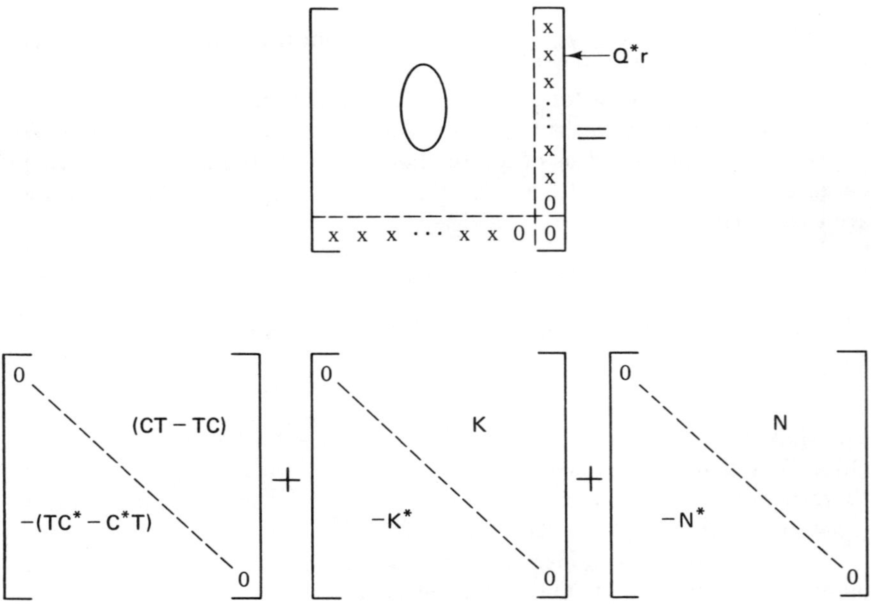

Figure 13-4-1 The structure of $\mathsf{Q}^*\mathsf{r}$

Proof. Drop the subscript j on which all quantities depend and pre-multiply (13-4-1) by Q^* to get

$$Q^*AQ - Q^*QT = Q^*re^* + Q^*F. \tag{13-4-8}$$

To eliminate A subtract from (13-4-8) its transpose and then apply (13-4-2) and (13-4-4) to find

$$\begin{aligned}(Q^*r)e^* - e(Q^*r)^* &= (I - Q^*Q)T - T(I - Q^*Q) + F^*Q - Q^*F \\ &= (C^*T - TC^*) + (CT - TC) \\ &\quad + (\Delta T - T\Delta) + F^*Q - Q^*F, \\ &= (C^*T - TC^*) + (CT - TC) \\ &\quad + (N - N^*) + (K - K^*).\end{aligned} \tag{13-4-9}$$

This important relation is illustrated in Fig. 13-4-1.

Since $\Delta T - T\Delta$ is skew symmetric it must be of the form $N - N^*$ and similarly $F^*Q - Q^*F = K - K^*$. By (13-4-4) $CT - TC$ is strictly upper triangular, as is the rank one matrix $(Q^*r)e^*$. The strictly upper triangular part of (13-4-9) is the key relation, i.e.,

$$(Q^*r)e^* = CT - TC + N + K. \tag{13-4-10}$$

Now s_i^*(13-4-10)s_i gives

$$\begin{aligned}y_i^*q_{j+1}\beta_{ji} &= (s_i^*Q^*)r(e^*s_i), && \text{since } r = q_{j+1}\beta_j, \\ &= s_i^*(CT - TC)s_i + s_i(N + K)s_i, && \text{by (13-4-10)}, \\ &= s_i^*Cs_i\theta_i - \theta_i s_i^*Cs_i + \gamma_{ii}, && \text{by (13-4-3)}, \\ &= \gamma_{ii},\end{aligned} \tag{13-4-11}$$

This gives (13-4-6).

To obtain (13-4-7) consider s_i^*(13-4-8)s_k, $i \neq k$,

$$y_i^*Ay_k - y_i^*y_k\theta_k = y_i^*q_{j+1}\beta_{jk} + s_i^*Q^*Fs_k. \tag{13-4-12}$$

To eliminate $y_i^*Ay_k$ form s_i^*(13-4-8)s_k − s_k^*(13-4-8)s_i and then use (13-4-6) to remove $y_i^*q_{j+1}$; finally (13-4-7) emerges. □

These results are of most use when $\|G_j\|$ is tiny, like $\varepsilon\|A\|$, and we know of no case where $\|G_j\| > \varepsilon\|A\|$ and ε is the unit roundoff. For the simple Lanczos algorithm $\|q_i\| = 1$ and so

$$\|Q_j\|^2 = \|Q_j^*Q_j\| \leq \text{trace}(Q_j^*Q_j) = j, \tag{13-4-13}$$

but this is too crude, especially when $j > n$. A good computable estimate is $\|Q\| = \sqrt{m}$, where m is the size of the biggest cluster of duplicate "Ritz vectors" (Ex. 13-4-1). In any case, when $\Delta_j = O$ in (13-4-2), then $N_j = O$ in

(13-4-5) and, Ex. 13-4-3,

$$\|G_j\| = \|K_j\| \leq \|K_j\|_F = (1/\sqrt{2})\|K_j - K_j^*\|_F \leq \sqrt{2}\,\|Q_j\|\|F_j\|_F; \tag{13-4-14}$$

and so $\|F_j\|_F$ determines whether or not $\|G_j\|$ is small. Recall that for any B, $\|B\|_F \equiv \sqrt{\text{trace}(B^*B)}$. In [Paige, 1976] a careful error analysis shows that

$$\|F_j\|_F \leq (7 + \alpha)\sqrt{j}\,\varepsilon\|A\|, \tag{13-4-15}$$

where α accounts for the error contributed by the user's program for computing Ax. However we know of no exception to the stronger assertion

$$\|F_j\| \leq \varepsilon\|A\|. \tag{13-4-16}$$

Another way in which the practical Lanczos algorithm deviates from the exact one is that the "Ritz vectors" y_i do not all remain of unit length as j increases. A complicated analysis of Paige, based on (13-4-6) above and some results from Chap. 7, shows that it is **only** the presence of other Ritz values which are extra copies of θ_i that permits $\|y_i\|$ to shrink to values like $\sqrt{\varepsilon}$. Thus a set of duplicate Ritz pairs can be expected to have associated y's of very different lengths.

Exercises on Sec. 13-4

13-4-1. Assume that, for each j, any two "Ritz vectors" y_i, y_k are either orthogonal or parallel. Deduce that $\|Q\| \leq \sqrt{m}$, where m is the size of the biggest cluster of Ritz values.

13-4-2. Derive a three-term recurrence dominating the quantities $\|Q_i^* q_{i+1}\|$, namely if $\|Q_i^* q_{i+1}\| \leq \xi_{i+1}$ for $i = 1, \ldots, j-1$, then $\|Q_j^* q_{j+1}\| \leq \xi_{j+1} \equiv \{\|T_j - \alpha_j\|\xi_j + \beta_{j-1}\xi_{j-1} + (1 + \|Q_j\|)\|F_j\|\}/\beta_j$.

13-4-3. Prove that $\|FG\|_F \leq \|F\|\|G\|_F$ whenever FG is defined. Then derive all the relations in (13-4-14). The monotonicity theorem in Chap. 10 may be useful.

13-5 AN ALTERNATIVE FORMULA FOR β_j

There is a variant of the Lanczos algorithm which produces a nonsymmetric tridiagonal matrix J_j instead of T_j. That version has been used extensively but this section shows why it is inferior to the simple implementation given above. The observation is due to Paige. The analysis [Scott, 1978] is a neat application of Paige's theorem and serves to make formulas (13-4-6) and (13-4-7) more familiar.

The Algorithm. Pick r_0 with $\beta_0' = \|r\| \neq 0$. For $j = 1, 2, \ldots$.

1. $q_j \longleftarrow r_{j-1}/\beta_{j-1}'$.
2. $u_j \longleftarrow Aq_j$.
3. $\eta_{j-1} \longleftarrow q_{j-1}^* u_j \qquad (q_0 = 0)$.
4. $\alpha_j \longleftarrow q_j^* u_j$.
5. $r_j \longleftarrow u_j - q_j\alpha_j - q_{j-1}\eta_{j-1}$.
6. $\beta_j' \longleftarrow \|r_j\|$.
7. Compute θ_i, y_i, β_{ji}' as desired.
8. If satisfied then stop.

The rationale behind this version is that **local orthogonality**, $q_{i+1}^* q_{i-1} = q_{i+1}^* q_i = 0$, is enforced to working accuracy.

A negative value of η_{j-1} should not be permitted but wildly disparate positive values of η_i and β_i' (they are equal in exact arithmetic) will aggravate the effect of roundoff errors. We suppose here that $\eta_i > 0$ for all i. The i-th row of J_j is $(\ldots \beta_{i-1}' \; \alpha_i \; \eta_i \ldots)$.

LEMMA There exists $\Omega_j = \operatorname{diag}(\omega_1, \ldots, \omega_{j-1}, 1)$ such that

$$\Omega_j^{-1} J_j \Omega_j = T_j \text{ and } \beta_i = \sqrt{\eta_i \beta_i'}\,, \; i = 1, \ldots, n-1. \qquad (13\text{-}5\text{-}1)$$

The proof constitutes Ex. 13-5-1.

The equation governing the algorithm can now be rewritten

$$\begin{aligned} AQ_j\Omega_j &= Q_j\Omega_j(\Omega_j^{-1}J_j\Omega_j) + r_je_j^*\Omega_j + F_j\Omega_j, \\ &= Q_j\Omega_jT_j + r_je_j^* + F_j\Omega_j, \end{aligned} \qquad (13\text{-}5\text{-}2)$$

since $\omega_j = 1$.

We want to apply Paige's theorem (13-4-5) as before but now the column lengths of $Q_j\Omega_j$ are not 1 but $\omega_1, \ldots, \omega_{j-1}, 1$ instead, i.e., $\Delta_j = \Omega_j^2$ in (13-4-2). Moreover N_j, the upper part of $\Delta_jT_j - T_j\Delta_j$ is zero except that, for $i = 1, \ldots, j-1$,

$$\begin{aligned} n_{i,i+1} &= \beta_i(\omega_{i+1}^2 - \omega_i^2), \\ &= \beta_i\omega_{i+1}^2(1 - \beta_i'/\eta_i), \qquad \text{by Ex. 13-5-1}, \\ &= \frac{\beta_{j-1}' \cdots \beta_{i+1}'}{\eta_{j-1} \cdots \eta_{i+1}}\left(1 - \frac{\beta_i'}{\eta_i}\right)\beta_i, \qquad \text{eliminating the } \omega\text{'s}. \end{aligned}$$

Thus the greater the asymmetry in J_j the greater is $(\|K_j\| + \|N_j\|)$ as well as the numerators γ_{ij} in (13-4-6) and (13-4-7). No such troubles plague the symmetric variant.

Exercise on 13-5

13-5-1. Prove lemma (13-5-1) and find the formulas for the ω_i, $i = 1, \ldots, j-1$.

13-6 CONVERGENCE ⇒ LOSS OF ORTHOGONALITY

Paige's formulas (13-4-6) and (13-4-7) substantiate the claims made earlier (Sec. 13-3) about the behavior of the simple Lanczos algorithm in finite precision arithmetic. The appearance of the ratio s_{ji}/s_{jk}, together with its reciprocal, in (13-4-7), clarifies the whole situation. Despite the worst case bounds (13-4-14) and (13-4-15) all the evidence suggests that the elements of the skew matrix $\mathsf{G}_j \equiv \mathsf{S}_j^*(\mathsf{K}_j + \mathsf{N}_j)\mathsf{S}_j$ which appear in these bounds satisfy

$$|\gamma_{ik}^{(j)}| \leqslant \varepsilon\|\mathsf{A}\|, \qquad \text{for all } i, k, j, \tag{13-6-1}$$

and so $|\mathsf{y}_i^*\mathsf{q}_{j+1}|$ is governed entirely by β_{ji} $(= \beta_j s_{ji})$.

We leave it as an important exercise for the reader to draw the following conclusions:

1. Orthogonality among the q_i, $i = 1, \ldots, j$, is well maintained until one of the Ritz vectors begins to converge ($\beta_{ji} \doteqdot \sqrt{\varepsilon}\ \|\mathsf{A}\|$).
2. Each new Lanczos vector q_{j+1} and each bad "Ritz vector" ($\beta_{ji} > \|\mathsf{A}\|/j$) has a significant component in the direction of each good "Ritz vector" ($\beta_{ji} < \sqrt{\varepsilon}\ \|\mathsf{A}\|$).
3. The emergence of (almost) duplicate copies of previously converged Ritz pairs is quite consistent with (13-4-7).

For example, consider the step at which a third Ritz value $\dddot{\theta}$ first joins the cluster formed by $\dot{\theta}$ and $\ddot{\theta}$ as a clearly recognizable new member. It will be found that $\dot{\theta}$ and $\ddot{\theta}$ are perturbed away from the common eigenvalue λ by the intrusion of $\dddot{\theta}$. The same phenomenon occurs to the "Ritz vectors" $\dot{\mathsf{y}}$ and $\ddot{\mathsf{y}}$. On subsequent Lanczos steps $\dot{\theta}$, $\ddot{\theta}$, and $\dddot{\theta}$ all converge back on λ until a fourth copy of λ "condenses out," barges in on the cluster, and elbows the others out of the way. And so it would continue indefinitely.

Paige's theorem says nothing about the frequency with which duplicate copies of eigenvalues may appear. The cycle time for each individual eigenvector seems to be fairly constant and to depend strongly on the gaps in the spectrum relative to the spread. The subject appears not to have been investigated.

The tables below illustrate (13-4-6) and (13-4-7) in the context of Example 13-3-1 considered above.

EXAMPLE 13-6-1

$\Lambda = \text{diag}(0, 1 \times 10^{-3}, 2 \times 10^{-3}, 3 \times 10^{-3}, 4 \times 10^{-3}, 1.0)$,
$\mathbf{q}_1 = (1, 1, 1, 1, 1, 1)/\sqrt{6}$

At $j = 3$ $\beta_3 = 0.0003094$, and

Table 13-6-1

i	θ_i	$\|\mathbf{y}_i\|$	β_{ji}	$\|\mathbf{y}_i^*\mathbf{q}_4\|$	
1	0.587×10^{-4}	0.9999745	0.2225×10^{-3}	0.4297×10^{-3}	
2	0.3415×10^{-3}	1.000025	0.2228×10^{-3}	0.4297×10^{-3}	
3	1.000000	1.000001	0.48×10^{-7}	*0.923981*	←

In each row the product of the entries in the last two columns is approximately $10^{-7} = \varepsilon\|\mathbf{A}\|$ as required by (13-4-6). In addition mutual orthogonality between good and bad Ritz vectors is poor as required by (13-4-7).

$Y_3^* Y_3$

1.0	0.15×10^{-6}	-0.2×10^{-3}	←
	1.0	0.2×10^{-3}	←
Symmetric		1.0	

At $j = 5$ $\beta_5 = 0.0001098$ and

Table 13-6-2

i	θ_i	$\|\mathbf{y}_i\|$	β_{ji}	$\|\mathbf{y}_i^*\mathbf{q}_6\|$
1	0.157×10^{-4}	1.000490	0.486×10^{-4}	0.33997×10^{-2}
2	0.2001×10^{-3}	1.000477	0.778×10^{-4}	0.21557×10^{-2}
3	0.3845×10^{-3}	0.999509	0.486×10^{-4}	0.35172×10^{-2}
4	0.9999996	*0.75043*	0.414×10^{-4}	0.70092×10^{-2}
5	1.000000	*1.148672*	0.681×10^{-5}	0.70535×10^{-2}
		↑	↑	

$Y_5^* Y_5$

1.0	-0.89×10^{-3}	0.10×10^{-5}	-0.40×10^{-6}	-0.35×10^{-6}
	1.0	0.85×10^{-3}	0.54×10^{-6}	0.60×10^{-6}
		1.0	-0.42×10^{-6}	-0.35×10^{-6}
			0.83	$\mathit{0.52 \times 10^{-1}}$ ←
Symmetric				1.3

As a result of his analysis and experience Paige suggested that the simple Lanczos process be continued as long as storage permitted. At the end of the run all the relevant Ritz pairs are computed and the duplicate copies are simply discarded. This scheme has been used with success but in the remaining sections we consider a modification of the Paige-style Lanczos algorithm which is more efficient and, more significantly, can be used automatically without expert judgment.

13-7 MAINTAINING ORTHOGONALITY

As we have seen, the Lanczos vectors $\mathbf{q}_1, \mathbf{q}_2, \mathbf{q}_3, \ldots$ lose mutual orthogonality as the number of steps increases. The original cure for this condition was proposed by Lanczos himself and required the explicit orthogonalization of $\mathbf{q}_{j+1}$ against all previous $\mathbf{q}$'s. The following step is added after 5 in the algorithm in Sec. 13-1.

$$5\tfrac{1}{2}.\quad \mathbf{r}_j \longleftarrow \mathbf{r}_j - \mathbf{q}_\nu(\mathbf{q}_\nu^* \mathbf{r}_j), \qquad \nu = j, j-1, \ldots, 2, 1.$$

(Note that $\mathbf{r}_j$ is explicitly orthogonalized against $\mathbf{q}_j$ and $\mathbf{q}_{j-1}$.) Consequently all the $\mathbf{q}$'s must be kept handy and the arithmetic cost of each step soars (Ex. 13-7-1). This variant of the algorithm is called Lanczos **with reorthogonalization.**

It is of some (mainly academic) interest to note that reorthogonalization of itself cannot guarantee to produce $\mathbf{q}$'s which are orthogonal to working accuracy. The reason is given in Sec. 6-9 and, as suggested there, the remedy is to test the decrease in norm of each vector and repeat an orthogonalization whenever necessary.

Another fact worth noting is that the cost of reorthogonalization can be halved by keeping the matrix $\mathbf{Q}_j$ in factored form $\mathbf{H}_1\mathbf{H}_2 \cdots \mathbf{H}_j\mathbf{E}_j$ where $\mathbf{E}_j = (\mathbf{e}_1, \mathbf{e}_2, \ldots, \mathbf{e}_j)$ and each $\mathbf{H}_i$ is a reflector matrix $\mathbf{H}(\mathbf{w}_i)$ as described in

Sec. 6-2. It is only necessary to keep the vectors $\mathbf{w}_i$, $i = 1, \ldots, j$, and the first $i - 1$ elements of $\mathbf{w}_i$ are zero. See [Golub, Underwood, and Wilkinson, 1972] for more details. Despite this improvement there is strong incentive to avoid the storage and arithmetic costs of reorthogonalization.

Paige's result (13-4-6) shows that in the simple Lanczos method $\mathbf{q}_{j+1}$ tilts most in the direction of those Ritz vectors $\mathbf{y}_i^{(j)}$, if any, which are fairly good approximations to eigenvectors; more precisely,

$$\mathbf{y}_i^{(j)*}\mathbf{q}_{j+1} = \gamma_{ii}^{(j)}/\beta_{ji}, \tag{13-7-1}$$

where $|\gamma_{ik}^{(j)}| \leqslant \varepsilon\|\mathbf{A}\|$ for all i, j, k, by (13-6-1), and $\beta_{ji} \equiv \beta_j s_{ji} \doteqdot \|\mathbf{A}\mathbf{y}_i^{(j)} - \mathbf{y}_i^{(j)}\theta_i^{(j)}\|$ by (13-2-1). Moreover the tilting can be monitored **without computing** $\mathbf{y}_i^{(j)}$ because each β_{ji} can be computed with approximately $4j$ ops once the Ritz values $\theta_i^{(j)}$ are known.

Formula (13-7-1) suggests a more discriminating way of maintaining a good measure of orthogonality than full reorthogonalization, namely orthogonalize $\mathbf{q}_{j+1}$ against $\mathbf{y}_i^{(j)}$ when and only when β_{ji} becomes small. To do this $\mathbf{y}_i^{(j)}$ must be computed and stored and this itself costs half as much as full reorthogonalization at one step. What makes the difference is that the new orthogonalizations do not occur very often and the ever growing $\mathbf{Q}_j$ does not have to be kept on hand for each step.

To see whether there is merit in the idea we examine two simple examples in each of which a vector $\bar{\mathbf{q}}$ is orthogonalized against $\mathbf{y}_2$ to produce

$$\mathbf{q} = (1 - \mathbf{y}_2\mathbf{y}_2^*)\bar{\mathbf{q}}/\|(1 - \mathbf{y}_2\mathbf{y}_2^*)\bar{\mathbf{q}}\|.$$

In general the orthogonality of the $\mathbf{q}_i$ and the $\mathbf{y}_i^{(j)}$, $i = 1, \ldots, j$ is measured by

$$\kappa_j \equiv \|1 - \mathbf{Q}_j^*\mathbf{Q}_j\| = \|1 - Y_j^* Y_j\|. \tag{13-7-2}$$

In the examples $\bar{\kappa}_3$ is the value before orthogonalization and κ_3 is the value afterward.

EXAMPLE 13-7-1

$$\mathbf{y}_1 = \begin{bmatrix} 1 \\ 0 \\ 0 \end{bmatrix}, \quad \mathbf{y}_2 = \begin{bmatrix} 1 \\ 1 \\ 0 \end{bmatrix}\frac{1}{\sqrt{2}}, \quad \bar{\mathbf{q}} = \begin{bmatrix} 0 \\ 1 \\ 1 \end{bmatrix}\frac{1}{\sqrt{2}}, \quad \mathbf{q} = \begin{bmatrix} -1 \\ 1 \\ 2 \end{bmatrix}\frac{1}{\sqrt{6}}.$$

Then

$$\mathbf{y}_1^*\mathbf{q} = -1/\sqrt{6}\,, \qquad \mathbf{y}_1^*\mathbf{y}_2 = 1/\sqrt{2}\,, \qquad \mathbf{y}_2^*\mathbf{q} = 0,$$

$$\kappa_2 = 1/\sqrt{2}\,, \qquad \bar{\kappa}_3 = 1/\sqrt{2}\,, \qquad \kappa_3 \geqslant 1/\sqrt{2}\,.$$

Nothing has been gained.

EXAMPLE 13-7-2

$$\mathbf{y}_1 = \begin{bmatrix} 1 \\ 0 \\ 0 \end{bmatrix}, \quad \mathbf{y}_2 = \begin{bmatrix} 10^{-4} \\ 1 \\ 0 \end{bmatrix}, \quad \bar{\mathbf{q}} = \begin{bmatrix} 0 \\ 10^{-2} \\ 1 \end{bmatrix}, \quad \mathbf{q} = \begin{bmatrix} -10^{-6} \\ 0 \\ 1 \end{bmatrix}.$$

Then

$$\mathbf{y}_1^*\mathbf{q} = -10^{-6}, \qquad \mathbf{y}_1^*\mathbf{y}_2 = 10^{-4}, \qquad \mathbf{y}_2^*\mathbf{q} = 10^{-10},$$

$$\kappa_2 = 10^{-4}, \qquad \bar{\kappa}_3 = 10^{-2}, \qquad \kappa_3 = 10^{-4}.$$

The previous level of orthogonality has been maintained.

Example 13-7-1 shows that orthogonalization, by itself, need not improve the situation. Example 13-7-2 is more realistic and suggests that the technique is beneficial whenever $|\mathbf{y}_i^*\mathbf{q}_{j+1}|$ exceeds κ_j significantly.

Rewards for maintaining strong linear independence among the Lanczos vectors, i.e., $\|\mathbf{I} - \mathbf{Q}_j^*\mathbf{Q}_j\| < 0.01$.

1. Troublesome, redundant Ritz pairs cannot be formed.
2. The number of Lanczos steps is kept to a minimum and, in any case, can never exceed n.
3. Multiple eigenvalues can be found, one by one. The reason is given in the next section.
4. The method can be used as a black box and requires no delicate parameters to be set by the user beyond indicating the desired accuracy and the amount of fast storage available.

On the debit side the program must compute and store any good "Ritz vectors" to be used for purging the new $\mathbf{q}$'s whether or not they are of interest to the user. For instance, if three (algebraically) small eigenvalues are wanted the algorithm may well be obliged to compute three or more "Ritz vectors" belonging to large Ritz values simply because some of them converge quickly. Scott is exploring the consequences of orthogonalizing the new $\mathbf{q}$'s against only those "Ritz vectors" which are wanted.

In what follows we shall sometimes say $\mathbf{q}$ is **purged of** $\mathbf{y}$ as a synonym for $\mathbf{q}$ is orthogonalized against $\mathbf{y}$.

Suppose that we permit κ_j [defined in (13-7-2)] to grow slowly, as j increases, from an initial value $\kappa_1 \doteqdot \varepsilon$ but never to exceed a certain value

κ. If $\kappa \doteqdot n\varepsilon$ then orthogonalization will be forced almost all the time and the cost may exceed that of full reorthogonalization. At the other extreme if $\kappa \geqslant 1$ then the Paige-style Lanczos process is recovered. Thus κ lets us interpolate between the two extreme versions of the Lanczos process.

Exercise on 13-7

13-7-1. Apart from the cost of computing $\mathbf{Aq}_j$ the op count for simple Lanczos is $5n$ and for full reorthogonalization is $(j + 3)n$, if it is assumed that $\mathbf{Q}_j$ is not held in factored form as described in [Golub, Underwood, Wilkinson, 1972]. Verify these counts.

*13-8 SELECTIVE ORTHOGONALIZATION

A small modification to the simple Lanczos algorithm ensures that the vectors $\mathbf{q}_1, \mathbf{q}_2, \ldots$ maintain a reasonable, preset level of linear independence. Let $\kappa_j \equiv \|\mathbf{I} - \mathbf{Q}_j^*\mathbf{Q}_j\|$ and suppose that it is required to keep $\kappa_j \leqslant \kappa$ for some κ in the interval $(n\varepsilon, 0.01)$. If $\varepsilon = 10^{-14}$ and $n = 10^4$ then the interval still spans eight orders of magnitude. Both experience and informal analysis suggest that κ should be chosen near $\sqrt{\varepsilon}$ (to within an order of magnitude) in order to reap the fruits of full reorthogonalization at a cost close to, and sometimes less than, that of the simple algorithm.

The new version modifies the vector $\mathbf{r}_j$ of the simple algorithm of Sec. 13-1 before normalizing it and some notation is needed to distinguish $\mathbf{r}_j$ before from $\mathbf{r}_j$ afterwards. We reserve $\mathbf{r}_j$ for the final form and use $\mathbf{r}'_j$ for the original. From (13-4-1)

$$\mathbf{r}'_j \equiv \mathbf{Aq}_j \equiv \mathbf{q}_j\alpha_j - \mathbf{q}_{j-1}\beta_{j-1} - \mathbf{f}_j,$$

where $\mathbf{f}_j$ accounts for roundoff errors during the j-th step and $\|\mathbf{f}_j\|$ remains below $n\varepsilon\|\mathbf{A}\|$ for all j. We are interested in $\angle(\mathbf{r}'_j, \mathbf{q}_i)$, $i = 1, \ldots, j$, or equivalently in $\angle(\mathbf{r}'_j, \mathbf{y}_i^{(j)})$, $i = 1, \ldots, j$. Recall that $\mathbf{y}_i^{(j)} = \mathbf{Q}_j\mathbf{s}_i$ and $(\theta_i, \mathbf{s}_i)$ is an eigenpair of $\mathbf{T}_j$. While the angles remain close to $\pi/2$ there is no need to depart from the simple algorithm and the new algorithm sets $\mathbf{r}_j \equiv \mathbf{r}'_j$. By Paige's theorem (13-4-6)

$$\cos \angle(\mathbf{r}'_j, \mathbf{y}_i^{(j)}) = \gamma_{ii}^{(j)} / \beta'_{ji}\|\mathbf{y}_i^{(j)}\|,$$

and $|\gamma_{ii}^{(j)}| \leqslant \varepsilon\|\mathbf{A}\|$ for all i, j, and $\beta'_{ji} \equiv \beta'_j s_{ji}$, $\beta'_j = \|\mathbf{r}'_j\|$, $s_{ji} = \mathbf{e}_j^*\mathbf{s}_i$. By the final remarks in Sec. 13-4 the restriction $\kappa \leqslant 0.01$ keeps $|1 - \|\mathbf{y}_i^{(j)}\|| \ll \kappa$ for all i, j, and we shall ignore the factor $\|\mathbf{y}_i^{(j)}\|$ throughout this section. The important consequence is that the angles can be monitored by keeping track of the easily computed quantities β'_{ji}.

As soon as the Lanczos vectors begin to lose orthogonality attention centers on the set of indices

$$\mathcal{L}(j) = \left\{ i: |\cos \angle(\mathbf{y}_i^{(j)}, \mathbf{r}_j')| \geq \kappa/\sqrt{j} \right\}.$$

Let $|\mathcal{L}(j)|$ denote the number of indices in $\mathcal{L}(j)$. For most values of j it turns out that $|\mathcal{L}(j)| = 0$ but when $|\mathcal{L}(j)| > 0$ then the idea is to purge $\mathbf{r}_j'$ of the associated $\mathbf{y}_i^{(j)}$, $i \in \mathcal{L}(j)$, which we call the **threshold** "Ritz vectors." The hope is that the resulting vector $\mathbf{r}_j$ will satisfy $|\cos \angle(\mathbf{y}_k^{(j)}, \mathbf{r}_j)| < \kappa/\sqrt{j}$ for **all** the values $k = 1, \ldots, j$. Then $\mathbf{r}_j$ is normalized to become $\mathbf{q}_{j+1}$ and we have

$$\|\mathbf{Q}_j^*\mathbf{q}_{j+1}\| = \|\mathbf{Y}_j^*\mathbf{q}_{j+1}\| < \sqrt{j}\left(\kappa/\sqrt{j}\right) = \kappa.$$

The factor $\sqrt{j}$ is a crude overbound and could not be attained. A more realistic bound is given in Ex. 13-8-1.

In order to purge $\mathbf{r}_j'$ it is necessary to compute $\mathbf{y}_i^{(j)}$ for $i \in \mathcal{L}(j)$. This involves bringing the old $\mathbf{q}$'s back from secondary storage and we say that the modified algorithm **pauses** whenever $|\mathcal{L}(j)| > 0$. On the other hand if $\kappa \geq \sqrt{\varepsilon}$ then $\beta_{ji}' \leq \varepsilon\|\mathbf{A}\|/(\kappa/\sqrt{j}) = \sqrt{j\varepsilon}\,\|\mathbf{A}\|$ and it follows that $\theta_i^{(j)}$, $i \in \mathcal{L}(j)$, will often agree with an eigenvalue of $\mathbf{A}$ to working accuracy. For some, but not all applications, these threshold "Ritz vectors" will already be acceptable and so there is no reluctance to compute them.

It is worth repeating that when κ exceeds $n\varepsilon$ then the $\mathbf{y}_i^{(j)}$ are neither orthonormal (to working accuracy) nor the true Ritz vectors from span $\mathbf{Q}_j$. That is why we persist in using quotes when referring to them. Before proceeding with the description of the algorithm we must make an observation about (true) Ritz vectors.

There is an annoying identification problem as j varies because the set of Ritz vectors changes completely at each step. In general there is no natural association between $\mathbf{y}_i^{(j)}$ and $\mathbf{y}_i^{(j+1)}$ for a given i. This is evident in Example 10-1-1 with $j = 1$ and $j = 2$. However as soon as a sequence $\{\theta_i^{(j)}\}$, for some fixed i such as ± 1 or ± 2, settles down in its first few digits, i.e., as soon as convergence becomes apparent, then it is meaningful to speak about $\mathbf{y}_i$ and **its** (vector) values at various steps.

With these remarks in mind we follow the history of a typical "Ritz vector" in the modified algorithm. At step 25, say, $\theta_i^{(25)}$ emerges from a crowd of undistinguished "Ritz values" in the middle of the spectrum and settles down, $|\theta_i^{(24)} - \theta_i^{(25)}|/|\theta_i^{25}| < 0.1$. Later $\mathbf{y}_i^{(40)}$ becomes a threshold vector; there is a pause wherein $\mathbf{y}_i^{(40)}$ is computed and possibly some other threshold vectors as well. Next it always happens that $\mathbf{y}_i^{(41)}$ is a threshold vector and agrees with $\mathbf{y}_i^{(40)}$ to several figures (depending on κ). It is tempting to forego the expense of pausing to compute $\mathbf{y}_i^{(41)}$ and to use $\mathbf{y}_i^{(40)}$ in its place for purging $\mathbf{r}_{41}'$. We will return to this point shortly.

The crucial fact, which is not immediately obvious, is that for $j = 42, 43, \ldots, i \notin \mathcal{L}(j)$. The effects of roundoff may or may not put i into $\mathcal{L}(j)$ again before the computation is over. In any case what matters is that $i \notin \mathcal{L}(j)$ for most values of j. That is part of the reason that the modified algorithm is as economical as the simple one.

Let us return to $\mathbf{y}_i^{(40)}$ and $\mathbf{y}_i^{(41)}$. True Ritz vectors are orthonormal. Moreover $\mathbf{y}_i^{(40)}$ will deviate most from a true Ritz vector on account of components of those $\mathbf{y}_\nu^{(40)}$ which crossed the threshold earlier and are therefore better converged; we call these the **good** "Ritz vectors" at step 40. Since the Lanczos vectors $\mathbf{q}_1, \mathbf{q}_2, \ldots, \mathbf{q}_j$ must be brought back from secondary storage there is little extra trouble in recomputing **all** the good $\mathbf{y}_\nu^{(40)}$ (more precisely those which have not yet converged to working accuracy). It is also possible to orthonormalize the good $\mathbf{y}_\nu^{(40)}$ and produce new vectors which we call $\mathbf{y}_1, \ldots, \mathbf{y}_i$ without a superscript. In that case $\mathbf{y}_i$ is used for purging both $\mathbf{r}'_{40}$ and $\mathbf{r}'_{41}$. There will not be a pause at $j = 41$ unless another "Ritz vector" crosses the threshold at that step. The only pairs $(\theta_\mu, \mathbf{s}_\mu)$ which need be computed at each step are for the extreme (outermost) θ's that are not yet good. There is no need to recompute good θ's at each step, even if they have not converged.

Note that the modification is independent of the user's accuracy requirements. Results to working accuracy can be obtained but so can rough approximations. There is no anomaly if a Ritz vector is acceptable before it is good because the adjectives good and bad pertain to the degree of orthogonality which is to be maintained. There are strong arguments for taking $\kappa = \sqrt{\varepsilon}$ (see the next section) and so no extra choice is thrust on the user.

Suppose that $\mathbf{A}$ has a multiple eigenvalue λ_1. At step 30, say, $\mathbf{y}_1^{(30)}$ will be one of λ_1's eigenvectors to within working accuracy. After step 30 the selective orthogonalization keeps $\mathbf{y}_1^*\mathbf{r}_j = 0(\varepsilon\|\mathbf{A}\|)$ for $j > 30$. Roundoff error will introduce nonzero components of other eigenvectors for λ_1 which are orthogonal to $\mathbf{y}_1$. In time a new Ritz vector, say $\mathbf{y}_8^{(70)}$, will converge to one of these eigenvectors. Then $\mathbf{y}_1^*\mathbf{r}_j = 0(\varepsilon\|\mathbf{A}\|)$, $\mathbf{y}_8^*\mathbf{r}_j = 0(\varepsilon\|\mathbf{A}\|)$ for $j > 70$ and so on until λ_1's eigenspace is spanned.

Section 13-8-1 describes the selective orthogonalization algorithm in more detail.

*13-8-1 **LANSO** FLOWCHART (**LAN**CZOS ALGORITHM WITH **S**ELECTIVE **O**RTHOGONALIZATION)*

The flowchart reflects the current implementation of *SO* but the algorithm is still under development.

Parameters

ℓc = index of last fully converged Ritz vector ($\beta_{ji} < j\varepsilon\|A\|$) from the left end of the spectrum

ℓg = index of last good Ritz vector [$\ell g \in \mathcal{L}(k)$ for some $k \leqslant j$] from the left

rc, rg as above for the right end of the spectrum.

Initialize

$\ell c = \ell g = rc = rg = |\mathcal{L}| = 0$. $\mathcal{L} = \varnothing$ (empty). $q_0 = o$. Pick $r'_0 \neq o$.

Loop

For $j = 1, 2, \ldots, n$ repeat steps 1 through 5.

1. If $|\mathcal{L}| > 0$ then purge r' of threshold vectors to get r and set $\beta_{j-1} \longleftarrow \|r\|$.
2. If $\beta_{j-1} = 0$ then stop or else normalize r to get q_j.
3. Take a Lanczos step to get α_j, r', β'_j.
4. $\theta_i^{(j)} \longleftarrow \lambda_i[T_j]$ for $i = \ell g + 1$, $\ell g + 2$ and $i = -(rg + 1), -(rg + 2)$. Compute associated s_{ji}. Set $|\mathcal{L}| = 0$.
5. If β'_{ji} $(= \beta'_j s_{ji}) < \sqrt{\varepsilon}\,\|T_j\|$ for any of these i then pause.

Pause

1. Form $\mathcal{L}$ $(\equiv \{i: \beta'_{ji} < \sqrt{\varepsilon}\,\|T_j\|)$. Update ℓg, rg.
2. Summon Q_j and compute $y_\ell^{(j)} = Q_j s_\ell$ for $\ell = \ell c, \ldots, \ell g$ and $\ell = -rc, \ldots, -rg$.
3. *Optional step*: Perform modified Gram-Schmidt on the new $y_\ell^{(j)}$; use the most accurate first. Update $s_{j\ell}$ accordingly.
4. If enough y's are acceptable then stop.
5. Compute $y_\ell^* r'$ for each good y_ℓ; if too big add ℓ to $\mathcal{L}$.

It is only necessary to keep the threshold vectors after the pause, the Lanczos vectors can be rewound.

EXAMPLE 13-8-1 Example of Selective Orthogonalization

$n = 6$.

$A = \text{diag}(0., 0.00025, 0.0005, 0.00075, 0.001, 10.)$.

$q_1 = 6^{-1/2}(1., 1., 1., 1., 1., 1.)^*$.

Unit roundoff $\doteqdot 10^{-14}$. Note that 0.75×10^{-6} is written $.75E-06$.

Simple Lanczos was run for six steps.

$$Q_6^* Q_6$$

$$\begin{bmatrix}
.10E+01 & .75E-14 & -.30E-10 & .25E-06 & .97E-02 & .41E+00 \\
.75E-14 & .10E+01 & .33E-10 & .55E-06 & .22E-01 & .91E+00 \\
-.30E-10 & .33E-10 & .10E+01 & -.97E-10 & .19E-05 & .79E-04 \\
.25E-06 & .55E-06 & -.97E-10 & .10E+01 & .11E-09 & .23E-08 \\
.97E-02 & .22E-01 & .19E-05 & .11E-09 & .10E+01 & -.12E-12 \\
.41E+00 & .91E+00 & .79E-04 & .23E-08 & -.12E-12 & .10E+01
\end{bmatrix}$$

The Lanczos algorithm with selective orthogonalization was run for six steps. It paused after four steps and computed a good Ritz vector for the eigenvalue 10. It then took two more steps orthogonalizing against this vector.

$Q_6^* Q_6$ for Selective Orthogonalization

$$\begin{bmatrix}
.10E-01 & .75E-14 & -.30E-10 & .25E-06 & -.11E-09 & .92E-10 \\
.75E-14 & .10E+01 & .33E-10 & .55E-06 & .51E-10 & -.36E-10 \\
-.30E-10 & .33E-10 & .10E+01 & -.97E-10 & -.44E-10 & -.37E-07 \\
.25E-06 & .55E-06 & -.97E-10 & .10E+01 & .24E-07 & -.64E-08 \\
-.11E-09 & .51E-10 & -.44E-10 & .24E-07 & .10E+01 & .10E-13 \\
.92E-10 & -.36E-10 & -.37E-07 & -.64E-08 & .10E-13 & .10E+01
\end{bmatrix}$$

Note that the leading 4 by 4 principal minor is the same in both matrices. Note that robust linear independence has been maintained by the selective orthogonalization scheme.

Exercises on Sec. 13-8

13-8-1. Assume that $y_k^{(j)*} r_j = y_k^{(j)*} r'_j$ for $k \notin \mathcal{L}(j)$. Use Paige's theorem (13-4-6) to obtain

$$\|Q_j^* q_{j+1}\| \leq (\varepsilon \|A\| / \beta_j) \left(\sum_{k \notin \mathcal{L}(j)} s_{jk}^{-2} \right)^{-1/2}.$$

Assume that $\|y_k^{(j)}\| = 1$ and $|\gamma_{ii}^{(j)}| \leq \varepsilon \|A\|$.

13-8-2. Let $q_{j+1}\beta_j = Aq_j - q_j\alpha_j - q_{j-1}\beta_{j-1}$ for all j. Let $Az = z\lambda$ and suppose that $z^* q_{40} = z^* q_{41} = 0$. Show that, in exact arithmetic, $z^* q_j = 0$ for $j > 41$.

*13-9 ANALYSIS OF SELECTIVE ORTHOGONALIZATION

This section discusses various aspects of the selective orthogonalization procedure described in Sec. 13-8.

The simple Lanczos algorithm will continue indefinitely in a finite precision environment and Paige was concerned with the question of convergence, as $j \longrightarrow \infty$, of certain "Ritz pairs" $(\theta_i^{(j)}, \mathbf{y}_i^{(j)})$ to eigenpairs of $\mathbf{A}$. The modified algorithm preserves a strong measure of linear independence among the $\mathbf{q}_i$ and so the procedure must terminate ($\beta_j \leqslant n\varepsilon\|\mathbf{A}\|$) with $j \leqslant n$. Consequently our attention turns away from convergence to the influence of the value of κ on the execution of the modified algorithm.

13-9-1 ORTHONORMALIZING THE GOOD RITZ VECTORS

This extra feature is not a necessary part of selective orthogonalization but it has merit when there is no subroutine available to refine crude output from a Lanczos process and supply careful a posteriori error bounds on the lines of Chap. 10.

For simplicity relabel the "Ritz vectors" so that $\mathbf{y}_1^{(j)}, \ldots, \mathbf{y}_i^{(j)}$ are the good ones and suppose that $\{i\} = \mathcal{L}(j)$. Without loss of generality we suppose that $\beta_{j1} \leqslant \beta_{j2} \leqslant \cdots \leqslant \beta_{ji}$. During the pause the algorithm computes, for $\ell = 1, 2, \ldots, i$,

$$\mathbf{y}_\ell^{(j)} \longleftarrow \mathbf{Q}_j \mathbf{s}_\ell,$$

$$\tilde{\mathbf{y}}_\ell \longleftarrow \mathbf{y}_\ell^{(j)} - \sum_{\nu=1}^{\ell-1} \mathbf{y}_\nu (\mathbf{y}_\nu^* \mathbf{y}_\ell^{(j)}),$$

$$\mathbf{y}_\ell \longleftarrow \tilde{\mathbf{y}}_\ell / \|\tilde{\mathbf{y}}_\ell\|.$$

What Ritz value should be associated with $\mathbf{y}_\ell$? Naturally the Rayleigh quotient $\rho(\mathbf{y}_\ell)$ would be best but we do not want to compute $\mathbf{A}\mathbf{y}_\ell$. The answer is suggested by the following result.

LEMMA.	$\rho(\mathbf{y}_\ell) = \theta_\ell^{(j)} + O(\kappa^2\|\mathbf{A}\|)$,	$l \leqslant i < j$.	(13-9-1)

The proof is left as Ex. 13-9-2. Note that the best bound we can put on the coefficients $\gamma_{\nu\ell} \equiv \mathbf{y}_\nu^* \mathbf{y}_\ell^{(j)}$ is $|\gamma_{\nu\ell}| \leqslant \kappa_j + O(\kappa_j^2)$ (Ex. 13-9-1). If $\kappa^2 \leqslant \varepsilon$ there is no loss in using $\theta_\ell^{(j)}$ as the Ritz value for $\mathbf{y}_\ell$.

In the remaining subsections we will not consider the use of the vectors $\mathbf{y}_\ell$, $\ell \leqslant i$ but will continue to work with the $\mathbf{y}_\ell^{(j)}$ for purging $\mathbf{r}_j'$.

13-9-2 THE EFFECT OF PURGING ON ANGLES

We drop the superscript j on each "Ritz vector" $\mathbf{y}_k$. We need to compare $\angle(\mathbf{y}_k, \mathbf{r}_j)$ and $\angle(\mathbf{y}_k, \mathbf{r}_j')$ for all $k \notin \mathcal{L}(j)$. Of course $\mathbf{y}_i^*\mathbf{r}_j = 0(\varepsilon\|\mathbf{A}\|)$

for $i \in \mathcal{L}(j)$, by construction. After the purge

$$\mathbf{r}_j \equiv \mathbf{r}_j' - \sum_{\nu \in \mathcal{L}(j)} \mathbf{y}_\nu \xi_\nu, \qquad \xi_\nu = \mathbf{y}_\nu^* \mathbf{r}_j',$$

and so

$$\mathbf{y}_k^* \mathbf{r}_j = \mathbf{y}_k^* \mathbf{r}_j' - \sum_\nu (\mathbf{y}_k^* \mathbf{y}_\nu)\xi_\nu. \tag{13-9-2}$$

In exact arithmetic both ξ_ν and $\mathbf{y}_k^*\mathbf{y}_\nu$ vanish but in practice it is only necessary that their **product** be tiny, like $\varepsilon\|\mathbf{A}\|$, to ensure that the purgings do not degrade the inner products $\mathbf{y}_k^*\mathbf{r}_j'$.

Now ξ_ν is definitely not small since

$$\begin{aligned} |\xi_\nu| &= |\mathbf{y}_\nu^* \mathbf{r}_j'| / \|\mathbf{y}_\nu\|, \\ &= |\cos \angle(\mathbf{y}_\nu, \mathbf{r}_j')|\, \beta_j', \qquad (\beta_j' \equiv \|\mathbf{r}'_j\|), \\ &\geq \beta_j' \kappa / \sqrt{j}\,, \qquad \text{by definition of } \mathcal{L}(j). \end{aligned}$$

The other factor, $\mathbf{y}_k^*\mathbf{y}_\nu$, in (13-9-2) is bounded by κ_j and this bound is realistic for some values of k. Some terms in the sum in (13-9-2) will be larger than $\beta_j'\kappa\kappa_j/j$. Moreover $\kappa_j \doteq \kappa$ eventually. In order to preserve $\angle(\mathbf{y}_k, \mathbf{r}_j')$ to working accuracy it seems necessary to have

$$\kappa^2 \leq \varepsilon. \tag{13-9-3}$$

On the other hand if $|\xi_\nu|$ can be much greater than $\kappa\|\mathbf{A}\|$ then (13-9-3) will not protect the configuration from a gradual acceleration in the loss of orthogonality. If Paige's theorem (13-4-6) continues to hold then the **sudden** convergence of one of the $\mathbf{y}_i$, indicated by $\beta_{ji} \ll \kappa\|\mathbf{A}\|$, might provoke such a large ξ_ν. The following result [Scott, 1978] shows that this fear is unfounded.

LEMMA If $\mathbf{r}_j' = \mathbf{A}\mathbf{q}_j - \mathbf{q}_j\alpha_j - \mathbf{q}_{j-1}\beta_{j-1} - \mathbf{f}_j$ then

$$|\mathbf{y}_i^*\mathbf{r}_j'| \leq \|\mathbf{A}\mathbf{y}_i - \mathbf{y}_i\theta_i\| + \kappa_j[(\alpha_j - \theta_i)^2 + \beta_{j-1}^2]^{1/2} + \|\mathbf{f}_j\|. \tag{13-9-4}$$

The proof is left as Ex. 13-9-3.

The effect of selective orthogonalization is to spoil the nice bound $\|\mathbf{A}\mathbf{y}_k - \mathbf{y}_k\theta_k\| \leq \beta_{jk} + O(\varepsilon\|\mathbf{A}\|)$ for the bad "Ritz vectors" $\mathbf{y}_k$. The next subsection suggests that for the good $\mathbf{y}_i$ ($i \in \mathcal{L}(\nu)$ for some $\nu \leq j$) the convenient bound still holds. Moreover, by the definition of $\mathcal{L}(j)$,

$\beta_{ji} \leq \varepsilon\sqrt{j}\ \|A\|/\kappa$. Thus lemma (13-9-4) yields the following interesting bound:

$$|y_i^* r_j'| \leq \varepsilon\sqrt{j}\ \|A\|/\kappa + \kappa(2\|A\|) + 0(\varepsilon\|A\|), \qquad i \in \mathfrak{L}(j).$$

The right-hand side is minimized (approximately) by the choice $\kappa = \sqrt{\varepsilon}$ and then $|\xi_i|$ can never rise much above $\sqrt{\varepsilon}\ \|A\|$.

13-9-3 THE GOVERNING FORMULA

Suppose that, for the first time, a "Ritz vector" y_1 crosses the threshold at step j, i.e., $|\cos \angle(y_1, r_j')| > \kappa/\sqrt{j}$. After the purge

$$r_j = r_j' - y_1\xi_1, \qquad \xi_1 = y_1^* r_j'/\|y_1\|. \tag{13-9-5}$$

The basic relation given in (13-4-1) is

$$AQ_j - Q_jT_j = r_j'e_j^* + F_j, \tag{13-9-6}$$

and when r_j' is eliminated from (13-9-5) and (13-9-6) we find

$$AQ_j - Q_jT_j - \xi_1 y_1 e_j^* = r_j e_j^* + F_j. \tag{13-9-7}$$

It is helpful to bring $y_1e_j^*$ to the left side because y_1 is in span Q_j and $y_1^* r_j = 0(\varepsilon\|A\|)$ by construction. Recall that $y_1 = Q_js_1$ and $T_js_1 = s_1\theta_1$; so (13-9-7) becomes

$$AQ_j - Q_j(T_j + \xi_1 s_1 e_j^*) = r_j e_j^* + F_j. \tag{13-9-8}$$

The rank one perturbation of T_j compensates for the fact that T_j is not the projections of A on span Q_j.

It may be verified (Ex. 13-9-4) that the eigenpairs of the perturbed matrix are:

$$\begin{cases} \theta_1 + \xi_1 s_{j1}, & s_1 \\ \theta_k, & s_k + s_1\left[\xi_1 s_{jk}/(\theta_k - \theta_1)\right] + O(\varepsilon), \qquad k > 1. \end{cases} \tag{13-9-9}$$

By Paige's theorem $\xi_1(y_1^* r_j')/\|y_1\| = \gamma_{11}/(\beta_{j1}'\|y_1\|)$, where $\gamma_{11} = 0(\varepsilon\|A\|)$. Consequently $\xi_1 s_{j1} = \gamma_{11}/(\beta_j'\|y_1\|)$, i.e., the quantity s_{j1} cancels out and there is no need to modify the pair (θ_1, y_1) unless β_j' is small. However a small β_j' is not consistent with our assumption that only one "Ritz vector" crossed the threshold at step j.

On the other hand the significant changes to the other eigenvectors s_k come as no surprise. The formulas indicate precisely how to remove the component of y_1 that has crept into the other "Ritz vectors" y_k. Since we are not interested in computing the true bad Ritz vectors from span Q_j there seems to be no point in recording the modifications to the s_k. This is a relief.

To see the general pattern more clearly we go on to the next step and assume that no new threshold vectors appear. The algorithm takes the lazy way out and uses $y_1^{(j)}$ instead of $y_1^{(j+1)}$ to purge r'_{j+1}. A good estimate for $\sin \angle(y_1^{(j)}, y_1^{(j+1)})$ is $s_{j+1,1}$, the $(j+1, 1)$ element of S_{j+1}, and we expect that $s_{j+1,1}^{(j+1)} < s_{j1}^{(j)} = 0(\sqrt{\varepsilon})$. In any case whether or not $y_1^{(j)}$ and $y_1^{(j+1)}$ are close, we find

$$AQ_{j+1} - Q_{j+1}\left[T_{j+1} + \begin{bmatrix} s_1^{(j)} \\ o \end{bmatrix}(0, \ldots, 0, \xi_1^{(j)}, \xi_1^{(j+1)})\right] = r_{j+1}e_{j+1}^* + F_{j+1}.$$

The vector $s_1^{(j)}$ must be endowed with an extra zero element at the bottom to be conformable with T_{j+1}. In pictures

$$\begin{bmatrix} x & x & & & \square & \triangle \\ x & x & x & & \square & \triangle \\ & x & x & x & \square & \triangle \\ & & x & x & \boxed{x} & \triangle \\ & & & x & x & x \\ & & & & x & x \end{bmatrix} \qquad \begin{aligned} \square &= \text{elements of } s_1\xi_1^{(j)}, \\ \triangle &= \text{elements of } s_1\xi_1^{(j+1)}. \end{aligned}$$

The bottom two elements of the perturbing vectors are $O(\varepsilon\|T_{j+1}\|)$.

It might be supposed that a similar pattern occurs at step $j + 2$ but that is not the case. After two purgings r'_{j+2} will be orthogonal to y_1 to working accuracy (thanks to the three-term recurrence):

$$\begin{aligned} y_1^* r'_{j+2} &= y_1^*(Aq_{j+2} - q_{j+2}\alpha_{j+2} - q_{j+1}\beta_{j+1} - f_{j+2}), \\ &= (Ay_1)^* q_{j+2} - O(\varepsilon) - O(\varepsilon) - O(\varepsilon), \\ &= (y_1\theta_1 + q_{j+1}\beta_{j1})^* q_{j+2} + O(\varepsilon), \\ &= O(\varepsilon). \end{aligned}$$

In fact y_1 will not become a threshold vector again unless roundoff boosts latent components of y_1 in the current r-vector up to the required level. There will be no more purgings until a new y_i, say y_2, crosses the threshold at step m. Since Q_m must be called in to compute $y_2^{(m)}$ ($\equiv Q_m s_2^{(m)}$) there is little extra cost to computing $y_1^{(m)}$ and storing it over $y_1^{(j)}$ at this time.

At this point we remember the perturbations to T_{j+1} associated with the purging at steps j and $j + 1$. Thus $y_2^{(m)}$ should be computed from $Q_m\hat{s}_2 = Q_m\left(s_2^{(m)} - \begin{bmatrix} s_1^{(j)} \\ o \end{bmatrix}\mu\right)$. The effect of this correction is to make $y_2^{(m)}$

more nearly orthogonal to $y_1^{(m)}$. Consequently it is simpler, and more effective, to **ignore the perturbations** and simply orthonormalize $Q_m s_2^{(m)}$ against normalized $y_1^{(m)}$ to get a new orthonormal pair when necessary. One possibility is to do this immediately as discussed at the beginning of this section; another possibility is to wait until the end.

A rigorous analysis of the selective orthogonalization process is beyond the scope of this book. Details concerning implementation of the method are given in [Parlett and Scott, 1979].

Exercises on 13-9

13-9-1. Prove, by induction or otherwise, that

$$|\gamma_{\nu\ell}| \equiv |y_\nu^* y_\ell^{(j)}| \leqslant \kappa_j + O(\kappa_j^2).$$

13-9-2. Assume that $y_\ell^{(j)*} r_j = 0$, use Ex. 13-9-1, and use $\rho(y_\ell; A) = \theta_\ell^{(j)} + \rho(y_\ell; A - \theta_\ell^{(j)})$ to prove lemma (13-9-1).

13-9-3. Prove Lemma (13-9-4).

13-9-4. Verify that the eigenvectors of $T_j + \xi_1 s_1 e_j^*$ are given by (13-9-9).

13-9-5. Define, for each computed Ritz pair (θ, y), the sequence $\{\tau_j\}$ by

$$\tau_{j+1} \equiv [\tau_j|\theta - \alpha_j| + \tau_{j-1}\beta_{j-1} + 3\varepsilon\|A\|]/\beta_j'.$$

Show that if $|y^* q_{j-1}| \leqslant \tau_{j-1}$ and $|y^* q_j| \leqslant \tau_j$ then $|y^* q_{j+1}| \leqslant \tau_{j+1}$. Assume that $Ay - y\theta = q\beta$ where $q = q_\ell$ for some $\ell < j - 1$ and $\beta < \sqrt{\varepsilon}\,\|A\|$.

13-9-6. Write the governing equation for selective orthogonalization as

$$AQ_j - Q_j(T_j + J_j) = r_j e_j^* + F_j$$

where J_j is strictly upper triangular and contains the appropriate multiples of eigenvectors s_ℓ used in computing threshold "Ritz vectors"; thus $J_j = \Sigma s_\ell \xi_\ell e_\ell^*$. Assume that $\kappa = \sqrt{\varepsilon}$, that $I - Q_j^* Q_j = C_j^* + C_j$ with $\|C_j\| \leqslant \kappa$, and that $\|J_j\| \leqslant \kappa\|A\|$. Imitate the proof of Paige's theorem, neglect all quantities which are $0(\varepsilon\|A\|)$, and deduce that, for $i = 1, \ldots, j$

$$\boxed{y_i^* q_{j+1}\beta_{ji} = \gamma_{ii}^{(j)} - s_i^* J_j s_i.}$$

Show that for threshold vectors $s_i^* J_i s_i$ is negligible.

*13-10 BAND (OR BLOCK) LANCZOS

Even when used with full reorthogonalization the basic Lanczos algorithm cannot detect the multiplicity of the eigenvalues which it computes. The reasons are given in Sec. 12-2. This limitation prompted the development of the block version of the Lanczos process which is capable of determining multiplicities up to the block size.

The idea is not to start with a single vector q_1 but with a set of mutually orthonormal vectors which we take as the columns of a starting n by ν **matrix** Q_1. Typical values for ν are 2, 3, and 4. The generalization of the algorithm of Sec. 13-1 to this new situation is straightforward and we shall describe it briefly. Associated with Q_1 is the big Krylov subspace

$$\hat{\mathcal{K}}^{\nu j}(Q_1) \equiv \text{span}(Q_1, AQ_1, \ldots, A^{j-1}Q_1).$$

(We assume, for simplicity, that $A^{j-1}Q_1$ has rank ν but, as we shall see, the failure of this assumption causes no difficulties. It merely complicates the description of the process.) The Rayleigh-Ritz procedure applied to $\hat{\mathcal{K}}^{\nu j}$ for $j = 1, 2, \ldots$ produces the distinguished orthonormal basis $\hat{Q}_j \equiv (Q_1, Q_2, \ldots, Q_j)$ and, in this basis, the projection of A is the block tridiagonal matrix

$$\hat{T}_j \equiv \begin{bmatrix} A_1 & B_1^* & & \\ B_1 & A_2 & \cdot & \\ & \cdot & \cdot & B_{j-1}^* \\ & & B_{j-1} & A_j \end{bmatrix}, \qquad A_i \text{ is } \nu \text{ by } \nu.$$

The B_i may be chosen to be upper triangular. By our assumption that $A^{j-1}Q_1$ has full rank it follows (Ex. 13-10-2) that each B_i is invertible.

13-10-1 BLOCK LANCZOS

R_0, n by ν, is given. For $j = 1, 2, \ldots$ repeat:

1. $R_{j-1} = Q_j B_{j-1}$, the QR factorization of R_{j-1} (Sec. 6-7).
2. $R_j \longleftarrow AQ_j - Q_{j-1}B_{j-1}^*$, $Q_0 = 0$.
3. $A_j \longleftarrow Q_j^* R_j$.
4. $R_j \longleftarrow R_j - Q_j A_j$.
5. Compute and test Ritz pairs. If satisfied then stop.

This algorithm is considerably more complicated than the simple Lanczos algorithm; in particular the computation of eigenpairs of $\hat{T}_j$ has a cost proportional to $(\nu^2)(\nu j)$. The reward is that $\hat{T}_j$ can have eigenvalues of multiplicity up to and including ν and thus can deliver approximations to multiple eigenvalues of A.

In practice, however, the Q_i's lose mutual orthogonality as soon as convergence sets in. The remedies described earlier are available: either full reorthogonalization or selective orthogonalization.

An attractive and successful alternative is to run the algorithm until the first Ritz pair converges and orthogonality is lost. Then start again with a new R_0 which is orthogonal to all known eigenvectors and iterate until all the wanted eigenvectors have been found. Much of the power of Lanczos is lost by restarting and the apt choice of a new R_0 which uses information from previous runs is not easy. Nevertheless the method has been useful.

At this point we go back and, following an idea in [Ruhe, 1979], reformulate block Lanczos in a way which puts it on the same footing as the simple Lanczos algorithm. The original A is directly reduced to **band** matrix form.

13-10-2 BAND LANCZOS

Pick orthonormal $q_1, \ldots, q_\nu$. Set $r = q_\nu$, $t_{\nu,0} = 1$, and for $j = 1, 2, \ldots,$ repeat:

1. $q_{j+\nu-1} \longleftarrow r/t_{j+\nu-1,j-1}$.
2. $r \longleftarrow Aq_j - \sum_{i=j-\nu}^{j-1} q_i t_{ij}$ $\quad (q_k = o \text{ if } k < 1)$.
3. For $i = j, \ldots, j+\nu-1$, $\begin{cases} t_{ij} \longleftarrow q_i^* r, \\ r \longleftarrow r - q_i t_{ij}. \end{cases}$
4. $t_{j+\nu,j} \longleftarrow \|r\|$. If $t_{j+\nu,j} = 0$ reduce ν.
5. Compute and test Ritz pairs. If satisfied then stop.

In exact arithmetic the band algorithm is identical to the block algorithm. In practice it has the virtue of requiring no special QR factorization subprograms.

At the end of step j the computed quantities satisfy

$$A\hat{Q}_j - \hat{Q}_j\hat{T}_j = R_j E_j^* + \hat{F}_j \tag{13-10-1}$$

where $E_j^* = (0, \ldots, 0, I_\nu)$ and $\hat{F}_j$ accounts for roundoff in executing step j. The Ritz vectors are of the form $y_i = \hat{Q}_j s_i$ where $\hat{T}_j s_i = s_i\theta_i$. The accuracy of a Ritz pair may be assessed in the usual way; multiplying (13-10-1) by s_i we find

$$\|Ay_i - y_i\theta_i\| \doteq \beta_{ji} \equiv \|B_j(E_j^* s_i)\|. \tag{13-10-2}$$

The band version is somewhat more complicated than the single vector algorithms. However if many of the wanted eigenvalues are known to have multiplicity ν then band Lanczos **with selective orthogonalization** (*SO*) will be more efficient than simple *SO* because all copies of a multiple eigenvalue will be found at the same step instead of one by one. On the other hand if all eigenvalues are simple then simple *SO* is preferable.

Exercises on Sec. 13-10

13-10-1. Do an operation count for one step of the band Lanczos algorithm. Assume that the band QL algorithm is used to compute 2ν eigenvalues of T (Sec. 8-16) at 2 QL transforms per eigenvalue. Each $\mathbf{s}_i$ requires $(\nu + 1)^2\nu j$ ops.

13-10-2. Assume that $t_{i+\nu,j} > 0, j = 1, \ldots, (j - 1)\nu$. Why does T have no eigenvalues of multiplicity greater than ν?

13-10-3. In exact arithmetic $Q_j^*Q_j = I_\nu$ and $A\hat{Q}_j - \hat{Q}_j\hat{T}_j = R_jE_j^*$. Prove directly from these relations that

$$\hat{Q}_{j-1}^*R_j = O$$

and,

$$\text{if } A_j = Q_j^*AQ_j \qquad \text{then } \hat{Q}_j^*R_j = O.$$

Notes and References

The paper [Lanczos, 1950] began it all. Various connections with other methods are given in [Householder, 1964] and [Wilkinson, 1965]. A deeper understanding of the simple Lanczos algorithm emerged from the pioneering unpublished Ph.D. thesis [Paige, 1971]. Some important facets of that work were published in [Paige, 1972 and 1976] but a proof of the theorem proved in Sec. 13-4 has not appeared in the open literature.

The use of the bottom element of an eigenvector of T_j in assessing the accuracy of a Ritz vector was shown by Paige and also in [Kahan and Parlett, 1976] but was not picked up by the community of users. Nevertheless engineers, chemists, and physicists have used the simple Lanczos process with success in large, difficult problems by incorporating a variety of checks on the progress of the computation. See [Whitehead, 1972], [Davidson, 1975], and [van Kats and van der Vorst, 1976]. The idea of using the algorithm iteratively, i.e., restarting periodically, goes back to some of the early attempts to use Lanczos on the computers available in the 1950s. The use of blocks is described in [Golub, 1973], [Cullum and Donath, 1974], and [Underwood, 1975]. Selective Orthogonalization was introduced in [Parlett and Scott, 1979] and Scott now advocates that *SO* be applied only at the wanted end of the spectrum. The Lanczos algorithm is now being adapted to compute the whole spectrum of large A; see [Cullum and Willoughby, 1979 and 1980].

14

Subspace Iteration

14-1 INTRODUCTION

Subspace iteration is a straightforward generalization of both the power method and inverse iteration which were presented in Chap. 4. Given $\mathbf{A}$ and a subspace $\mathbb{S}$ of $\mathcal{E}^n$ there is no difficulty in **defining** a new subspace

$$\mathbf{A}\mathbb{S} \equiv \{\mathbf{As}: \mathbf{s} \in \mathbb{S}\}.$$

Repetition of the idea produces the Krylov sequence of subspaces

$$\mathcal{K}(\mathbb{S}) = \{\mathbb{S}, \mathbf{A}\mathbb{S}, \mathbf{A}^2\mathbb{S}, \ldots\}$$

and the questions are: How can the sequence be represented and is it worthwhile?

Before taking up these problems we must dispose of a very reasonable objection. In practice $\mathbb{S}$ is given indirectly via an orthonormal basis $\mathbf{S} = (\mathbf{x}_1, \mathbf{x}_2, \mathbf{x}_3)$ say. Then $\mathbf{A}^k\mathbb{S}$ is spanned by $\mathbf{A}^k\mathbf{S} = (\mathbf{A}^k\mathbf{x}_1, \mathbf{A}^k\mathbf{x}_2, \mathbf{A}^k\mathbf{x}_3)$. Even if these columns are normalized they are simply the k-th terms in three separate power sequences each of which will converge (slowly) to the dominant eigenvector $\mathbf{z}_n$, where $\mathbf{Az}_n = \mathbf{z}_n\lambda_n$ and $\lambda_n = \|\mathbf{A}\|$. How can three slowly convergent sequences be preferable to one?

The answer is that $\mathbf{A}^k\mathbf{S}$ is a bad basis for a good subspace $\mathbf{A}^k\mathbb{S}$. In Sec. 14-4 we shall see that $\mathbf{A}^k\mathbb{S}^p$ does converge to span $(\mathbf{z}_n, \mathbf{z}_{n-1}, \ldots, \mathbf{z}_{n-p+1})$, the dominant invariant subspace of dimension p, and, for large enough k, one application of the Rayleigh-Ritz procedure (Chap. 11) produces good approximations to the individual eigenvectors.

If it is feasible to solve linear systems such as $(\mathsf{A} - \sigma)\mathsf{x} = \mathsf{b}$, either by factoring $\mathsf{A} - \sigma$ or by iteration, then subspace iteration can be effected with $(\mathsf{A} - \sigma)^{-1}$ to obtain the p eigenvalues closest to σ together with their eigenvectors. This is the most common way in which the technique is used.

The main point is that the reward for working with several columns at once, and orthonormalizing them from time to time, is an improved factor in the linear convergence of successive subspaces. When several clustered eigenvalues are wanted the improvement is dramatic and more than makes up for the extra work incurred by working with a bigger subspace than is really wanted. In other configurations the method can be very slow. The difficulty is that the distribution of eigenvalues is usually unknown. Consequently it is not clear how large p, the dimension of $\mathcal{S}$, should be taken. Moreover the efficiency of the method often depends strongly on the value of p.

Sophisticated versions of subspace iteration were developed during the 1960s and 1970s when the Lanczos algorithm was under a cloud. Now that easy-to-use, reliable Lanczos programs are available it is pertinent to ask whether subspace iteration should be laid to rest. The answer is no because there are a few circumstances in which its use is warranted.

1. If no secondary storage is available and the fast store can hold only a few n-vectors at a time then there seems to be no choice but to discard previous vectors in the Krylov sequence and employ subspace iteration.
2. If the relative gap between the wanted eigenvalues and the others is enormous, as in inverse iteration with a good shift, then only one power step, or a few, will be needed for convergence. The situation is so good that the advantages of the Lanczos method are not needed. However Lanczos also works very well in these favorable circumstances.

The finest example of how far subspace iteration can be taken toward the goal of **automatic** computation is Rutishauser's program **ritzit** which is presented in Contribution II/9 of the Handbook. The program is complicated enough to be efficient in a wide variety of applications, but it is intelligible and every effort was made to keep down the number of decisions thrust on the user. The next three sections describe some of that work.

14-2 IMPLEMENTATIONS

The exposition is simplified by supposing that A is positive definite and that its dominant p eigenpairs (α_i, z_i), $i = n - p + 1, \ldots, n$ are to be found. The matrix A is not modified and need not be known explicitly because its role in the program is that of an operator receiving vectors u and returning vectors Au. Any special features of A which permit economies in storage or arithmetic operations should be exploited by the user in coding the matrix-times-vector subprogram which Rutishauser calls OP (for operator).

The ν-th step in each implementation transforms an orthonormal basis $\mathsf{S}_{\nu-1}$ of $\mathsf{A}^{\nu-1}\mathcal{S}$ into an orthonormal basis S_ν of $\mathsf{A}^\nu\mathcal{S}$. In addition there must be a test for convergence. (The rectangular matrix S_ν bears no relation to T_j's eigenvector matrix S_j of Chap. 13.)

14-2-1 IMPLEMENTATION NO. 1: SIMPLE SUBSPACE ITERATION

Table 14-2-1

	Action	*Cost*
(a)	Compute $\mathsf{C}_\nu = \mathsf{AS}_{\nu-1}$.	p calls on OP.
(b)	Test each column for convergence.	pn ops.
(c)	Orthonormalize $\mathsf{C}_\nu = \mathsf{Q}_\nu\mathsf{R}_\nu$ (by modified Gram-Schmidt, Sec. 6-7).	$p(p+1)n$ ops.
(d)	Set $\mathsf{S}_\nu = \mathsf{Q}_\nu$.	0.

Remarks on Table 14-2-1. The columns of S_ν are **not** optimal approximations to the target eigenvectors from span S_ν. Even if $\mathcal{S} = \text{span}(\mathsf{z}_n, \mathsf{z}_{n-1})$ the columns of S_ν will converge only linearly to z_n and z_{n-1} as $\nu \longrightarrow \infty$, despite the fact that they are already in $\mathcal{S} = \text{span}\ \mathsf{S}_0$ (Ex. 14-2-1). This defect suggests that the RR procedure (Sec. 11-3) should be applied frequently to S_ν.

14-2-2 IMPLEMENTATION NO. 2: SUBSPACE ITERATION + RAYLEIGH-RITZ

Table 14-2-2

	Action	*Cost*
(a)	Compute $\mathsf{C}_\nu = \mathsf{AS}_{\nu-1}$.	p calls on OP, C overwrites S.
(b)	Orthonormalize $\mathsf{C}_\nu = \mathsf{Q}_\nu\mathsf{R}_\nu$ by modified Gram-Schmidt.	$p(p+1)n$ ops, Q overwrites C.
(c)	Form $\hat{\mathsf{H}}_\nu = \mathsf{Q}_\nu^*(\mathsf{AQ}_\nu)$.	p calls on OP, $\frac{1}{2}p(p+1)n$ ops for $\hat{\mathsf{H}}$
(d)	Factor $\hat{\mathsf{H}}_\nu = \mathsf{G}_\nu\Theta_\nu\mathsf{G}_\nu^*$,	κp^3 ($\kappa \doteqdot 5$) (κ depends on the spectral decomposition method).
(e)	Form $\mathsf{S}_\nu = \mathsf{Q}_\nu\mathsf{G}_\nu$, the Ritz vectors for $\mathsf{A}^\nu\mathcal{S}$.	kp^2n ops, S overwrites Q.

1. Compute $\quad C_\nu = A\, S_{\nu-1}$

2. Orthonormalize $\quad C_\nu = Q_\nu\, R_\nu$

3. Form Rayleigh quotient $\quad \hat{H}_\nu = Q_\nu^*\, AQ_\nu$

4. Factor $\quad \hat{H}_\nu = G_\nu\, \theta_\nu\, G_\nu^*$ — Spectral decomposition

5. Form $\quad S_\nu = Q_\nu\, G_\nu$

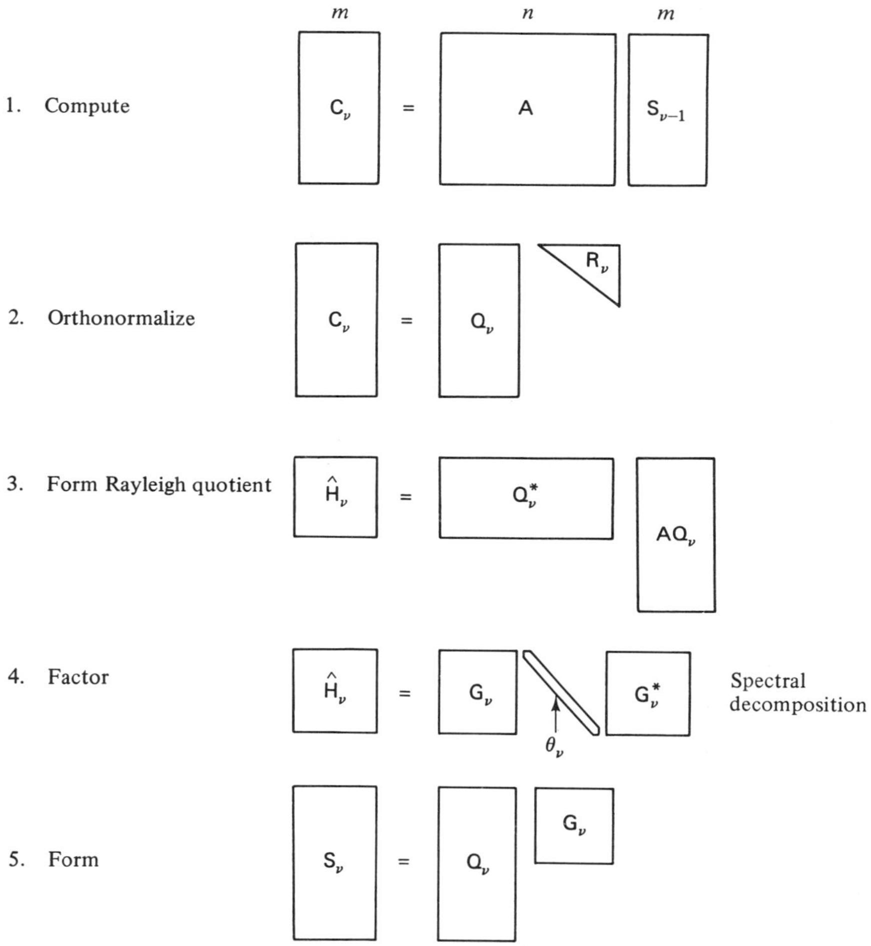

Figure 14-2-1 One step of Implementation No. 2

Fig. 14-2-1 gives a picture of one step of Implementation No. 2.

Remarks on Table 14-2-2. If G_ν is found as a product of orthonormal matrices, say $G_\nu = P_1 \cdots P_k$ then (e) can be carried out at the same time as (d) by the following algorithm: $S_\nu = Q_\nu$, then for $i = 1, \ldots, k$, $S_\nu \leftarrow S_\nu P_i$. There is no need for G_ν to be formed explicitly and a single n by p array suffices for S, C, and Q.

S_ν is the best basis in $A^\nu \mathbb{S}$ and its columns converge to the z's. However the price is high; in particular the p extra calls to subroutine OP in (c) help to double the cost of each step as compared with Implementation No. 1. Fortunately there is a clever way to avoid these extra calls based on the fact that $\phi(\mathsf{A})$ has the same eigenvectors as A for any analytic function ϕ that does not coalesce distinct eigenvalues.

For the moment leave the function ϕ unspecified and seek the Rayleigh-Ritz approximations to $\phi(\mathsf{A})$ from $\mathsf{A}^{\nu}\mathcal{S}$. As in the previous implementations let

$$\mathsf{C}_{\nu} = \mathsf{AS}_{\nu-1}. \qquad (14\text{-}2\text{-}1)$$

We seek a new basis, say $\mathsf{C}_{\nu}\mathsf{F}_{\nu}$ with the p by p matrix F_{ν} satisfying two conditions, namely

$$(\mathsf{C}_{\nu}\mathsf{F}_{\nu})^*(\mathsf{C}_{\nu}\mathsf{F}_{\nu}) = \mathsf{I}_p \qquad \text{(orthonormality)}, \qquad (14\text{-}2\text{-}2)$$

and

$$(\mathsf{C}_{\nu}\mathsf{F}_{\nu})^*\phi(\mathsf{A})(\mathsf{C}_{\nu}\mathsf{F}_{\nu}) \equiv \Delta_{\nu}^{-2} = \text{diagonal} \qquad \text{(giving Ritz vectors)}. \qquad (14\text{-}2\text{-}3)$$

The right choice is $\phi(\zeta) = \zeta^{-2}$ so that (14-2-3) collapses into

$$(\mathsf{F}_{\nu}^*\mathsf{S}_{\nu-1}^*\mathsf{A})\mathsf{A}^{-2}(\mathsf{AS}_{\nu-1}\mathsf{F}_{\nu}) = \mathsf{F}_{\nu}^*\mathsf{F}_{\nu} = \Delta_{\nu}^{-2}. \qquad (14\text{-}2\text{-}4)$$

Thus $\mathsf{F}_{\nu}\Delta_{\nu}$ will be orthonormal and (14-2-2) becomes

$$\begin{aligned}\mathsf{C}_{\nu}^*\mathsf{C}_{\nu} &= (\mathsf{F}_{\nu}^{-*}\Delta_{\nu}^{-1})\Delta_{\nu}^2(\Delta_{\nu}^{-1}\mathsf{F}_{\nu}^{-1}),\\ &= (\mathsf{F}_{\nu}\Delta_{\nu})\Delta_{\nu}^2(\mathsf{F}_{\nu}\Delta_{\nu})^*, \qquad \text{by (14-2-4)}. \qquad (14\text{-}2\text{-}5)\end{aligned}$$

Thus F_{ν} and Δ_{ν} are determined by the spectral decomposition of $\mathsf{C}_{\nu}^*\mathsf{C}_{\nu}$.

14-2-3 IMPLEMENTATION NO. 3: NO. 2 APPLIED TO A^{-2}

Table 14-2-3

Action	*Cost*
(a) Compute $\mathsf{C}_{\nu} = \mathsf{AS}_{\nu-1}$.	p calls on OP, C overwrites S.
(b) Compute $\mathring{\mathsf{H}}_{\nu} = \mathsf{C}_{\nu}^*\mathsf{C}_{\nu}$.	$\frac{1}{2}p(p+1)n$ ops.
(c) Factor $\mathring{\mathsf{H}}_{\nu} = \mathsf{B}_{\nu}\Delta_{\nu}^2\mathsf{B}_{\nu}^*$, the spectral decomposition.	κp^3 ops (κ depends on the method).
(d) Form $\mathsf{S}_{\nu} = \mathsf{C}_{\nu}\mathsf{B}_{\nu}\Delta_{\nu}^{-1}$ $(= \mathsf{C}_{\nu}\mathsf{F}_{\nu})$.	kp^2n ops, S overwrites C.

Remarks on Table 14-2-3. $\mathring{\mathsf{H}}_{\nu}$ is the projection of A^2 onto span $\mathsf{S}_{\nu-1}$ (Ex. 14-2-2). The new S_{ν} will differ from the S_{ν} of the previous implementation. Moreover it is not expressed as a product of orthonormal matrices. In 1971 Reinsch found a clever way to remove this blemish as follows. Let the QR factorization of C_{ν} be $\mathsf{Q}_{\nu}\mathsf{R}_{\nu}$ as in No. 2. Then, from (b) in No. 3,

$$\mathring{\mathsf{H}}_{\nu} = \mathsf{C}_{\nu}^*\mathsf{C}_{\nu} = \mathsf{R}_{\nu}^*\mathsf{Q}_{\nu}^*\mathsf{Q}_{\nu}\mathsf{R}_{\nu} = \mathsf{R}_{\nu}^*\mathsf{R}_{\nu}.$$

The LR transformation was the forerunner of the QR transformation and when applied to $\mathring{H}_\nu$ it produces a more nearly diagonal matrix H_ν defined by reversing the factors of $\mathring{H}_\nu$,

$$H_\nu \equiv R_\nu R_\nu^* = R_\nu \mathring{H}_\nu R_\nu^{-1},$$
$$= R_\nu B_\nu \Delta_\nu^2 B_\nu^* R_\nu^{-1}, \qquad \text{by (c) of No. 3,}$$
$$= (R_\nu B_\nu \Delta_\nu^{-1})\Delta_\nu^2(\Delta_\nu B_\nu^* R_\nu^{-1}).$$

It may be verified (Ex. 14-2-3) that $P_\nu \equiv R_\nu B_\nu \Delta_\nu^{-1}$ is the orthonormal matrix of eigenvectors of H_ν. So, from (d) in No. 3,

$$S_\nu = C_\nu B_\nu \Delta_\nu^{-1} = Q_\nu R_\nu B_\nu \Delta_\nu^{-1} = Q_\nu P_\nu.$$

*14-2-4 IMPLEMENTATION NO. 4 [IN **RITZIT**]*

Table 14-2-4

Action	*Cost*
(a) Compute $C_\nu = AS_{\nu-1}$.	p calls on OP, C over S.
(b) Factor $C_\nu = Q_\nu R_\nu$.	$p(p+1)n$ ops, Q over C.
(c) Form $H_\nu = R_\nu R_\nu^*$.	$\frac{1}{3}p^3$ ops.
(d) Factor $H_\nu = P_\nu \Delta_\nu^2 P_\nu^*$.	κp^3 ops ($\kappa \doteq 5$).
(e) Form $S_\nu = Q_\nu P_\nu$.	kp^2n ops, S over Q.

Note: Steps (d) and (e) should be done together.

Exercises on Sec. 14-2

14-2-1. Take $S_0 = (s_1, s_2)$ with $s_i = (z_n \pm z_{n-1})/\sqrt{2}$ and verify that $S_\nu e_1$ converges linearly to z_n. What is the convergence factor?

14-2-2. Show that $\mathring{H}_\nu$ is the projection of A^2 onto span $S_{\nu-1}$. See Secs. 1-4 and 11-4 for definitions and discussion of projections.

14-2-3. Show that $P_\nu \equiv R_\nu B_\nu \Delta_\nu^{-1}$ is actually orthogonal.

14-2-4. Since $\phi(A)$'s eigenvectors are the same as A's why are the Ritz vectors from S_ν for $\phi(A)$ not the same as the Ritz vectors from S_ν for A? *Hint*: See Sec. 11-4.

14-2-5. Make a table showing the total op count and storage requirement for each of the four implementations.

14-3 IMPROVEMENTS

14-3-1 CHEBYSHEV ACCELERATION

The technique described in this section is used in many branches of numerical analysis and so warrants more than a brief mention.

Even with Implementation No. 4 the Rayleigh-Ritz procedure is expensive and so it is tempting to take a few steps of the basic power method between each invocation of RR. To be specific suppose that the algorithm computes, in turn, $S_{\nu+j} = AS_{\nu+j-1}, j = 1, \ldots, m$. However it is no more trouble to incorporate shifts σ_j and compute instead $S_{\nu+j} = (A - \sigma_j)S_{\nu+j-1}, j = 1, \ldots, m$. Thus the algorithm could compute

$$S_{\nu+m} = \phi(A)S_\nu$$

for **any** monic polynomial ϕ of the degree m. The problem is to select a helpful ϕ.

After the first application of the RR procedure the program possesses approximations $\theta_{-1}, \ldots, \theta_{-p}$ to the dominant eigenvalues. The θ_{-i} are the eigenvalues of H_ν or $\mathring{H}_\nu$ or $\hat{H}_\nu$. A little reflection shows that a useful, but by no means optimal choice is a ϕ which is as small as possible on the interval $[\theta_1, \theta_{-p}]$. Since θ_1 is unknown it is customary to use $[0, \theta_{-p}]$ if A is positive semi-definite and $[-\theta_{-p}, \theta_{-p}]$ otherwise. This problem in approximation theory is solved by the Chebyshev polynomials adapted to the appropriate interval. These polynomials are described briefly in Appendix B. The beautiful fact is that there is no need to know and then use the zeros of the appropriate Chebyshev polynomial $T_m(\xi)$ as shifts σ_j because a simple three-term recurrence with constant coefficients permits the calculation of $T_j(A)$ from $T_{j-1}(A)$ and $T_{j-2}(A)$. Specifically if T_m is adapted to the interval $[-e, e]$ then

$$\left\{\begin{aligned} T_{\nu+j}(\xi) &= \frac{2\xi}{e} T_{\nu+j-1}(\xi) - T_{\nu+j-2}(\xi), \\ S_{\nu+j} &= \frac{2}{e} AS_{\nu+j-1} - S_{\nu+j-2}, \end{aligned}\right\} \qquad j = 2, \ldots, m.$$

Moreover there is no need to save $S_{\nu+j-2}$ provided that each column is updated from S_ν all the way to $S_{\nu+m}$ in turn. This means that only two extra n-vectors of storage are needed to effect this acceleration of the convergence of S_ν.

How should m be chosen? An important advantage of the recurrence is that it is independent of m and so m can be freely changed during the computation. In Sec. 14-4 it will be shown that the convergence factor for the dominant Ritz vector is $\theta_{-p-1}/\theta_{-1}$. If this ratio is small, like 0.1, then there is little need for acceleration and m can be held at 1. However when the ratio is close to 1, like 0.98, then there is much to be gained from a large m. The only constraint is that the columns of $S_{\nu+m}$ must be kept fully independent. In his program **ritzit** Rutishauser requires

$$\|S_{\nu+m}\| \doteqdot T_m(\theta_{-1}/\theta_{-p}) \leqslant \cosh 8 < 1500.$$

The program which embodies this powerful device is remarkably simple and even elegant. However to keep a proper perspective it must be recalled that the Lanczos algorithm, when run for m steps, gives even more than Chebyshev acceleration using the unknown optimal interval $[\alpha_1, \alpha_{-p-1}]$. This observation is based on the theory of Sec. 12-4.

14-3-2 RANDOMIZATION

This is a device to protect the algorithm from an unhappy choice of an initial subspace $\mathcal{S}$ which is effectively orthogonal to one of the desired eigenvectors.

After each orthogonalization of the basis vectors the one with the innermost Rayleigh quotient θ_{-p} is replaced by a random vector orthogonal to the rest of the basis. This is a simple way of making it **most** unlikely that any wanted eigenvectors will be missed.

It is advisable to wait until the Rayleigh quotients have settled down and Rutishauser waits until 3 Rayleigh-Ritz approximations have been made. Here is an example of one of the rare ad hoc parameters that occur in the program. The alternative of monitoring the Rayleigh quotients for stabilization of θ_{-m} seems to be more complicated than is warranted.

Note that this device impinges on Chebyshev acceleration because the quantity θ_{-p}, the p-th Rayleigh quotient, determines the interval for the Chebyshev polynomial. Some care is needed in the precise designation of the ends of the interval to keep them free from spurious current values of θ_{-p}. The reader is referred to Contribution II/9 in the Handbook for the details.

14-3-3 TERMINATION CRITERIA

The Rayleigh-Ritz approximations θ_i should move monotonically to their limits α_i ($i = \pm 1, \pm 2, \ldots$), in the absence of roundoff, as subspace iteration continues. In **ritzit** Rutishauser accepts a θ_i as soon as it stagnates (thus achieving accuracy close to working precision). No test is made on the Ritz vectors y_i until θ_i has been accepted.

Recall from Sec. 11-7 that

$$|\sin \angle(\mathsf{y}_i, \mathsf{z}_i)| \leq \|\mathsf{r}_i\|/\text{gap}$$

where $\mathsf{r}_i = (\mathsf{A} - \theta_i)\mathsf{y}_i$ is known but the gap, namely $\min\{|\alpha_{i-1} - \alpha_i|, |\alpha_i - \alpha_{i+1}|\}$ is not. Rutishauser uses the θ's to approximate the gap and thus obtains a computable estimate of the error angle which can be tested for the required accuracy.

Those experienced in numerical computation know that preset tolerances can occasionally fail to be met. Hence Rutishauser builds into the simple error measure ($\|\mathsf{r}_i\|/\text{gap}$), a slow but sure decay which ensures that the program will terminate even in those cases in which the desired accuracy is greater than the program can achieve. However, Rutishauser did not make use of the residual norm for clustered eigenvalues (see Sec. 11-5) nor of the refined error bounds in Chap. 10.

*14-4 CONVERGENCE

Let $\mathsf{Z} \equiv (\mathsf{z}_1, \ldots, \mathsf{z}_m)$, where $\mathsf{A}\mathsf{z}_i = \mathsf{z}_i\alpha_i$, $i = 1, \ldots, m$ be the matrix of wanted eigenvectors. To be specific we assume that

$$0 < \alpha_1 \leqslant \alpha_2 \leqslant \cdots \leqslant \alpha_m < \alpha_{m+1} \leqslant \cdots, \tag{14-4-1}$$

so that $\mathcal{Z} \equiv \mathcal{Z}^m \equiv \text{span } \mathsf{Z}$ is the dominant invariant subspace under A^{-1}. Let $\mathcal{S}$ be any m-dimensional subspace of $\mathcal{E}^n$ and $\{\mathsf{A}^{-k}\mathcal{S} : k = 0, 1, 2, \ldots\}$ the associated sequence generated by subspace iteration. The quantity $\angle(\mathcal{Z}, \mathsf{A}^{-k}\mathcal{S})$, defined in Sec. 11-7, is too crude a measure of how well $\mathsf{A}^{-k}\mathcal{S}$ approximates $\mathcal{Z}$ because the user wants to know how well specific vectors, such as z_1, can be approximated from $\mathsf{A}^{-k}\mathcal{S}$. So the natural objects to study are

$$\psi_j^{(k)} \equiv \angle\left(\mathsf{z}_j, \mathsf{A}^{-k}\mathcal{S}\right) \equiv \min \angle(\mathsf{z}_i, \mathsf{x}) \qquad \text{over } \mathsf{x} \in \mathsf{A}^{-k}\mathcal{S}.$$

We want to see how fast $\psi_i^{(k)} \longrightarrow 0$ and, at the same time, make the proof as similar to the one-dimensional case as possible. The reader might refer to Sec. 4-2.

In order to have convergence of $\{\mathsf{A}^{-k}\mathcal{S}\}$ to $\mathcal{Z}$, and not some other invariant subspace, as $k \longrightarrow \infty$ it is necessary to assume that $\angle(\mathcal{Z}, \mathcal{S}) < \pi/2$, or, equivalently, that for any orthonormal basis S of $\mathcal{S}$

$$\mathsf{Z}^*\mathsf{S} \text{ is invertible.} \tag{14-4-2}$$

A useful notion in the analysis is the (matrix) angle Ψ between $\mathcal{Z}$ and $\mathcal{S}$ defined by

$$\Psi = \cos^{-1}(\mathsf{Z}^*\mathsf{S}\mathsf{S}^*\mathsf{Z})^{1/2}. \tag{14-4-3}$$

Functions of matrices can be defined in various ways but in our case Ψ is only needed to give meaning to matrices such as $\sin \Psi$ and $\tan \Psi$ which are messy when expressed in terms of S and Z. Assumption (14-4-2) ensures that Ψ is well defined. It is not diagonal.

In analogy with the one-dimensional case there exists an orthonormal basis S such that $\mathsf{Z}^*\mathsf{S} = \mathsf{S}^*\mathsf{Z} = \cos \Psi$ (Ex. 14-4-1). This basis S can be expressed in terms of Z as follows:

$$\mathsf{S} = \mathsf{Z}\,\boxed{\cos \Psi} + \mathsf{J}\,\boxed{\sin \Psi} \tag{14-4-4}$$

where (Ex. 14-4-2)

$$\mathsf{J}^*\mathsf{J} = \mathsf{I}_m, \qquad \mathsf{Z}^*\mathsf{J} = \mathsf{O}. \tag{14-4-5}$$

> THEOREM Under assumptions (14-4-1) and (14-4-2) on $\mathcal{S}$ and $\mathcal{Z}$ each eigenvector z_i, $i \leqslant m$, satisfies
>
> $$\tan \angle(\mathsf{z}_i, \mathsf{A}^{-k}\mathcal{S}) \leqslant (\alpha_i/\alpha_{m+1})^k \tan \angle(\mathcal{Z}, \mathcal{S}). \tag{14-4-6}$$

Proof. Pre-multiply (14-4-4) by A^{-k} and post-multiply by $(\sec \Psi)\Lambda^k$ to find

$$\begin{aligned}\mathsf{A}^{-k}\mathsf{S}(\sec \Psi)\Lambda^k &= \mathsf{A}^{-k}\mathsf{Z}\Lambda^k + \mathsf{A}^{-k}\mathsf{J}(\tan \Psi)\Lambda^k,\\ &= (\mathsf{Z}\Lambda^{-k})\Lambda^k + \mathsf{A}^{-k}\mathsf{J}(\tan \Psi)\Lambda^k, \end{aligned} \tag{14-4-7}$$

where $\Lambda = \operatorname{diag}(\alpha_1, \ldots, \alpha_m)$. The key fact is that Z is orthogonal to $\mathsf{A}^{-k}\mathsf{J}$ (Ex. 14-4-3) and it is convenient to rewrite $\mathsf{A}^{-k}\mathsf{J}$ as

$$\mathsf{A}^{-k}\mathsf{J} = \mathsf{J}_k\Omega_k, \qquad \Omega_k = (\mathsf{J}^*\mathsf{A}^{-2k}\mathsf{J})^{1/2}, \tag{14-4-8}$$

so that J_k is orthonormal. To bound Ω_k note that

$$\|\Omega_k\|^2 = \max_{\|v\|=1} \mathsf{v}^*\mathsf{J}^*\mathsf{A}^{-2k}\mathsf{J}\mathsf{v} \leqslant \alpha_{m+1}^{-2k}, \tag{14-4-9}$$

since $\mathsf{Z}^*\mathsf{J} = \mathsf{O}$. Now consider the j-th columns in (14-4-7) to find

$$\begin{aligned}\mathsf{x}_j^{(k)} &\equiv \mathsf{A}^{-k}\mathsf{S}(\sec \Psi)\mathsf{e}_j\lambda_j^k = \mathsf{z}_j + \mathsf{u}_j, \qquad \mathsf{u}_j = \mathsf{J}_k\Omega_k(\tan \Psi)\mathsf{e}_j\alpha_j^k,\\
\tan \angle(\mathsf{z}_j, \mathsf{A}^{-k}\mathcal{S}) &\leqslant \tan \angle(\mathsf{z}_j, \mathsf{x}_j^{(k)}) = \|\mathsf{u}_j\|/1, \qquad (14\text{-}4\text{-}10)\\
&= \|\Omega_k(\tan \Psi)\mathsf{e}_j\|\alpha_j^k, \qquad \text{using (14-4-8)},\\
&\leqslant \alpha_j^k\|\Omega_k\|\|(\tan \Psi)\mathsf{e}_j\|,\\
&\leqslant (\alpha_j/\alpha_{m+1})^k \tan \angle(\mathsf{z}_j, \mathsf{S}(\sec \Psi)\mathsf{e}_j),\\
&\qquad\qquad \text{by (14-4-9) and Ex. 14-4-4},\\
&\leqslant (\alpha_j/\alpha_{m+1})^k \tan \angle(\mathcal{Z}, \mathcal{S}). \qquad \square\end{aligned}$$

One way to appreciate this result is to contrast the computation of z_1 by subspace iteration and inverse iteration (Ex. 14-4-5).

Theorem (14-4-6) shows that a certain sequence $\{\mathsf{x}_j^{(k)}: \mathsf{x}_j^{(k)} = \mathsf{A}^{-k}\mathsf{S}(\sec \Psi)\mathsf{e}_j\} \longrightarrow \mathsf{z}_j$ as $k \longrightarrow \infty$, but it does not address the behavior of the sequence $\{\mathsf{y}_j^{(k)}\}$ actually computed by say Implementation No. 4. Some authors invoke the "optimality" of Ritz vectors to conclude that $\{\mathsf{y}_j^{(k)}\}$ must converge at least as quickly as does $\{\mathsf{x}_j^{(k)}\}$. Such an argument is mistaken because $\mathsf{y}_j^{(k)}$ is **not** the closest unit vector in $\mathsf{A}^{-k}\mathcal{S}$ to z_j. Thus the x's might converge quicker than the y's.

Rutishauser shows that $\mathbf{y}_j^{(k)} \longrightarrow \mathbf{x}_j^{(k)}$ at the same asymptotic rate as $\mathbf{x}_j^{(k)} \longrightarrow \mathbf{z}_j$ and consequently the same improved factor (α_j/α_{m+1}) governs the linear convergence of $\{\mathbf{y}_j^{(k)}\}$ to $\mathbf{z}_j$. The difficulty lies in the complicated nature of Ritz vectors and a brief digression is needed before establishing these claims. If $\mathbf{B}$ is a nonorthonormal basis for span $\mathbf{B}$ then each Ritz approximation for $\mathbf{A}^2$ from span $\mathbf{B}$ is of the form $\mathbf{Bt}$ where $\mathbf{t}$ satisfies

$$(\mathbf{B}^*\mathbf{A}^2\mathbf{B} - \mu^2\mathbf{B}^*\mathbf{B})\mathbf{t} = \mathbf{o}, \qquad (14\text{-}4\text{-}11)$$

and μ is the associated Ritz value (Ex. 14-4-6).

THEOREM When subspace iteration uses Implementation No. 4 then each Ritz vector $\mathbf{y}_i^{(k)}$ is related to the vector $\mathbf{x}_i^{(k)}$ of Theorem (14-4-6) as $k \longrightarrow \infty$, by

$$\sin \angle(\mathbf{y}_i^{(k)}, \mathbf{x}_i^{(k)}) = O[(\alpha_i/\alpha_{m+1})^k], \qquad i = 1, \ldots, m. \qquad (14\text{-}4\text{-}12)$$

Proof. In the basis $\mathbf{B} \equiv \mathbf{A}^{-k}\mathbf{S}(\sec \Psi)\Lambda^k$, given in (14-4-7), $\mathbf{x}_i^{(k)}$ is represented by $\mathbf{e}_i$ and $\mathbf{y}_i^{(k)}$ is represented by a solution $\mathbf{t}_i$ of (14-4-11). On substituting the right side of (14-4-7) for $\mathbf{B}$ formula (14-4-11) becomes

$$\left[\Lambda^2 - \mu^2 + (\alpha_{m+1}^{-1}\Lambda)^k\mathbf{H}_k(\alpha_{m+1}^{-1}\Lambda)^k\right]\mathbf{t} = \mathbf{o} \qquad (14\text{-}4\text{-}13)$$

where, by Ex. 14-4-7,

$$\|\mathbf{H}_k\| \leq (\alpha_{m+1}\|\tan \Psi\|)^2. \qquad (14\text{-}4\text{-}14)$$

The perturbation to $\Lambda^2 - \mu^2$ in (14-4-13) vanishes as $k \longrightarrow \infty$ and, for large enough k, there is a μ_i close to α_i and a $\mathbf{t}_i$ close to $\mathbf{e}_i$. Next consider k so large that

$$|\mu_i - \alpha_i| \leq \delta \equiv \frac{1}{2}\min|\alpha_i - \alpha_j| \qquad \text{over } \alpha_j \neq \alpha_i.$$

For such k $(\alpha_i^2, \mathbf{e}_i)$ is a good approximate eigenpair to $(\mu_i^2, \mathbf{t}_i)$ and the residual vector $\mathbf{r}_i$ for (14-4-13) satisfies

$$\begin{aligned}\|\mathbf{r}_i\| &= \left\|\left[\Lambda^2 + (\alpha_{m+1}^{-1}\Lambda)^k\mathbf{H}_k(\alpha_{m+1}^{-1}\Lambda)^k - \alpha_i^2\right]\mathbf{e}_i\right\|,\\ &= \|(\alpha_{m+1}^{-1}\Lambda)^k\mathbf{H}_k\mathbf{e}_i\|(\alpha_i/\alpha_{m+1})^k,\\ &\leq (\alpha_m/\alpha_{m+1})^k(\alpha_i/\alpha_{m+1})^k(\alpha_{m+1}\|\tan \Psi\|)^2, \qquad (14\text{-}4\text{-}15)\end{aligned}$$

The last inequality uses (14-4-14).

By Theorem 11-7-1 (the gap results)

$$\sin \angle(\mathbf{t}_i, \mathbf{e}_i) \leq \|\mathbf{r}_i\|/\delta. \qquad (14\text{-}4\text{-}16)$$

For large k the basis $\mathbf{B}$ is almost orthonormal,

$$|\mathbf{b}_i^*\mathbf{b}_j| \leqslant \left(\frac{\alpha_i}{\alpha_{m+1}}\right)^k\left(\frac{\alpha_j}{\alpha_{m+1}}\right)^k \|\tan \Psi\|^2, \qquad i \neq j. \qquad (14\text{-}4\text{-}17)$$

When the effects of (14-4-16) and (14-4-17) are combined (Ex. 14-4-8) it turns out that (14-4-17) dominates in the limit and

$$\sin \angle(\mathbf{y}_i^{(k)}, \mathbf{x}_i^{(k)}) \leqslant \left(\frac{\alpha_i}{\alpha_{m+1}}\right)^k \|\tan \Psi\| \qquad \text{as } k \longrightarrow \infty, \qquad (14\text{-}4\text{-}18)$$

which completes the proof. □

Exercises on Sec. 14-4

14-4-1. Start with any orthonormal basis $\mathbf{S}$ of $\mathcal{S}$ and then find m by m orthogonal $\mathbf{G}$ such that $\hat{\mathbf{S}} = \mathbf{SG}$ satisfies $\mathbf{Z}^*\hat{\mathbf{S}} = \hat{\mathbf{S}}^*\mathbf{Z} = \cos \Psi$, a positive definite matrix.

14-4-2. Verify that $\mathbf{J} = \hat{\mathbf{S}} \operatorname{cosec} \Psi - \mathbf{Z} \cot \Psi$.

14-4-3. Show that $\mathbf{Z} \perp \mathbf{A}^{-k}\hat{\mathbf{S}}$.

14-4-4. Show that $\tan \angle(\mathbf{z}_i, \mathbf{S} \sec \Psi \mathbf{e}_j) = \|(\tan \Psi)\mathbf{e}_j\|$.

14-4-5. Find expressions for the number of steps required to reduce the error angle by a factor of 1000 by subspace iteration. Divide your answer by m and compare with the result for inverse iteration. Is this comparison fair?

14-4-6. An orthonormal basis for span $\mathbf{B}$ is $\mathbf{B}(\mathbf{B}^*\mathbf{B})^{-1/2}$. Use this to establish (14-4-11).

14-4-7. Show that $\mathbf{H}_k = \tan \Psi(\Omega_{k-1}^2 - \mu^2\Omega_k^2) \tan \Psi \alpha_{m+1}^{2k}$ and then confirm (14-4-14).

14-4-8. Write

$$\mathbf{t}_i = \mathbf{e}_i + \eta_i\mathbf{f}_i, \qquad \|\mathbf{f}_i\| = 1, \qquad |\eta_i| = O\left[\left(\frac{\alpha_i}{\alpha_{m+1}}\frac{\alpha_m}{\alpha_{m+1}}\right)^k\right].$$

Also

$$\cos \angle(\mathbf{y}_i, \mathbf{x}_i) = \mathbf{t}_i^*\mathbf{BBe}_i/\|\mathbf{Bt}_i\| \cdot \|\mathbf{Be}_i\|.$$

Use these two facts to establish (14-4-18).

14-5 SECTIONING

There is a variant of the implementations given in Secs. 14-2 and 14-3 which is aimed at the subclass of large problems in which all the eigenvalues in a given interval (α, β) must be computed, however many there may be. The interval may be quite wide. Of course the spectrum could be sliced at α and β to determine this number but what makes the problem interesting is the desire to keep the number of factorizations to a minimum, preferably to one.

The method uses only one application of the Rayleigh-Ritz procedure. The iterative part is transferred to determining starting vectors which actually span the invariant subspace associated with (α, β). The expensive step is factoring $A - \mu$ for some μ at, or close to, the midpoint $(\alpha + \beta)/2$, but the factors permit a relatively swift execution of inverse iteration. The goal is to distinguish rapidly between three possible configurations: (a) no eigenvalues in (α, β), (b) one or more eigenvalues close to μ, (c) eigenvalues bunched near α and β but not near μ. The distinction is made by monitoring carefully the rate of convergence of inverse iteration on a single vector which is kept orthogonal to directions already found to be in the invariant subspace associated with (α, β). Inverse iteration is efficient for cases (a) and (b) but not for (c). By detecting (c) early the program can abandon the current shift and, **only when warranted**, start again with a well-chosen shift near α and another near β. This brings the total to 3 factorizations for difficult cases.

In case (b) the iteration is stopped as soon as the vector is seen to be lying in the desired subspace; there is no waiting about for convergence. Since the program is given α and β it can make use of Chebyshev acceleration to further reduce the components in those eigenvectors belonging to eigenvalues outside (α, β). In this way columns are added to the starting matrix until case (a) obtains, signalling that the invariant subspace has been captured. One application of the RR procedure, using A, yields the wanted eigenvectors. The all important details are given in [Jensen, 1972] where sectioning was first presented.

A different type of sectioning has been proposed recently in [Wilson, 1978], apparently without knowledge of Jensen's work. By minimizing an operation count and making certain simplifying assumptions Wilson concludes that a good choice for block size in subspace iteration is $\sqrt{m/2}$, where m is the half-bandwidth of A. To be specific, suppose that the optimal block size for the given matrix is declared to be 10.

Wilson's strategy makes freer use of factorizations than does Jensen's and uses spectrum slicing to find subintervals of (α, β) each containing about 7 eigenvalues. Wilson then begins with the subinterval nearest β (i.e., the innermost end of the big interval) and uses subspace iteration to find the eigenvalues and eigenvectors in it. After that he works down toward α using some vectors from the previous iteration as starting vectors and making sure that all starting vectors are orthogonal to the eigenvectors found previously.

The usual practice had been to work from α to β.

When the bandwidth w is very small relative to n then factorization is comparable in cost to other vector operations such as orthogonalization. In that case it pays to use spectrum slicing and the techniques described in Sec. 3-5 to locate an eigenvalue accurately and then find the eigenvector with one or two steps of inverse iteration. When the eigenpairs are found one by one there is no need to use blocks which are larger than the number of wanted eigenvectors.

Notes and References

An early work analyzing block methods was [Bauer, 1957] but the most important references on subspace iteration are [Rutishauser, 1969 and 1971]. The geometrical aspects are discussed in [Parlett and Poole, 1973].

The civil engineers used and developed the method somewhat independently of the numerical analysts and the language used is therefore different. For example, Jennings uses the term "interaction matrix" for what we call the projection of A onto a subspace. Recent references which describe tests of several variants are [Bathé and Wilson, 1976] and [Jennings, 1977].

The development of sectioning, [Jensen, 1972], was very valuable for large problems. See [Jennings and Agar, 1978] for recent developments.

15

The General Linear Eigenvalue Problem

15-1 INTRODUCTION

This chapter takes up the task of computing some, or all, of the pairs $(\lambda, \mathbf{z})$ such that $(\mathsf{A} - \lambda\mathsf{M})\mathbf{z} = \mathbf{o}$, $\mathbf{z} \neq \mathbf{o}$ given two symmetric matrices A and M. The scalar λ is called an **eigenvalue** (or **root**) of the pair (A, M) and $\mathbf{z}$ is an eigenvector. In [Gantmacher, 1959] the matrix $\mathsf{A} - \lambda\mathsf{M}$ is called a matrix **pencil**. The rather strange use of the word "pencil" comes from optics and geometry. An aggregate of (light) rays converging to a point does suggest the sharp end of a pencil and, by a natural extension, the term came to be used for any **one parameter family** of curves, spaces, matrices, or other mathematical objects. In structural analysis A is the stiffness matrix (usually written as K) and M is the mass matrix.

Two pencils $(\mathsf{A}_1, \mathsf{M}_1)$ and $(\mathsf{A}_2, \mathsf{M}_2)$ are said to be **equivalent** if there exist invertible matrices E and F such that

$$\mathsf{A}_2 = \mathsf{E}\mathsf{A}_1\mathsf{F}, \qquad \mathsf{M}_2 = \mathsf{E}\mathsf{M}_1\mathsf{F}. \tag{15-1-1}$$

The roots of two equivalent pencils are the same and the eigenvectors are simply related (Ex. 15-1-1). Moreover the roots $\lambda_1, \lambda_2, \ldots$ are zeros of the **characteristic polynomial**

$$\chi(\tau) \equiv \det[\tau\mathsf{M} - \mathsf{A}]. \tag{15-1-2}$$

Symmetry is too precious a property to surrender and so we shall consider only **congruent** pencils, that is, equivalent pencils in which $E = F^*$. It is not necessary that $F^* = F^{-1}$ in order to preserve eigenvalues but in practice orthonormal F are popular because by Fact 1-10 in Chap. 1, $\|A_1\| = \|A_2\|$, $\|M_1\| = \|M_2\|$ and so no dangerous element growth can occur in carrying out the congruence transformation explicitly.

It is natural to seek the analogue to the spectral theorem, Fact 1-4 in Chap. 1, and hence to find the canonical form (i.e., simplest pair) in each class of congruent pencils. The answer is known as the simultaneous reduction to diagonal form.

For some pencils (A, M) there is an invertible F such that

$$F^*AF = \Phi = \text{diag}(\phi_1, \ldots, \phi_n),$$

$$F^*MF = \Psi = \text{diag}(\psi_1, \ldots, \psi_n).$$

There are two important departures from the spectral theorem: (1) although the ratios ϕ_i/ψ_i, $i = 1, \ldots, n$ are unique the matrices Φ and Ψ are not and (2) the reduction is not always possible.

If $\psi_i \neq 0$, $i = 1, \ldots, n$ then it is possible to normalize the pair Φ, Ψ by making $\Psi = I$ and $\Phi = \Lambda = \text{diag}(\lambda_1, \ldots, \lambda_n)$, but this is not always advisable. This point is elaborated in Secs. 15-2 and 15-3 which present basic material on matrix pencils or, equivalently, on pairs of quadratic forms.

With that preparation in hand Secs. 15-4, 15-5, 15-6, and 15-7 discuss the numerical reduction of (A, M) to (Λ, I) while the rest of the chapter is concerned with extending the methods of earlier chapters to large pencils (A, M).

Exercise on Sec. 15-1

15-1-1. Let $A_2 = EA_1F$, $M_2 = EM_1F$. Show that the roots of (A_1, M_1) and (A_2, M_2) are the same and show how the eigenvectors are related.

15-2 SYMMETRY IS NOT ENOUGH

The generalized eigenvalue problem is, in principle, more difficult than the standard one because of three new phenomena which can occur. Fortunately they can be illustrated on 2 by 2 pencils.

Eigenpairs

I. $\mathsf{A} = \begin{bmatrix} 1 & 0 \\ 0 & 0 \end{bmatrix}, \quad \mathsf{M} = \begin{bmatrix} 1 & 0 \\ 0 & 0 \end{bmatrix}. \qquad (1, \mathbf{e}_1); \quad (\tfrac{0}{0}, \mathbf{e}_2).$

II. $\mathsf{A} = \begin{bmatrix} 1 & 0 \\ 0 & 0 \end{bmatrix}, \quad \mathsf{M} = \begin{bmatrix} 0 & 0 \\ 0 & 1 \end{bmatrix}. \qquad (\tfrac{1}{0}, \mathbf{e}_1); \quad (\tfrac{0}{1}, \mathbf{e}_2).$

III. $\mathsf{A} = \begin{bmatrix} 0 & 1 \\ 1 & 0 \end{bmatrix}, \quad \mathsf{M} = \begin{bmatrix} 1 & 0 \\ 0 & -1 \end{bmatrix}. \qquad \left[i, \begin{pmatrix} i \\ -1 \end{pmatrix}\right]; \quad \left[i, \begin{pmatrix} i \\ 1 \end{pmatrix}\right].$

(Here $i^2 = -1$).

In (I) all scalars are eigenvalues for $\mathbf{e}_2$. In (II) ∞ is an eigenvalue for a well-defined eigenvector $\mathbf{e}_1$. In (III) there are complex eigenvalues even though A and M are symmetric. This last phenomenon is clarified by the following surprising result.

> THEOREM Any real square matrix B can be written as $\mathsf{B} = \mathsf{AM}^{-1}$ or $\mathsf{B} = \mathsf{M}^{-1}\mathsf{A}$ where A and M are suitable symmetric matrices. (15-2-1)

The proof is the subject of Ex. 15-2-3. It follows that any difficulty arising in the computation of B's eigenvalues can afflict the task of solving $(\mathsf{A} - \lambda\mathsf{M})\mathbf{x} = (\mathsf{AM}^{-1} - \lambda)\mathsf{M}\mathbf{x} = \mathbf{o}$.

Pencils, like (I), which have $\chi(t) = 0$ for **all** t are called **singular**. Often, but not always pencils are singular because A and M have some null vectors in common, $\mathsf{A}\mathbf{x} = \mathbf{o}$, $\mathsf{M}\mathbf{x} = \mathbf{o}$, $\mathbf{x} \neq \mathbf{o}$. Such x are, strictly speaking, eigenvectors and any number whatsoever is a matching eigenvalue. This behavior is unorthodox to say the least and the first part of any analysis of a pencil should be to find any common null space and get rid of it. Theoretically such a subspace is removed by **deflation**, that is, by restricting A and M to the **complementary** invariant subspace in $\mathcal{E}^n$.

For practical work the danger which suggests itself is that there may be vectors x which are nearly annihilated by both A and M. In such a case an eigenvalue program may compute some innocent looking eigenvalues which are not only hypersensitive to perturbations in A and M but whose very presence degrades the stability of the other eigenvalues. An example will give substance to these remarks.

$$\mathsf{A} = \begin{bmatrix} 1 & 0 \\ 0 & 10^{-8} \end{bmatrix}, \quad \mathsf{M} = \begin{bmatrix} 1 & 0 \\ 0 & 2 \times 10^{-8} \end{bmatrix}: \qquad \text{Solution } (1, \mathbf{e}_1), \left(\tfrac{1}{2}, \mathbf{e}_2\right). \tag{15-2-2}$$

Now perturb A by elements of the order 10^{-8}. Let

$$\mathsf{A}' = \begin{bmatrix} 1 & \sqrt{2} \times 10^{-8} \\ \sqrt{2} \times 10^{-8} & 2 \times 10^{-8} \end{bmatrix}. \tag{15-2-3}$$

The roots of $(\mathsf{A}', \mathsf{M})$ are approximately 1 ± 10^{-4} with eigenvectors approximately $\mathbf{e}_2 \mp 10^{-4}\mathbf{e}_1$. Thus a change of 10^{-8} in A has changed the root 1 by 10^{-4}, a magnification of 10,000, while the root $\frac{1}{2}$ changes completely. Eigenvalues such as $\frac{1}{2}$ are **ill-disposed**, an apt term coined by G. W. Stewart. Now (15-2-2) is equivalent to

$$\mathsf{A} = \begin{bmatrix} 1 & 0 \\ 0 & 1 \end{bmatrix}, \quad \mathsf{M} = \begin{bmatrix} 1 & 0 \\ 0 & 2 \end{bmatrix}: \quad \text{Solution } (1, \mathbf{e}_1), \left(\tfrac{1}{2}, \mathbf{e}_2\right) \tag{15-2-4}$$

which is perfectly conditioned.

Mathematically, in exact arithmetic, we cannot distinguish between (15-2-2) and (15-2-4). It may be vital for certain applications to recognize that (15-2-2) and (15-2-4) are not equally permissible representations of a problem, but such knowledge is external to the standard theory of matrix pencils and must be supplied as an extra element of the problem in hand. It is best if this can be done by designating the scaling which is appropriate for the data.

It is worth emphasizing that infinite eigenvalues, Case II, are not necessarily ill-disposed. In fact, the infinite eigenvalues of (A, M) are the zero eigenvalues of (M, A) and this is one more hint that each eigenvalue of a pencil should be presented as a quotient and not reduced to a single number. Our normal measure of the separation of two eigenvalues, namely $|\lambda_i - \lambda_j|$, is also called into question. Since the roles of A and M are interchangeable it would seem natural to use a measure of the separation of λ_i and λ_j which is invariant under reciprocation. The **chordal metric**,

$$\chi(\lambda, \mu) \equiv |\lambda - \mu| / \sqrt{1 + \lambda^2}\sqrt{1 + \mu^2},$$

enjoys this property (Exs. 15-2-4 and 15-2-5) and crops up in the study of non-Euclidean geometry.

With this tool in hand it should be possible to generalize the error bounds in Chaps. 4, 10, and 11 to cover the pencil (A, M) in a properly invariant manner. See [Crawford, 1976] and [Stewart, 1979] for some perturbation results of this kind. However the error bounds in this book do not require that the matrices being compared be close in any sense.

Exercises on Sec. 15-2

15-2-1. Verify the given solutions of (I), (II), and (III). Give a 2 by 2 symmetric pencil with only one eigenvector.

15-2-2. Find a singular 2 by 2 pencil in which A and M do not share a common null space.

15-2-3. Note that

$$\left[\begin{array}{cc|cc} \rho & 1 & \mu & 0 \\ 0 & \rho & 0 & \mu \\ \hline -\mu & 0 & \rho & 1 \\ 0 & -\mu & 0 & \rho \end{array}\right] = \left[\begin{array}{cc|cc} 0 & \mu & 1 & \rho \\ \mu & 0 & \rho & 0 \\ \hline 1 & \rho & 0 & -\mu \\ \rho & 0 & \mu & 0 \end{array}\right] \begin{bmatrix} 0 & 0 & 0 & 1 \\ 0 & 0 & 1 & 0 \\ 0 & 1 & 0 & 0 \\ 1 & 0 & 0 & 0 \end{bmatrix}$$

Let $B = FJ_rF^{-1}$ where J_r is the real Jordan form. Show that J_r can be written as $J_r = \tilde{A}\tilde{M}^{-1}$. Then conclude that $B = AM^{-1}$ where $A = F\tilde{A}F^*$, $M = F\tilde{M}F^*$.

15-2-4. Show that $\chi(\lambda, \mu) = \chi(1/\lambda, 1/\mu)$. Let $\lambda = \alpha_1/\beta_1$, $\mu = \alpha_2/\beta_2$, and express $\chi(\lambda, \mu)$ in terms of the α's and β's. Note that χ is invariant under orthogonal transformations

$$\begin{bmatrix} \gamma \\ \delta \end{bmatrix} = \begin{bmatrix} c & -s \\ s & c \end{bmatrix} \begin{bmatrix} \alpha \\ \beta \end{bmatrix}.$$

15-2-5. Show that χ is the length of the chord joining $\tilde{\lambda}$ and $\tilde{\mu}$ in the figure below.

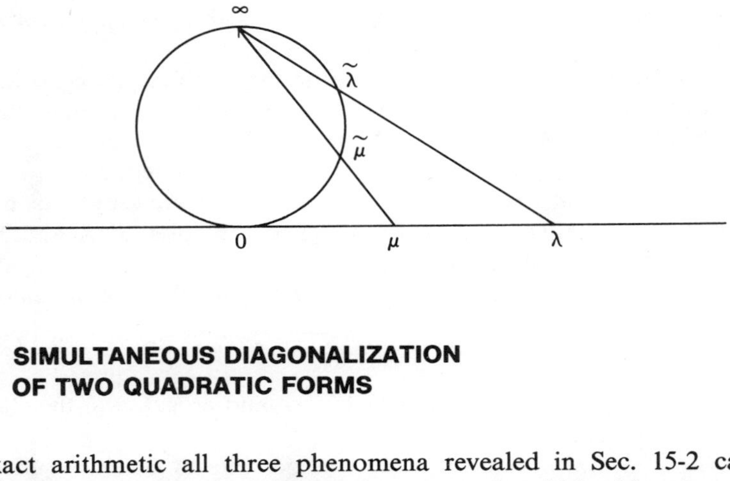

15-3 SIMULTANEOUS DIAGONALIZATION OF TWO QUADRATIC FORMS

In exact arithmetic all three phenomena revealed in Sec. 15-2 can be banished by confining attention to those cases in which either A or M or some combination $\alpha A + \mu M$ is positive definite. This restriction seems natural when one notes that the standard eigenvalue problem corresponds to $M = I$ and I is the prototype of all positive definite matrices.

In fact when given the pencil (A, M) there is no loss in considering the equivalent pencil $(\hat{A}, \hat{M}) = (\gamma A + \sigma M, -\sigma A + \gamma M)$ for any pair (γ, σ) such that $\gamma^2 + \sigma^2 = 1$. What is the best pair to choose? To answer this consider

$$\mu(A, M) \equiv \inf_{\|x\|=1} \left\{(x^*Ax)^2 + (x^*Mx)^2\right\}^{1/2}. \qquad (15\text{-}3\text{-}1)$$

The following interesting result is the subject of [Uhlig, 1979].

THEOREM Iff $\mu(\mathsf{A}, \mathsf{M}) > 0$ then there is a pair (γ, σ) such that $\hat{\mathsf{M}} = \gamma\mathsf{M} - \sigma\mathsf{A}$ is positive definite provided $n > 2$. (15-3-2)

Even when M is positive definite it might be preferable to work with $(\hat{\mathsf{A}}, \hat{\mathsf{M}})$ if $\lambda_1(\hat{\mathsf{M}}) \gg \lambda_1(\mathsf{M})$.

At present there is no economical way of finding the best pair and exploiting theorem (15-3-2). Pencils with $\mu > 0$ are called definite and in theory the remainder of this chapter applies to definite pencils but we shall confine ourselves to those in which definiteness is explicit.

It is the definiteness of the pencil (A, M) which guarantees the simultaneous reduction to diagonal form.

THEOREM If M is positive definite then there are many invertible matrices F such that F*AF and F*MF are both diagonal and real. (15-3-3)

Proof. By the spectral theorem and the hypothesis,

$$\mathsf{M} = \mathsf{G}\Delta^2\mathsf{G}^*$$

where G is orthonormal and $\Delta = \mathrm{diag}(\delta_1, \ldots, \delta_n)$ is real. The only, but crucial use of positive definiteness is that M's eigenvalues are positive and allow the definition of a real Δ. Now reduce the pencil to standard form as follows:

$$\mathsf{A} \longrightarrow \Delta^{-1}\mathsf{G}^*\mathsf{A}\mathsf{G}\Delta^{-1} = \mathsf{H}, \qquad \mathsf{M} \longrightarrow \Delta^{-1}\mathsf{G}^*\mathsf{M}\mathsf{G}\Delta^{-1} = \mathsf{I}.$$

By the spectral theorem for H there is an orthonormal P so that

$$\mathsf{H} = \mathsf{P}\Lambda\mathsf{P}^*$$

and so, with $\mathsf{F} \equiv \mathsf{G}\Delta^{-1}\mathsf{P}$, a product of invertible matrices,

$$\mathsf{A} \longrightarrow \mathsf{F}^*\mathsf{A}\mathsf{F} = \Lambda, \qquad \mathsf{M} \longrightarrow \mathsf{F}^*\mathsf{M}\mathsf{F} = \mathsf{I}.$$

For any nonsingular diagonal matrix Ω, reduction by $\mathsf{F}\Omega$ also yields a diagonal pencil. □

The proof suggests a simple way to compute the roots of (A, M) but that topic is postponed until the next section. Another form of theorem

(15-3-3) is

> THEOREM If M is positive definite then the symmetric pencil (A, M) has n real roots $\lambda_1, \lambda_2, \ldots, \lambda_n$ in the interval $[-\|\mathsf{M}^{-1}\mathsf{A}\|, \|\mathsf{M}^{-1}\mathsf{A}\|]$ and, to match them, n linearly independent eigenvectors $\mathsf{z}_1, \ldots, \mathsf{z}_n$. Moreover z_i and z_j are M-orthogonal if $\lambda_i \neq \lambda_j$, i.e.,
>
> $$\mathsf{z}_i^* \mathsf{M} \mathsf{z}_j = 0. \tag{15-3-4}$$
>
> If $\lambda_i = \lambda_j$ then z_i and z_j may be **chosen** to be M-orthogonal.

The proof is left as Ex. 15-3-3.

Another bonus from the positive definiteness of M is that both bilinear forms

$$(\mathsf{x}, \mathsf{y})_M \equiv \mathsf{y}^* \mathsf{M} \mathsf{x}, \qquad (\mathsf{x}, \mathsf{y})_{M^{-1}} \equiv \mathsf{y}^* \mathsf{M}^{-1} \mathsf{x}, \tag{15-3-5}$$

are genuine **inner product** functions. The vector space $\mathcal{R}^n$ supplemented with either inner product becomes a genuine inner product space $\mathcal{M}^n$ and almost all properties of $\mathcal{E}^n$, such as the Cauchy-Schwarz inequality, carry over to $\mathcal{M}^n$. Exs. 15-3-3 and 15-3-4 explore some of this material.

A pencil is positive definite only if both A and M are positive definite.

Exercises on Sec. 15-3

15-3-1. Prove theorem (15-3-4) by imitating the proofs of analogous results for the standard problem.

15-3-2. Show that (15-3-5) defines an inner product by verifying the following:
(a) $(\mathsf{x}, \mathsf{y})_M = (\mathsf{y}, \mathsf{x})_M$,
(b) $(\alpha \mathsf{x} + \beta \mathsf{y}, \mathsf{z})_M = \alpha(\mathsf{x}, \mathsf{z})_M + \beta(\mathsf{y}, \mathsf{z})_M$,
(c) $(\mathsf{x}, \mathsf{x})_M > 0$ if $\mathsf{x} \neq \mathsf{o}$.

15-3-3. Verify that $\|\mathsf{u}\|_M \equiv \sqrt{\mathsf{u}^* \mathsf{M} \mathsf{u}}$ satisfies the axioms for a norm:
(a) $\|\mathsf{u}_M\| > 0$ if $\mathsf{u} \neq \mathsf{o}$,
(b) $\|\alpha \mathsf{u}\|_M = |\alpha| \|\mathsf{u}\|_M$,
(c) $\|\mathsf{u} + \mathsf{v}\|_M \leqslant \|\mathsf{u}\|_M + \|\mathsf{v}\|_M$.
For which properties is it essential that M be positive definite?

15-3-4. Prove the Cauchy-Schwarz inequality

$$(\mathsf{x}, \mathsf{y})_M^2 \leqslant (\mathsf{x}, \mathsf{x})_M (\mathsf{y}, \mathsf{y})_M$$

by considering $(\mathsf{x} + \xi \mathsf{y}, \mathsf{x} + \xi \mathsf{y})_M$ for all real ξ.

15-3-5. Find θ such that $(\cos\theta)\begin{bmatrix} 1 & 0 \\ 0 & 0 \end{bmatrix} - (\sin\theta)\begin{bmatrix} 0 & 0 \\ 0 & 1 \end{bmatrix}$ is positive definite. What is the best choice for θ?

15-4 EXPLICIT REDUCTION TO STANDARD FORM

Thanks to the Handbook and EISPACK (see Sec. 2-8) high-quality programs are available for small standard eigenvalue problems. So there is much to be said for reducing the general problem $(A - \lambda M)x = o$ to an equivalent standard form $(A - \lambda I)y = o$. There are several ways to do it.

1. Form $M^{-1}A$ and lose both symmetry and sparsity.
2. Solve the standard eigenvalue problem for M to get orthonormal G and diagonal Δ such that $M = G\Delta^2G^*$. Then
$$A - \lambda M = G\Delta(\hat{A} - \lambda I)\Delta G^*,$$
where
$$\hat{A} = \Delta^{-1}G^*AG\Delta^{-1}.$$
3. Compute the Choleski decomposition $M = LL^*$. Then
$$A - \lambda M = L(\mathring{A} - \lambda I)L^*,$$
and
$$\mathring{A} = L^{-1}AL^{-*}.$$

See Contribution II/10 in the Handbook or the EISPACK guide for more details.

15-4-1 REMARKS

1. If eigenvectors are not wanted then the transform matrices L or G need not be preserved. If eigenvectors are wanted then either the transformations or the original pair must be saved.
2. Some of the transformations on A and M can be done simultaneously, for example, when G and L are expressed as products of simple matrices (Ex. 15-4-1).
3. An extension and refinement of Method 2 is described in Sec. 15-5.
4. Wilkinson gives an elegant way to compute $\mathring{A}$. Initially n by n arrays hold A and M. Copy the diagonals of A and M into one-dimensional arrays DA and DM. Then compute L and store its lower triangular part in the lower triangle of M. Finally compute $(L^{-1}A)L^{-*}$ in two stages by using only the lower triangle of A which, on completion, holds the lower triangular part of $\mathring{A}$. The details are left as Ex. 15-4-2.
5. In order to simplify discussion assume that both A and M are positive

definite and that the eigenvalues of the pair (A, M) are ordered by

$$0 < \lambda_1 \leqslant \cdots \leqslant \lambda_n.$$

Note that

$$\|\mathring{A}\| = \lambda_n \leqslant \|M^{-1}\|\|A\|.$$

In many cases some elements of $\mathring{A}$ are of the same magnitude as $\|M^{-1}\| \cdot \|A\|$ and much greater than $\|A\|$. This happens when M is ill-conditioned for inversion (nearly singular) and then the eigenvalues of $\mathring{A}$ are spread over many orders of magnitude, for example from 10^3 to 10^{20}. It is likely that the small eigenvalues will be computed with far less relative accuracy than the large ones. This is because the computed eigenvalues are exact for a matrix $\tilde{A}$ and often $\|\tilde{A} - \mathring{A}\| \doteqdot \varepsilon\|\mathring{A}\|$, where ε is the unit roundoff. A crude bound on the eigenvalue change is

$$\frac{|\tilde{\lambda}_i - \mathring{\lambda}_i|}{|\mathring{\lambda}_i|} \leqslant \frac{\|\tilde{A} - \mathring{A}\|}{|\mathring{\lambda}_i|} \doteqdot \varepsilon\left(\frac{\|\mathring{A}\|}{\mathring{\lambda}_i}\right), \qquad i = 1, \ldots, n.$$

For $i = 1, 2, 3$ it is quite possible to have $\|\mathring{A}\|/\mathring{\lambda}_i \doteqdot 1/\varepsilon$ and hence the possibility that $\mathring{\lambda}_i$ will have no correct figures.

This is the flaw in explicit reduction when M is nearly singular and the arithmetic is done with numbers of limited precision.

6. If M, and therefore L, is moderately ill-conditioned for inversion and if, in addition, the columns of L decrease monotonically then $\mathring{A}$ will usually be a graded matrix whose rows increase in norm (last row is biggest). Graded matrices usually define their smallest eigenvalues with greater relative accuracy than could be expected from standard norm estimates (absolute error $< \varepsilon\|\mathring{A}\|$). An example is given in Fig. 15-4-1.

$$A = \begin{bmatrix} 4 & 1 & 0 & 1 \\ 1 & 4 & 2 & 0 \\ 0 & 2 & 4 & 2 \\ 1 & 0 & 2 & 3 \end{bmatrix} \qquad M = \begin{bmatrix} 10000 & 4000 & 1000 & 100 \\ 4000 & 1700 & 430 & 60 \\ 1000 & 430 & 110 & 16.5 \\ 100 & 60 & 16.5 & 5.26 \end{bmatrix}$$

$$L = \begin{bmatrix} 100 & & & \\ 40 & 10 & & \\ 10 & 3 & 1 & \\ 1 & 2 & .5 & .1 \end{bmatrix}$$

$$\mathring{A} = \begin{bmatrix} .0004 & -.0006 & -.0022 & .119 \\ -.0006 & -.0384 & .0908 & -1.616 \\ -.0022 & .0908 & 3.150 & 2.658 \\ .119 & -1.616 & 2.658 & 223.8 \end{bmatrix}$$

Figure 15-4-1 Reduction of (A, M) to ($\mathring{A}$, I)

15-4-2 REDUCTION OF BANDED PENCILS

Section 7-5 presented an ingenious algorithm for reducing a banded matrix A to tridiagonal form T without temporarily enlarging the bandwidth in the process. In [Crawford, 1973] this approach is extended to transform banded pencils (A, M) into (T, I) without the use of extra storage.

We shall not describe the algorithm in detail for the following reasons. When A and M are small the storage advantage is not very significant. When A and M are large it is unlikely that all, or even a majority, of the eigenvalues will be wanted and methods which avoid explicit reduction are preferable. Moreover large A and M of narrow bandwidth are ideally suited to spectrum slicing (Sec. 3-3) or the Lanczos algorithm (Sec. 15-11).

Exercises on Sec. 15-4

15-4-1. Find the operation counts for the spectral decomposition and the Cholesky factorization of M. Exploit symmetry. Give op counts for reductions (15-4-2) and (15-4-3).

15-4-2. Let $B = L^{-1}A$. Find an algorithm which computes and writes the lower part of BL^{-*} over the lower part of A, losing A's diagonal in the process. Exploit symmetry where possible.

15-4-3. Consider a 2 by 2 pencil. If M is ill-conditioned and if $\mathring{A}$ is ill-conditioned, then $\mathring{A}$ must be strongly graded. True or false?

*15-5 THE FIX-HEIBERGER REDUCTION

This section describes a careful reduction to standard form designed specifically to cope with those cases in which M is, for practical purposes, positive **semi**-definite (at least one eigenvalue is zero). In other words M is either singular or so close to a singular matrix that it is preferable to work with the latter and face up to the consequences. The aim is to find any infinite or ill-disposed eigenvalues (see Sec. 15-2) before computing the respectable ones.

In what follows the notation

$$A, M \xrightarrow{P} \bar{A}, \bar{M}$$

means

$$\bar{A} = P^*AP, \qquad \bar{M} = P^*MP.$$

Also $P \oplus Q$ stands for diag(P, Q). Let η be a user-given criterion for negligibility.

There are three steps of increasing complexity.

Step 1. Find the spectral decomposition of M, $M = G(\Delta_1^2 \oplus \Delta_2^2)G^*$ where Δ_2^2 has the tiny eigenvalues of M, $\|\Delta_2\|^2 < \eta\|\Delta_1\|^2$. Replace Δ_2^2 by O. Partition G^*AG to match Δ_1 to find

$$A, M \xrightarrow{G} \begin{bmatrix} \tilde{A}_{11} & \tilde{A}_{12} \\ \tilde{A}_{12}^* & \tilde{A}_{22} \end{bmatrix}, \quad \begin{bmatrix} \Delta_1^2 & O \\ O & O \end{bmatrix}.$$

Step 2. Reduce Δ_1^2 to I and find the spectral decomposition of $\tilde{A}_{22}$, $\tilde{A}_{22} = F(\Phi \oplus \Psi)F^*$ where $\Phi \oplus \Psi$ is diagonal and $\|\Psi\| < \eta\|\Phi\|$. Put $\Psi = O$. Write $\Delta_1\tilde{A}_{12}F$ as $(\mathring{A}_{12}, \mathring{A}_{13})$ to find

$$(\cdot, \cdot) \xrightarrow{\Delta_1^{-1} \oplus F} \left[\begin{array}{c|cc} \mathring{A}_{11} & \mathring{A}_{12} & \mathring{A}_{13} \\ \hline \mathring{A}_{12}^* & \Phi & O \\ \mathring{A}_{13}^* & O & O \end{array}\right], \quad \left[\begin{array}{c|cc} I & O & O \\ \hline O & O & O \\ O & O & O \end{array}\right].$$

When M is well-conditioned the last two rows and columns will be empty and the result is $(\mathring{A}_{11}, I)$ which is Reduction No. 2 given in Sec. 15-4. If $\tilde{A}_{22}$ is ill-conditioned then the final form is

$$\begin{bmatrix} \mathring{A}_{11} & \mathring{A}_{12} \\ \mathring{A}_{12}^* & \Phi \end{bmatrix}, \quad \begin{bmatrix} I & O \\ O & O \end{bmatrix}.$$

The further reduction of this pencil to obtain the finite eigenvalues is left as Ex. 15-5-1.

In the general case $\mathring{A}_{13}$ needs further analysis.

Step 3. Trouble comes when $\mathring{A}_{13}$ does not have full rank but this is one of the cases when Fix-Heiberger is warranted and so we allow for this possibility. At this point it is necessary to compute the singular value decomposition of $\mathring{A}_{13}$, namely

$$\mathring{A}_{13} = Q\begin{bmatrix} \Sigma \\ O \end{bmatrix}P^*$$

where Q and P are orthonormal (of different sizes) and Σ is diagonal with nonnegative elements. The matrices Q and P are not unique, Q is an eigenvector matrix for $\mathring{A}_{13}\mathring{A}_{13}^*$ while P is an eigenvector matrix for $\mathring{A}_{13}^*\mathring{A}_{13}$. The diagonal elements σ_i of Σ are given by

$$\sigma_i = \sqrt{\lambda_{-i}[\mathring{A}_{13}^*\mathring{A}_{13}]}$$

and are called the singular values of $\mathring{A}_{13}$.

Next Σ is divided into $\Theta \oplus \Omega$ where Θ has the larger singular values and $\|\Omega\| \leqslant \eta\|\Theta\|$. Then put $\Omega = \mathsf{O}$ and define

$$\begin{aligned} n_1 &\equiv \text{the number of rows in } \Theta. \\ n_1 + n_2 &\equiv \text{the number of rows in } \mathring{\mathsf{A}}_{13}. \\ n_3 &\equiv \text{the number of rows in } \Phi. \end{aligned}$$

The submatrices in the Fix-Heiberger form are given by

$$\mathsf{Q}^*\mathring{\mathsf{A}}_{11}\mathsf{Q} = \begin{array}{c} \begin{matrix} n_1 & n_2 \end{matrix} \\ \begin{bmatrix} \mathsf{A}_{11} & \mathsf{A}_{12} \\ \mathsf{A}_{12}^* & \mathsf{A}_{22} \end{bmatrix} \end{array}, \qquad \mathsf{Q}^*\mathring{\mathsf{A}}_{12} = \begin{bmatrix} \mathsf{A}_{13} \\ \mathsf{A}_{23} \end{bmatrix}.$$

Then finally

$$(\cdot, \cdot) \xrightarrow{\mathsf{Q}\oplus\mathsf{I}\oplus\mathsf{P}} \begin{bmatrix} \mathsf{A}_{11} & \mathsf{A}_{12} & \mathsf{A}_{13} & \Theta & \mathsf{O} \\ & \mathsf{A}_{22} & \mathsf{A}_{23} & \mathsf{O} & \mathsf{O} \\ & & \Phi & \mathsf{O} & \mathsf{O} \\ & \text{sym} & & \mathsf{O} & \mathsf{O} \\ & & & & \mathsf{O} \end{bmatrix}, \qquad \mathsf{I} \oplus \mathsf{I} \oplus \mathsf{O} \oplus \mathsf{O} \oplus \mathsf{O}.$$

If $\Theta \neq \Sigma$ then the fifth (block) row and column of O's is actually present and the pencil (A, M) is singular. A warning should be given that any number is an eigenvalue. The fifth row and column may then be dropped and this deflates the troublesome null space. The genuine eigenvalues and eigenvectors can be found from the remaining four rows and columns as in the case when $\mathring{\mathsf{A}}_{13}$ has full rank in Step 2.

Let $(\mathsf{x}_1^*, \mathsf{x}_2^*, \mathsf{x}_3^*, \mathsf{x}_4^*)$ be a partitioned eigenvector of the 4 by 4 leading submatrix of the canonical form shown above. Solving these equations from the bottom row up yields

$$\begin{aligned} \mathsf{x}_1 &= \mathsf{o}, \\ \mathsf{x}_3 &= -\Phi^{-1}\mathsf{A}_{23}^*\mathsf{x}_2, \\ (\mathsf{A}_{22} - \mathsf{A}_{23}\Phi^{-1}\mathsf{A}_{23}^* - \lambda)\mathsf{x}_2 &= \mathsf{o}, \\ \mathsf{x}_4 &= -\Theta^{-1}(\mathsf{A}_{12}\mathsf{x}_2 + \mathsf{A}_{13}\mathsf{x}_3). \end{aligned}$$

Thus the n_2 finite eigenvalues come from the standard eigenvalue problem for x_2 in line 3. Derivation of the eigenvectors with infinite eigenvalue is left as Ex. 15-5-2. The formulas for the eigenvectors of the original problem are the subject of Ex. 15-5-3.

The purpose of the extra care exercised in the above reduction is the accurate determination of those small eigenvalues which happen to be well determined by the original data. If a suitable value of η is not apparent the calculation should be made with two different values of η (say $\eta = n\varepsilon$ and $\eta = \sqrt{\varepsilon}$) and the results compared.

In [Fix and Heiberger, 1972] bounds are given on the error caused by annihilating Δ_2, Ψ, and Ω.

Exercises on Sec. 15-5

15-5-1. Find a standard eigenvalue problem which gives the finite eigenvalues of

$$\begin{bmatrix} \mathring{A}_{11} & \mathring{A}_{12} \\ \mathring{A}_{12}^{*} & \Phi \end{bmatrix}, \quad \begin{bmatrix} I & O \\ O & O \end{bmatrix}.$$

Recall that Φ is diagonal and invertible.

15-5-2. Describe the eigenvectors of the Fix-Heiberger form which have infinite eigenvalues using the partition x_1, x_2, x_3, and x_4 employed in Sec. 15-5.

15-5-3. Give the formulas which turn the eigenvectors of the canonical form into those of the original pencil assuming that $\mathring{A}_{13}$ has full rank.

15-5-4. At the end of Step 1 we have

$$\left[\begin{array}{cccc|cc} 6 & 5 & 4 & 3 & 2 & 1 \\ & 4 & 3 & 2 & 1 & 1 \\ & & 2 & 1 & 2 & 3 \\ & & & 2 & 3 & 4 \\ \hline & & & & 10^{-4} & 0 \\ \text{sym} & & & & 0 & 0 \end{array}\right], \quad \left[\begin{array}{cccc|cc} 1 & & & & & \\ & 1 & & & \multicolumn{2}{c}{O} \\ & & 10^{-2} & & & \\ & & & 10^{-4} & & \\ \hline & & O & & 0 & \\ & & & & & 0 \end{array}\right]$$

Find the Fix-Heiberger form and, if a calculator is available, find all eigenvectors and eigenvalues.

15-6 THE QZ ALGORITHM

A stable generalization of the QR algorithm to arbitrary square pencils (B, N) was presented in [Moler and Stewart, 1973]. The method destroys symmetry and changes (A, M) to the equivalent pencil (J, K), where J and K are upper triangular. Orthogonal transformations are employed exclusively so that $\|A\| = \|J\|$, $\|M\| = \|K\|$, and the final accuracy is impervious to hazards such as M being singular or even indefinite. Ill-disposed eigenvalues are revealed as the quotients of small diagonal elements of J and K.

For small pencils QZ is an alternative to the Fix-Heiberger reduction and has the advantage of being generally available in computer center program libraries. Approximately $20n^3$ ops are required to find all the eigenvalues by QZ. The time penalty is less irksome than the need for two auxiliary n by n arrays. However for small n neither penalty is too serious unless a very large number of pencils are to be processed.

The loss of symmetry is displeasing and we turn next to a method which employs congruencies rather than equivalence transformations.

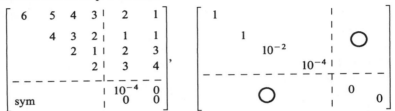

15-7 JACOBI GENERALIZED

The natural extension of Jacobi's idea (Chap. 9) is to find a congruence in the (i, j) plane to annihilate the (i, j) elements of both A and M. If M is positive definite this can always be done and there is considerable freedom in the choice of congruence.

The problem is essentially two-dimensional. Restricting attention to the typical (i, j) plane and using the simplest congruences yields

$$\begin{bmatrix} 1 & -\alpha \\ \beta & 1 \end{bmatrix}\begin{bmatrix} a_{ii} & a_{ij} \\ a_{ij} & a_{jj} \end{bmatrix}\begin{bmatrix} 1 & \beta \\ -\alpha & 1 \end{bmatrix} =$$

$$\begin{bmatrix} a_{ii} - 2\alpha a_{ij} + \alpha^2 a_{jj}, & \beta a_{ii} + (1 - \alpha\beta)a_{ij} - \alpha a_{jj} \\ \beta a_{ii} + (1 - \alpha\beta)a_{ij} - \alpha a_{jj}, & \beta^2 a_{ii} + 2\beta a_{ij} + a_{jj} \end{bmatrix}, \tag{15-7-1}$$

and similarly for M. In order to diagonalize both matrices a special pair of quadratic equations must be solved for α and β. Fortunately there is a closed form solution (see Ex. 15-7-1); the nonlinear term can be eliminated to reveal $\alpha = \delta_i/\nu$, $\beta = \delta_j/\nu$, where

$$\delta_i \equiv \det\begin{bmatrix} a_{ii} & a_{ij} \\ m_{ii} & m_{ij} \end{bmatrix}, \qquad \delta_j \equiv \det\begin{bmatrix} a_{jj} & a_{ij} \\ m_{jj} & m_{ij} \end{bmatrix}, \tag{15-7-2}$$

and ν satisfies the quadratic equation

$$\nu^2 - \delta_{ij}\nu - \delta_i\delta_j = 0, \tag{15-7-3}$$

$$\delta_{ij} \equiv \det\begin{bmatrix} a_{ii} & a_{jj} \\ m_{ii} & m_{jj} \end{bmatrix}. \tag{15-7-4}$$

When M is positive definite, then there is a nonzero solution to the quadratic (Ex. 15-7-2) and the congruence is proper, i.e., the determinant $1 + \alpha\beta \neq 0$. To keep α and β small it is essential to choose as ν the root of (15-7-3) which is farthest from 0.

The asymptotic quadratic convergence of the regular Jacobi process extends to this case [Zimmerman, 1969] provided that the process does converge. However convergence has not been proven.

These techniques have been used with some success on small A and M which are diagonally dominant, that is $a_{ii} > \sum_{j=1}^{n} |a_{ij}|$, $j \neq i$, for each i. See [Bathé and Wilson, 1976].

Exercises on Sec. 15-7

15-7-1. Find a closed form solution (α, β) to

$$\beta a_{ii} + (1 - \alpha\beta)a_{ij} - \alpha a_{jj} = 0,$$
$$\beta m_{ii} + (1 - \alpha\beta)m_{ij} - \alpha m_{jj} = 0.$$

15-7-2. Show that (15-7-3) has a nonzero solution when M is positive definite.

15-7-3. Show that with the proper choice of ν in (15-7-3) the parameters α and β satisfy $0 \leqslant \alpha\beta \leqslant 1$.

15-8 IMPLICIT REDUCTION TO STANDARD FORM

The three reductions described in Sec. 15-4 can be used implicitly. This observation is important in the treatment of large sparse A and M. Let us reconsider the three techniques.

1. The matrix $\mathsf{M}^{-1}\mathsf{A}$ is neither symmetric nor sparse but it is self-adjoint with respect to the M inner product;

$$(\mathsf{M}^{-1}\mathsf{Ax}, \mathsf{y})_M = (\mathsf{x}, \mathsf{M}^{-1}\mathsf{Ay})_M = \mathsf{y}^*\mathsf{Ax}.$$

The vector $\mathsf{w} = \mathsf{M}^{-1}\mathsf{Au}$ is computed in two steps.

(a) Form $\mathsf{v} = \mathsf{Au}$, exploiting sparsity.

(b) Either solve $\mathsf{Mw} = \mathsf{v}$ iteratively, if M cannot be factored, or solve $\mathsf{Lx} = \mathsf{v}$; then $\mathsf{L}^*\mathsf{w} = \mathsf{x}$, if $\mathsf{M} = \mathsf{LL}^*$.

Each technique in (b) can exploit sparsity.

2. In principle the matrix $\hat{\mathsf{A}} = \Delta^{-1}\mathsf{G}^*\mathsf{AG}\Delta^{-1}$ can be used without forming $\hat{\mathsf{A}}$. However the matrix G rarely inherits any sparse structure in M (except when $\mathsf{M} = \Delta^2$ and $\mathsf{G} = \mathsf{I}$!) and so this decomposition is not used implicitly to our knowledge.
3. The product $\mathsf{x} = \mathring{\mathsf{A}}\mathsf{u}$ where $\mathring{\mathsf{A}} = \mathsf{L}^{-1}\mathsf{AL}^{-*}$ can be formed in three steps:

Solve $\mathsf{L}^*\mathsf{v} = \mathsf{u}$ for v,

Form $\mathsf{w} = \mathsf{Av}$,

Solve $\mathsf{Lx} = \mathsf{w}$ for x.

Any band or profile structure in M is inherited by its Choleski factor L. The sparse matrix L need not be inverted. If A and M have bandwidth $2m + 1$, $m \ll n$, then the method given above requires approximately $(4m + 3)n$ ops to form x as against n^3 using the full form of Å. So both time and storage are saved by the implicit reduction when only a few eigenvalues and eigenvectors are required.

Fig. 15-8-1 shows how L inherits A's sparsity pattern.

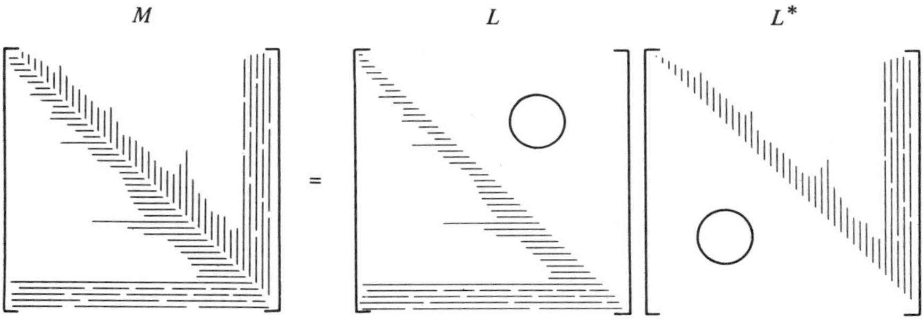

Figure 15-8-1 Preservation of profile in triangular factorization

15-9 SIMPLE VECTOR ITERATIONS

The previous section may have given the impression that if M can be factored into LL* then there is nothing better to do than apply the most appropriate method to the standard problem (Å, I) with Å either in explicit or implicit form. This approach is certainly in the mathematical tradition of reducing a new problem to one previously solved. However in the present context the standard reduction may not be warranted.

In order to examine this question we reconsider the power method and inverse iteration of Chap. 4. The governing equations of the two methods, with shift σ, are now

$$\text{PM:} \qquad \mathsf{M}\mathsf{v}_{k+1} = (\mathsf{A} - \sigma_k \mathsf{M})\mathsf{v}_k \tau_k, \tag{15-9-1}$$

$$\text{INVIT:} \qquad (\mathsf{A} - \sigma_k \mathsf{M})\mathsf{u}_{k+1} = \mathsf{M}\mathsf{u}_k \nu_k, \tag{15-9-2}$$

where τ_k and ν_k are normalizing constants. The new fact is that both iterations require the solution of a system of equations.

Case 1. If A and M are so large that factorization is infeasible then some inner iteration must be used to solve (15-9-1) or (15-9-2) and the two techniques are equal in cost—in the absence of special qualities in M.

Case 2. If A and M can be factored then we can compare two versions of INVIT, namely

1. Factor $A - \sigma_k M$ in order to solve (15-9-2) for u_{k+1}.
2. Solve $(\mathring{A} - \sigma_k I)w_{k+1} = w_k \pi_k$ for w_{k+1} in three steps:

$$x_k = Lw_k, \qquad (A - \sigma_k M)y_k = x_k, \qquad w_{k+1} = (L^* y_k)\pi_k. \tag{15-9-3}$$

If u_{k+1} is normalized in the same way as w_{k+1} then the two versions have the same cost and the first version is certainly more natural than the second version.

The preceding remarks show that for large matrices there is merit in retaining the original form of the problem, namely

$$(A - \lambda M)x = o.$$

1. The spectrum slicing technique and secant iterations of Chap. 3 extend directly and are well-suited to those cases in which A and M have narrow bandwidth and no eigenvectors are wanted. See [Bathé and Wilson, 1976, Chap. 11] for more details.
2. Most of the results of Chap. 4 extend to the pencil (A, M) provided that an appropriate norm is used in place of $\| \cdot \|$. Theoretically, the extension is fine but in practice the new norm increases the cost considerably. Perhaps we have been spoilt by the extreme simplicity of the Euclidean norm. The proofs of the results given below are close in spirit to the corresponding proofs in Chap. 4 and will be omitted. Some of them are quite straightforward, others less so. Recall that $\|x\|_{M^{-1}} \equiv \sqrt{x^* M^{-1} x}$.

THEOREM For arbitrary $u \neq o$ and σ there is an eigenvalue λ of (A, M) such that

$$|\lambda - \sigma| \leqslant \|(A - \sigma M)u\|_{M^{-1}} / \|Mu\|_{M^{-1}}. \tag{15-9-4}$$

As in the standard case an excellent choice for σ in INVIT is Rayleigh's quotient

$$\rho(u) = \rho(u; A, M) \equiv u^* A u / u^* M u, \qquad u \neq o.$$

> THEOREM When M is positive definite the Rayleigh quotient enjoys the following properties:
>
> *Homogeneity*: $\rho(\alpha \mathsf{u}) = \rho(\mathsf{u}), \alpha \neq 0$ (degree 0).
>
> *Boundedness*: $\rho(\mathsf{u})$ ranges over $[\lambda_1, \lambda_{-1}]$ as u ranges over the unit sphere.
>
> *Stationarity*: $\text{grad}\ \rho(\mathsf{u}) \equiv \nabla\rho(\mathsf{u}) = 2(\mathsf{Au} - \rho(\mathsf{u})\mathsf{Mu})^*/\mathsf{u}^*\mathsf{Mu}$. Thus ρ is stationary at, and only at, the eigenvectors of (A, M).
>
> *Minimum Residual*: $\|(\mathsf{A} - \sigma\mathsf{M})\mathsf{u}\|^2_{M^{-1}} \geq \|\mathsf{Au}\|^2_{M^{-1}} - |\rho(\mathsf{u})|^2\|\mathsf{Mu}\|^2_{M^{-1}}$, with equality when, and only when $\sigma = \rho(\mathsf{u})$.
>
> (15-9-5)

Numerical methods have been proposed to remedy the "defect" that $\rho(\mathsf{u})$ does not minimize $\|(\mathsf{A} - \sigma\mathsf{M})\mathsf{u}\|$ over all σ. This is rather like forgetting to change currency when visiting a foreign country.

As before we define the residual vector $\mathsf{r}(\mathsf{u})$ by

$$\mathsf{Au} = \rho(\mathsf{u})\mathsf{Mu} + \mathsf{r}(\mathsf{u})$$

and observe that this is again an orthogonal decomposition of Au in the M^{-1} inner product; $(\mathsf{r}(\mathsf{u}), \mathsf{Mu})_{M^{-1}} = 0$. Algebraically it happens that $\mathsf{r}^*\mathsf{u} = 0$ but that is not so significant in the present context.

The Rayleigh quotient iteration (RQI) produces the sequence $\{\mathsf{x}_k\}$ using

$$(\mathsf{A} - \rho_k\mathsf{M})\mathsf{x}_{k+1} = \mathsf{Mx}_k\tau_k, \qquad (15\text{-}9\text{-}6)$$

where τ_k is chosen so that $\|\mathsf{Mx}_k\|_{M^{-1}} = 1$ for all k and $\rho_k = \rho(\mathsf{x}_k)$. The convergence theory of RQI is given in the following three theorems.

> THEOREM Suppose that $\{\mathsf{x}_k\} \longrightarrow \mathsf{z}$, an eigenvector, as $k \longrightarrow \infty$. Let $\psi_k = \angle(\mathsf{Mx}_k, \mathsf{Mz})$ in the M^{-1} inner product. Then
>
> $$\lim_{k \to \infty} |\tan \psi_{k+1}/\tan^3 \psi_k| \leq 1. \qquad (15\text{-}9\text{-}7)$$

> THEOREM $\|(\mathsf{A} - \rho_{k+1}\mathsf{M})\mathsf{x}_{k+1}\|_{M^{-1}} \leq \|(\mathsf{A} - \rho_k\mathsf{M})\mathsf{x}_k\|_{M^{-1}}$ for all k.
>
> (15-9-8)

THEOREM For any nonzero starting vector x_0

(a) $\rho_k \longrightarrow \rho$, as $k \longrightarrow \infty$, and either

(b) $(\rho_k, x_k) \longrightarrow (\lambda, z)$, an eigenpair, cubically, or

(c) $\{x_{2k}\} \longrightarrow x_+$, $\{x_{2k+1}\} \longrightarrow x_-$, linearly, and $x_\pm$ are bisectors of eigenvectors belonging to eigenvalues $\rho \pm \tau$,

$$\tau = \lim_k \|(A - \rho_k M)x_k\|_{M^{-1}}.$$

The regime in (c) is unstable in the face of perturbations. (15-9-9)

15-9-1 OTHER ITERATIVE TECHNIQUES

The power method and RQI are techniques for solving the homogeneous system of equations $(A - \lambda M)z = o$. Of course λ is unknown and so the problem is not linear. Nevertheless almost every known technique for solving linear systems yields an analogous iteration for the eigenvalue problem. For example, there is an SOR (successive overrelaxation) method which can be very effective for special problems when triangular factorization is not possible. The reader is referred to [Ruhe, 1975 and 1977] for a full treatment of these ideas. It is our belief that no iterations of the form $x_{k+1} = \phi_k(x_k)$ can compete with the Lanczos algorithm which discards no previous information.

For the same reason we have not described those methods which seek to minimize $\|(A - \mu M)x\|$ by various gradient methods which have proved useful in the more general problem of minimizing nonlinear functions. In our case the Euclidean norm $\| \cdot \|$ is not natural to the problem and so there is no reason why such iterations should converge rapidly nor any evidence that they do.

One of the simplest iterative techniques is coordinate overrelaxation. The idea is simply to minimize $\rho(x + \alpha e_j; A, M)$ over α for each coordinate vector e_j in turn. Here x is the current approximate eigenvector whose j-th element is to be changed. In practice it turns out advantageous to overrelax and replace x by $x + \omega\alpha e_j$ for some ω in $(1, 2)$. Each step requires two matrix-vector multiplications and, inevitably, the convergence properties are not very satisfactory.

The usual argument in favor of these simple iterative techniques is that they require no factorization and consequently are the **only** methods available for those problems in which A and M are so huge that triangular factorization is supposedly not possible. This sentiment flouts the old adage, "Where there's a will there's a way." The example of the structural engineers suggests that there is no limit on the size of matrices which can be factored. Of course secondary storage must be used heavily when $n > 5000$ and the proper control of transfers between the various storage hierarchies is neither easy nor within the scope of this book.

Even when factorization is rejected the equation $\mathsf{Mx} = \mathsf{b}$ can be solved, either by iteration or by the conjugate gradient technique, and so the Lanczos algorithm and subspace iteration can be used as described in the remaining sections.

Exercises on Sec. 15-9

15-9-1. Prove (15-9-4).

15-9-2. Prove (15-9-5).

15-9-3. Do an operation count for one step of RQI assuming that A and M have half-bandwidth m. Assume that $\mathsf{A} - \rho_k \mathsf{M}$ permits triangular factorization and use the table in Sec. 3-1 for relevant operation counts.

15-9-4. Prove (15-9-7).

15-9-5. Prove (15-9-8).

15-10 RAYLEIGH-RITZ APPROXIMATIONS

Let S be an n by m matrix of full rank $m \leqslant n$. We want to find formulas for those linear combinations of S's columns which are collectively the best approximations to some eigenvectors of (A, M). In other words we want the best approximations from span S. Our criterion for "best" will be imported directly from the standard problem discussed in Chap. 11. This is somewhat unsatisfactory as regards an independent development of the theory of matrix pencils but it does avoid a rather complicated discussion of the proper geometric setting for (A, M).

Every symmetric positive definite matrix has a unique positive definite square root and, for theoretical purposes, it is convenient to examine the reduction of (A, M) to standard form by means of M's square root $\mathsf{M}^{\frac{1}{2}}$. The given basic equation

$$\mathsf{Az} - \mathsf{Mz}\lambda = \mathsf{o} \tag{15-10-1}$$

is rewritten as

$$(\mathsf{M}^{-\frac{1}{2}}\mathsf{AM}^{-\frac{1}{2}} - \lambda)\mathsf{M}^{\frac{1}{2}}\mathsf{z} = \mathsf{M}^{-\frac{1}{2}}\mathsf{o} = \mathsf{o}. \tag{15-10-2}$$

Consequently the objects of interest are $\hat{\mathsf{A}} \equiv \mathsf{M}^{-\frac{1}{2}}\mathsf{AM}^{-\frac{1}{2}}$, $\hat{\mathsf{z}} = \mathsf{M}^{\frac{1}{2}}\mathsf{z}$, and the modified trial vectors in the columns of $\mathsf{M}^{\frac{1}{2}}\mathsf{S}$.

For the reduced problem we can invoke the Rayleigh-Ritz approximations discussed in Sec. 11-4. The approximations are easiest to describe when the basis is orthonormal, i.e., when $(S^*M^{\frac{1}{2}})(M^{\frac{1}{2}}S) = I_m$. For theoretical work the easiest way to normalize S is to use

$$\hat{S} = M^{\frac{1}{2}}S(S^*MS)^{-\frac{1}{2}}. \tag{15-10-3}$$

Thus $\hat{S}^*\hat{S} = I_m$ and the classical Rayleigh-Ritz theory says that

$$\min\|\hat{A}\hat{S} - \hat{S}\hat{H}\|_F, \qquad \text{over } m \text{ by } m \quad \hat{H}, \tag{15-10-4}$$

is achieved uniquely by

$$\hat{H} \equiv \rho(\hat{S}) \equiv \hat{S}^*\hat{A}\hat{S}. \tag{15-10-5}$$

Note the use of $\|\cdot\|_F$ instead of $\|\cdot\|$. Furthermore the best approximations to eigenvectors of $\hat{A}$ from span $\hat{S}$ are $\hat{S}\hat{g}_i$, $i = 1, \ldots, m$, where $\hat{H}\hat{g}_i = \hat{g}_i\theta_i$ and $\|\hat{g}_i\| = 1$. In brief, the Ritz approximations are $\hat{S}\hat{G}$ where $\hat{G}\Theta\hat{G}^*$ is the spectral factorization of $\hat{H}$.

Now let us translate this characterization into expressions involving the original matrices A and M. From (15-10-5) and (15-10-3)

$$\hat{H} = (S^*MS)^{-\frac{1}{2}}(S^*AS)(S^*MS)^{-\frac{1}{2}}. \tag{15-10-6}$$

Fortunately $\hat{H}$ is not needed explicitly; its eigenpairs $(\theta_i, \hat{g}_i)$ yield the Ritz approximations. From (15-10-6) we derive

$$(S^*AS - \theta_i S^*MS)(S^*MS)^{-\frac{1}{2}}\hat{g}_i = o. \tag{15-10-7}$$

This equation is usually written as

$$(A_S - \theta_i M_S)g_i = o, \qquad i = 1, \ldots, m. \tag{15-10-8}$$

The best approximations to $\hat{z}_i$ is $\hat{S}\hat{g}_i$ and so, by (15-10-2), the best approximations to the original z's are the vectors

$$M^{-\frac{1}{2}}\hat{S}\hat{g}_i = S(S^*MS)^{-\frac{1}{2}}\hat{g}_i = Sg_i, \qquad i = 1, \ldots, m. \tag{15-10-9}$$

To summarize,

> The Rayleigh-Ritz approximations to the pencil (A, M) from span S are the pairs (θ_i, Sg_i), $i = 1, \ldots, m$, where (θ_i, g_i) satisfy (15-10-8). They are optimal in the sense of minimizing
>
> $$\|M^{-\frac{1}{2}}R\|_F^2 = \text{trace}(R^*M^{-1}R)$$
>
> over all residual matrices R where $r_i = Ax_i - Mx_i\mu_i$, $i = 1, \ldots, m$, and $x_i^*Mx_j = \delta_{ij}$ (Kronecker symbol), $x_i \in$ span S.

Exercises on Sec. 15-10

15-10-1. Let $\langle \cdot , \cdot \rangle$ be an inner product function on $\mathcal{R}^n$. Find the scalar μ which minimizes, for given x, $\langle \mathbf{Ax} - \mathbf{Mx}\mu, \mathbf{Ax} - \mathbf{Mx}\mu \rangle$.

15-10-2. Derive (15-10-7).

15-11 LANCZOS ALGORITHMS

Familiarity with the contents of Chap. 13 is strongly recommended.

Given A, M (positive definite), and M's Choleski factor L then the Lanczos algorithm, with or without selective orthogonalization, can be applied to the implicitly defined matrix $\mathring{\mathbf{A}} = \mathbf{L}^{-1}\mathbf{A}\mathbf{L}^{-*}$ as described in Sec. 15-7. The user supplied matrix-vector product program OP must incorporate L but the actual Lanczos program need not be modified at all. At the end each computed eigenvector y must be mapped back to an eigenvector z of (A, M) by solving $\mathbf{L}^*\mathbf{z} = \mathbf{y}$ for z. This usage is very satisfactory and standard working practice when M has narrow bandwidth.

On the other hand it is possible to use Lanczos on $\mathbf{M}^{-1}\mathbf{A}$, again defined implicitly, and this option is valuable when M cannot be factored conveniently. In this case the algorithm must be reformulated to take account of M and so the algorithms of Chap. 13 must be modified a little.

Mathematically the extension is immediate. The basic three-term recurrence becomes

$$\mathbf{M}\mathbf{q}_{j+1}\beta_{j+1} = \mathbf{A}\mathbf{q}_j - \mathbf{M}\mathbf{q}_j\alpha_j - \mathbf{M}\mathbf{q}_{j-1}\beta_j.$$

The indices of the β's have been shifted up by one. The Lanczos vectors $\mathbf{q}_1, \mathbf{q}_2, \ldots$ are not mutually orthonormal but they are M-orthonormal, i.e., $\mathbf{q}_i^*\mathbf{M}\mathbf{q}_j = \delta_{ij}$. The tridiagonal matrix $\mathbf{T}_j$ is the same as before and if the algorithm is continued for n steps it produces an invertible matrix Q which reduces (A, M) to (T, I). By the end of step j the algorithm has produced $\mathbf{Q}_j = (\mathbf{q}_1, \ldots, \mathbf{q}_j)$ and $\mathbf{T}_j$, the leading principal j by j submatrix of T. If $(\theta, \mathbf{s})$ is an eigenpair of $\mathbf{T}_j$ then $(\theta, \mathbf{Q}_j\mathbf{s})$ is the corresponding Ritz pair for the pencil (A, M) (see Sec. 15-10).

For theoretical and computational purposes it is helpful to introduce the auxiliary vectors $\mathbf{p}_1, \mathbf{p}_2, \ldots$ defined by $\mathbf{p}_i = \mathbf{M}\mathbf{q}_i$. Note that the sequences $\{\mathbf{q}_i\}$ and $\{\mathbf{p}_i\}$ are bi-orthonormal in $\mathcal{E}^n$, i.e., $\mathbf{p}_k^*\mathbf{q}_j = \delta_{kj}$. Selective orthogonalization generalizes in a straightforward manner. The simple algorithm is recovered by omitting the first step of the algorithm given below.

15-11-1 SELECTIVE LANCZOS ALGORITHM FOR (A, M)

Pick $\mathsf{u}_1 \neq \mathsf{o}$. Compute $\mathsf{r}_1 \longleftarrow \mathsf{M}\mathsf{u}_1$, $\beta_1 \longleftarrow \sqrt{\mathsf{u}_1^*\mathsf{r}_1} > 0$. For $j = 1, 2, \ldots$ repeat steps 1 through 9:

1. If there are any threshold vectors, then purge u_j of threshold vectors, and recompute $\mathsf{r}_j \longleftarrow \mathsf{M}\mathsf{u}_j$, $\beta_j \longleftarrow \sqrt{\mathsf{u}_j^*\mathsf{r}_j}$.
2. $\mathsf{q}_j \longleftarrow \mathsf{u}_j/\beta_j$.
3. $\bar{\mathsf{u}}_j \longleftarrow \mathsf{A}\mathsf{q}_j - \mathsf{p}_{j-1}\beta_j \qquad (\mathsf{p}_0 = \mathsf{o})$.
4. $\alpha_j \longleftarrow \mathsf{q}_j^*\bar{\mathsf{u}}_j$.
5. $\mathsf{p}_j \longleftarrow \mathsf{r}_j/\beta_j$.
6. $\mathsf{r}_{j+1} \longleftarrow \bar{\mathsf{u}}_j - \mathsf{p}_j\alpha_j$.
7. Solve $\mathsf{M}\mathsf{u}_{j+1} = \mathsf{r}_{j+1}$ for u_{j+1} $(= \mathsf{q}_{j+1}\beta_{j+1})$.
8. $\beta_{j+1} \longleftarrow \sqrt{\mathsf{u}_{j+1}^*\mathsf{r}_{j+1}}$.
9. Compute eigenpairs (θ_i, s_i) of T_j as needed. Test for threshold vectors and for a pause. Test for convergence. If satisfied then stop.

15-11-2 REMARKS

1. It has been said that Lanczos cannot be used unless M can be factored. However, iterative methods in Step 7 can make maximal use of sparsity. A simple starting vector is $\mathrm{diag}(\mathsf{M})^{-1}\mathsf{r}_{j+1}$.
2. Steps 5 and 7 are additions to the algorithms of Chap. 13. Step 5 increases the inner product count from five to six. Only four n-vectors of storage are needed; u, r, q, p, an increase of one.
3. q_j can be put into sequential storage after Step 4. There is no need to save p_j.
4. The quantities $\beta_{ji} = \beta_j|\mathsf{e}_j^*\mathsf{s}_i|$ still measure convergence but in the new norm. See Ex. 15-11-2.
5. The algorithm given above can be used to advantage even when M can be factored as indicated below.

15-11-3 HOW TO USE A LANCZOS PROGRAM

A task which occurs frequently is to compute the eigenvectors belonging to all the eigenvalues in an interval $[\sigma - \omega, \sigma + \omega]$. Both A and M will be large and sparse but one or two triangular factorizations may be feasible. In such a case it will be advantageous to apply the Lanczos algorithm to the pair (M, A $-$ σM) instead of (A, M); see [Ruhe and Ericsson, 1980].

If the computation of Aq or the solution of Mu = r involves heavy use of secondary storage then there may be no extra cost in computing AQ or solving MU = R for n by m matrices Q, U, and R for some $m > 1$. In these circumstances the block Lanczos algorithm with selective orthogonalization would seem to be the most effective procedure for computing eigenvalues and eigenvectors. Work on this project is still "in progress."

Exercises on Sec. 15-11

15-11-1. Assume that A and M have half-bandwidth m and that M has been factored into LL*. Do an operation count for one step of the algorithm assuming that Aq requires $(2m + 1)n$ ops.

15-11-2. Of what residual is β_{ji} the norm?

15-11-3. Verify that the same matrix T_j is produced, in exact arithmetic, by the algorithms of Chap. 13 acting on $\mathring{A}$ ($\equiv L^{-1}AL^{-*}$) and by the algorithm of this section applied to (A, M).

15-12 SUBSPACE ITERATION

The methods described in Chap. 14 extend readily to the problem $(A - \lambda M)x = o$ when M is positive definite. As pointed out in Sec. 15-9 the cost of one step of the power method (or direct iteration) is approximately the same as one step of inverse iteration (unless M is diagonal) and the improved convergence rate of the latter makes it the preferred version for computing the small eigenvalues of large, sparse pencils (A, M). The basic idea of subspace iteration is to combine block inverse iteration, using occasional shifts of origin, with the Rayleigh-Ritz approximations at each step.

Sophisticated implementations of the method have become the standard tool in the dynamic analysis of structures of all kinds. Typical problems have $n > 1000$ but fortunately A and M usually have a half-bandwidth around 100 and the pencil (A, M) is positive definite. Perhaps the 10 smallest eigenvalues are wanted together with their eigenvectors or, for earthquake calculations, all the eigenvalues less than some designated value.

At each step k the algorithm computes an n by m matrix $S^{(k)}$ whose columns are the current eigenvector approximations. As shown in Chap. 14 some columns will converge faster than others and in practice it is important to take advantage of this phenomenon although we shall discuss it no further. One disadvantage of subspace iteration is that m should be chosen larger than the number of eigenvectors actually wanted. The improved convergence rate more than makes up for the extra work but there is no really satisfying way to decide a priori on how many extra columns to carry. What is more the efficiency of the program depends quite sensitively on the choice of block size.

The algorithm is set out in Table 15-12-1.

Table 15-12-1 Summary of subspace iteration†

DATA: A, M — n by n, average half-bandwidth ω for both A and M. $S^{(k)}$ — n by m, approximate eigenvectors at step k. $\Theta^{(k)}$ — m by m, diagonal matrix of Ritz values. $1 \leqslant m < \omega < n$.	
INITIAL CALCULATIONS:	OPERATION COUNT
I. Select origin shift σ.	
II. Factor $A - \sigma M$ into $L\Delta L^*$.	$n(\omega + 1)(\omega + 2)/2$
III. Select m starting vectors to be the columns of $S^{(0)}$. Set $\Theta^{(0)} = I_m$	$\approx nm$
ITERATION: For $k = 1, 2, \ldots$	
1. Orthogonalize $S^{(k-1)}$ against known eigenvectors, when necessary.	
2. Form right-hand side, scaled by $\Theta^{(k-1)}$, $R^{(k)} = MS^{(k-1)}\Theta^{(k-1)}$.	$2nm(\omega + 1)$
3. Solve $LL^*\bar{S}^{(k)} = R^{(k)}$.	$2nm\omega$
4. "Project" A and M onto span $\bar{S}^{(k)}$ to get $A^{(k)} = \bar{S}^{(k)*}R^{(k)}$, $M^{(k)} = \bar{S}^{(k)}M\bar{S}^{(k)}$.	$2nm^2 + nm(2\omega + 1)$
5. Solve the m by m eigenvalue problem $(A^{(k)} - \sigma M^{(k)})G^{(k)} = M^{(k)}G^{(k)}\Theta^{(k)}$ for shifted Ritz values $\Theta^{(k)}$ and eigenvector matrix $G^{(k)}$ which is orthonormal with respect to $M^{(k)}$.	$10m^3$
6. Form new basis $S^{(k)} = \bar{S}^{(k)}G^{(k)}$.	nm^2
7. Test for convergence.	
SUBTOTAL	$3nm(2\omega + m + 1) + \cdots$

†Adapted from [Wilson, 1978]

15-12-1 COMMENTS ON TABLE 15-12-1

(I). The selection of σ is discussed in Sec. 14-5. In many applications there is extra information available to guide the user. Here is an instance of the difficulty in making subspace iteration into a black box program. The efficiency depends strongly on the choice of this parameter.

(II). See Chap. 3 for more details. Check $\nu(\Delta)$ to see where σ actually slices the spectrum.

(III). Sometimes knowledge of the application can guide the selection of starting vectors. The columns of $S^{(0)}$ need not be orthonormal with respect to M but they should be strongly linearly independent. In the absence of adscititious information a reasonable choice is to take column 1 as the sum of the first $[n/m]$ columns of I_n, column 2 as the sum of the second $[n/m]$ columns, and so on. Vectors with random elements do not seem to be very effective. The starting vectors must be orthogonalized against any known eigenvectors.

1. When eigenvectors are computed in batches (ℓ at a time) it is important to keep the current subspace orthonormal (with respect to M) to all known eigenvectors. However it is not necessary to orthogonalize at each iteration. The frequency with which a known eigenvector z_i should be purged depends on the ratio $\max_j |\theta_j^{(k-1)}|/|\lambda_i - \sigma|$. Thus if λ_1 is very close to σ then z_1 will converge in one or two steps and thereafter it can be deflated by restricting the columns of $S^{(k)}$ to be orthogonal to z_1. Of course m is reduced with resulting gains in efficiency.

4. As k increases the full m by m matrices $A^{(k)}$ and $M^{(k)}$ are increasingly dominated by their diagonal elements. This makes the Jacobi method described in Sec. 15-7 very attractive, despite the absence of any proof of convergence, because the number of sweeps required to make the off-diagonal elements negligible is usually between 3 and 6, depending on the working precision.

15-13 PRACTICAL CONSIDERATIONS

Those who are concerned with the computation of eigenvalues and eigenvectors of large pencils (A, M) are well aware that success depends on the happy combination of techniques for (a) generating A and M, (b) moving information in and out of the fast store, (c) computing the results, and (d) handling input/output. For large problems it becomes increasingly difficult to isolate the numerical method from the rest of the environment. An important consequence is the difficulty of measuring, a priori, the cost of a method and hence its efficiency.

When it is possible to execute m matrix-vector products Ax_1, $Ax_2, \ldots, Ax_m$ as quickly as a single one then the relative speeds of contending methods may be strongly affected and our habitual judgments are upset. At a lower level we observe that the times required for a fetch, a store, an add, a multiply, and a divide are all getting closer to each other and this trend will make many conventional operation counts misleading. On the other hand the importance of the divide/multiply ratio has simply moved to a higher level of abstraction: the problem-dependent ratio of the cost of solving $\mathbf{Ax} = \mathbf{b}$ to the cost of forming $\mathbf{Ab}$. This is where the sparsity structure and the computer system enter the picture in a crucial way.

This book has concentrated on those aspects of the numerical methods which are independent of the vagaries of the computing system but there is no intention of belittling the others. As numerical methods become better adapted to the current facilities so will the bottleneck in these routine calculations move to the data management facets of the task.

Notes and References

Matrix pencils are discussed in some detail in [Gantmacher, 1959, vol. II, chap. 2]. Most of the material in Sec. 15-2 comes from [Moler and Stewart, 1973] and [Stewart, 1979]. Definite quadratic forms are discussed in many books: [Strang, 1976], [Stewart, 1973], and [Franklin, 1968] to name a few.

The explicit reduction to standard form is covered in [Wilkinson, 1965] but the various options are not usually placed in close proximity. Sec. 15-5 comes from [Fix and Heiberger, 1972], Sec. 15-6 from [Moler and Stewart, 1973], and Sec. 15-7 from [Zimmermann, 1969] via [Bathé and Wilson, 1976].

The implicit reduction to standard form is well known but is not usually presented in this way. The extension of the Rayleigh quotient iteration and the Rayleigh-Ritz procedure to pencils is given in some detail because the material did not seem to be readily available elsewhere. It was tempting to present the Rayleigh-Ritz procedure from a more abstract point of view, letting it emerge naturally from the proper geometric setting. However the necessary preparation seems to exceed the reward.

References for coordinate relaxation are [Schwarz, 1977] for theory, and [Shavitt, Bender, and Pipano, 1973] on the practical aspects. This book does not treat relaxation methods and the reader is referred to the timely and comprehensive survey [Ruhe, 1977].

References for the Lanczos algorithm applied to $(\mathbf{A}, \mathbf{M})$ are [Weaver and Yoshida, 1971], [Golub, Underwood, and Wilkinson, 1972], and [Cullum and Donath, 1974]. The idea of using $\mathbf{M}^{-1}\mathbf{A}$ implicitly can be found in [Wiberg, 1973] and [McCormick and Noe, 1977]. A termination criterion based on watching the behavior of the Ritz values $\theta_i^{(j)}$ is given in [van Kats and van der Vorst, 1976]. The idea of the band Lanczos comes from [Ruhe, 1979].

The structural engineers have done such extensive work with subspace iteration that the tale of all the gimmicks and modifications they have tried would be a long one. Two recent references [Bathé and Wilson, 1976] and [Jennings, 1977] seem to synthesize all that experience. The whole field of large eigenvalue computations is surveyed in admirable style in [Stewart, 1976].

Appendix A

Rank One and Elementary Matrices

The elementary row operations on a matrix B are: (a) interchange two rows, (b) multiply a row by a nonzero scalar, and (c) add a multiple of one row to another. They can be effected by pre-multiplying B by appropriate rather simple matrices which are traditionally called elementary. The concept of an elementary matrix gains significance in the light of the important result that any invertible matrix can be written as a product of elementary matrices (in many ways).

In the 1950s Householder revised the definition to make it both simpler and more general.

Definition. An **elementary matrix** is any square matrix of the form

$$\mathsf{I} + \text{a rank one matrix.} \tag{A-1}$$

Elementary matrices can be stored in compact form and their inverses are easy to write down as we now show.

All **rank one** matrices are of the form xy^*, since every column is a multiple of x and every row is a multiple of y^*. It is standard practice to normalize x and y and write a rank one matrix as

$$\mathsf{u}\sigma\mathsf{v}^* \tag{A-2}$$

where $\|\mathsf{u}\| = \|\mathsf{v}\| = 1$. Consequently (A-1) can be rewritten as

$$\mathsf{E} = \mathsf{I} - \mathsf{u}\sigma\mathsf{v}^*. \tag{A-3}$$

It is easy to verify that E^{-1} exists if, and only if, $\sigma \mathsf{v}^*\mathsf{u} \neq 1$, in which case

$$\mathsf{E}^{-1} = \mathsf{I} - \mathsf{u}\tau\mathsf{v}^* \tag{A-4}$$

where $\tau^{-1} + \sigma^{-1} = \mathsf{v}^*\mathsf{u}$. Thus (A-4) confirms that E^{-1} is also elementary. Observe that u, v, σ, and τ define both E and E^{-1} and, in practical work, the product EB can be formed with only $2n^2$ ops as against n^3 if the structure of E is ignored. Observe that, in general, E would be a full matrix if it were to be formed explicitly.

Elementary matrices are indeed useful tools.

Exercise on Appendix A

A-1. Find the values of u, v, and σ which yield the traditional elementary matrices described in the first sentence.

Appendix B

Chebyshev Polynomials

For a given positive integer k the function $\cos k\phi$ is **not** a polynomial in ϕ but it does happen to be a polynomial of degree k in $\cos \phi$. This polynomial is, of course, well defined for values of its argument outside the interval $[-1, 1]$ but it cannot be expressed in terms of cosines. If $\cos \phi = \xi$ then $\phi = \cos^{-1} \xi = \operatorname{arc\,cos} \xi$ and

$$T_k(\xi) = \begin{cases} \cos(k \operatorname{arc\,cos} \xi) = \cos k\phi, & -1 \leqslant \xi \leqslant 1, \\ \cosh(k \operatorname{arc\,cosh} \xi), & |\xi| > 1, \end{cases}$$

$$= \tfrac{1}{2}\left[\left(\xi + \sqrt{\xi^2 - 1}\right)^k + \left(\xi + \sqrt{\xi^2 - 1}\right)^{-k}\right], \qquad |\xi| > 1,$$

$$\sim \tfrac{1}{2}(2\xi)^k \text{ as } \xi \longrightarrow \infty, \qquad k \text{ fixed.}$$

Moreover

$$T_k(1) = 1, \qquad \text{for all } k,$$

$$T_k(1 + 2\epsilon) \sim \tfrac{1}{2}(1 + 2\sqrt{\epsilon} + 2\epsilon)^k, \qquad \text{for } 0 \leqslant \epsilon < 0.1,$$

$$\sim \tfrac{1}{2}\exp(2k\sqrt{\epsilon}\,) \qquad \text{for } k\sqrt{\epsilon} > 1.$$

It is the rapid growth of $T_k(1 + \epsilon)$ for small ϵ and large k that makes Chebyshev acceleration so useful. See Table B-1.

Table B-1 Representative values of $T_m(1 + 2\gamma)$

m \ γ	10^{-4}	10^{-3}	10^{-2}	10^{-1}
10	1.0201	1.2067	3.7502	2.5227×10^2
10^2	3.7621	2.7876×10^2	2.3466×10^8	5.3436×10^{26}
2×10^2	2.7306×10^1	1.5542×10^5	1.1014×10^{17}	5.7107×10^{53}
10^3	2.4250×10^8	1.4507×10^{27}	2.5927×10^{86}	9.7179×10^{269}

A simple three-term recurrence permits the evaluation of $T_k(\xi)$ without knowledge of T_k's coefficients. The recurrence is a disguised form of

$$\cos(k + 1)\phi + \cos(k - 1)\phi = 2 \cos \phi \cos k\phi,$$

namely

$$T_{k+1}(\xi) = 2\xi T_k(\xi) - T_{k-1}(\xi).$$

The polynomial $T_k(\xi)/2^{k-1}$ is the smallest monic polynomial of degree k provided that the size of a function is taken as its maximal absolute value on $[-1, 1]$. Of more importance is the related fact that of all polynomials p of degree $\leqslant k$ which satisfy $p(\gamma) = \delta$ for some $|\gamma| > 1$ the smallest on $[-1, 1]$ is

$$\bar{p}(\xi) = \delta T_k(\xi)/T_k(\gamma)$$

and

$$\|\bar{p}\|_\infty = \max_{-1 \leqslant \xi \leqslant 1} |\bar{p}(\xi)| = |\delta|/|T_k(\gamma)|.$$

In order to get the analogous results for an interval $[\alpha, \beta]$ use the adapted Chebyshev polynomial

$$T_k(\xi : \alpha, \beta) \equiv T_k[(2\xi - \alpha - \beta)/(\beta - \alpha)],$$

$$= T_k\left(1 + 2\frac{\xi - \beta}{\beta - \alpha}\right).$$

The leading coefficient of $T_k(\xi; \alpha, \beta)$ is $\frac{1}{2}[4/(\beta - \alpha)]^k$.

The optimal characteristics of T_k stem from its famous **equioscillation** property: on $[-1, 1]$ T_k switches between its maximum absolute value with alternating signs exactly k times. No other polynomial of degree k or less can equioscillate more apart from multiples of T_k.

Extrema of T_7 are $\cos(j\pi/7), j = 0, \ldots, 7$

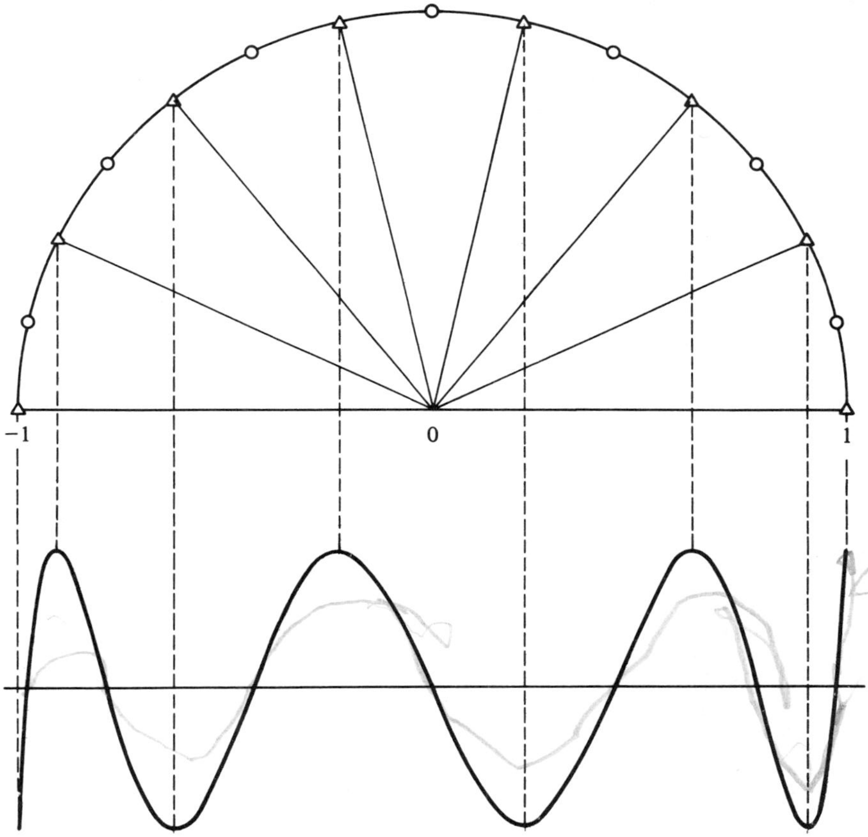

Key: △ Extrema
○ Zeros

Figure B-1 Graph of T_7

References

Bargmann, V.; C. Montgomery; and J. von Neumann, 1946. "Solution of Linear Systems of High Order," Princeton: Institute for Advanced Study.

Bathé, K. – J., and E. Wilson, 1976. *Numerical Methods in Finite Element Analysis*. Englewood Cliffs, N.J.: Prentice-Hall.

Bauer, F. L., 1957. "Das Verfahren der Treppeniteration und Verwandte Verfahren zur Lösung Algebraischer Eigenwertprobleme." *Zamp*, 8:214–235.

Brent, R. P., 1973. *Algorithms for Minimization without Derivatives*. Englewood Cliffs, N.J.: Prentice-Hall.

Browne, E. T., 1930. "On the Separation Property of the Roots of the Secular Equation." *Amer. J. Math.*, 52:841–850.

Bunch, J. R., and D. J. Rose, eds., 1976. *Sparse Matrix Computations*. New York: Academic Press.

Bus, J. C., and T. J. Dekker, 1975. "Two Efficient Algorithms with Guaranteed Convergence for Finding a Zero of a Function." *Trans. on Mathematical Software*, 1:330–345.

Cauchy, A., 1821. "*Cours D'Analyse de L'Ecole Polytechnique*." *Oeuvres Complètes*, vols. 2 and 3.

Chatelin, F., and J. Lemordant, 1978. "Error Bounds in the Approximation of the Eigenvalues of Differential and Integral Operators." *J. Math. Anal. Appl.*, 22:257–271.

Cline, A. K.; G. H. Golub; and G. W. Platzman, 1976. "Calculation of Normal Modes of Oceans Using a Lanczos Method." In *Sparse Matrix Computations*, eds. J. R. Bunch, and D. J. Rose, pp. 409–426. New York: Academic Press.

Corneil, D., 1965. *Eigenvalues and Orthogonal Eigenvectors of Real Symmetric Matrices*. Master's thesis, Dept. of Computer Science, University of Toronto.

Crandall, S. H., 1951. "Iterative Procedures Related to Relaxation Methods for Eigenvalue Problems." *Proc. Roy. Soc. London*, Ser. A, 207:416–423.

Crawford, C. R., 1973. "Reduction of a Band-Symmetric Generalized Eigenvalue Problem." *Comm. A.C.M.*, 16:41–44.

______, 1976. "A Stable Generalized Eigenvalue Problem." *Soc. Ind. Appl. Math. J. Num. Anal.*, 6:854–860.

Cullum, J., and W. E. Donath, 1974. "A Block Generalization of the Symmetric *S*-step Lanczos Algorithm." Report #RC 4845 (#21570). Yorktown Heights, New York; IBM Thomas J. Watson Research Center.

Cullum, J., and R. A. Willoughby, 1979. "Lanczos and the Computation in Specified Intervals of the Spectrum of Large, Sparse, Real Symmetric Matrices." In *Sparse Matrix Proceedings 1978*, eds. I. Duff and G. W. Stewart. Philadelphia: SIAM Publ.

______,1980. "Computing Eigenvectors (and Eigenvalues) of Large, Symmetric Matrices using Lanczos Tridiagonalization." *Proc. Biennial Conf. on Num. Anal.*, Univ. of Dundee, Scotland, ed. G. A. Watson. New York: Springer-Verlag.

Daniel, J. W.; W. B. Gragg; L. Kaufman; and G. W. Stewart, 1976. "Reorthogonalization and Stable Algorithms for Updating the Gram-Schmidt QR Factorization." *Math. Comp.*, 30:772–795.

Davidson, E. R., 1975. "The Iterative Calculation of a Few of the Lowest Eigenvalues of Corresponding Eigenvectors of Large Real Symmetric Matrices." *Jour. Comp. Phys.*, 17:87–94.

Davis, C., and W. Kahan, 1970. "The Rotation of Eigenvectors by a Perturbation–III." *Soc. Ind. Appl. Math. J. Num. Anal.*, 7:1–46.

De Boor, C., and G. H. Golub, 1978. "The Numerically Stable Reconstruction of a Jacobi Matrix from Spectral Data." *Lin. Alg. and its Appl.*, 21:245–260.

Dekker, T. J., and J. F. Traub, 1971. "The Shifted QR Algorithm for Hermitian Matrices." *Lin. Alg. and its Appl.*, 11:137–154.

Faddeev, D. K., and V. N. Faddeeva, 1963. *Computational Methods of Linear Algebra*. Translated by R. C. Williams. San Francisco: W H. Freeman.

Fischer, E., 1905. "Concerning Quadratic Forms with Real Coefficients." *Monatsh. Math Phys.*, 16:234–249.

Fix, G., and R. Heiberger, 1972. "An Algorithm for the Illconditioned Generalized Eigenvalue Problem." *Soc. Ind. Appl. Math. J. Num. Anal.*, 9:788–88.

Forsythe, G., and C. B. Moler, 1967. *Computer Solutions of Linear Algebraic Systems*. Englewood Cliffs, N.J.: Prentice-Hall.

Fox, L., 1964. *An Introduction to Numerical Linear Algebra*. New York: Oxford Univ. Press.

Francis, J. G. F., 1961 and 1962. "The QR Transformation, Parts I and II." *Computer J.*, 4:265–271, 332–345.

Franklin, J. N., 1968. *Matrix Theory*. Englewood Cliffs, N.J.: Prentice-Hall.

Gantmacher, F. R., 1959. *The Theory of Matrices*, vol. II. New York: Chelsea Publ. Co..

Gentleman, W. M., 1973. "Least Squares Computations by Givens Transformations without Square Roots." *J. Inst. Math. Appl.*, 12:329–336.

______, 1975. "Error Analysis of QR Decompositions by Givens Transformations." *Lin. Alg. and its Appl.*, 10:189–197.

Gill, P. E.; G. H. Golub; W. Murray; and M. A. Saunders, 1974. "Methods for Modifying Matrix Factorizations." *Math. Comp.*, 28:505–535.

Gill, P. E.; W. Murray; and M. A. Saunders, 1975. "Methods for Computing and Modifying the LDU Factors of a Matrix." *Math. Comp.*, 29:1051–1077.

Givens, W., 1954. "Numerical Computation of the Characteristic Values of a Real Symmetric Matrix." ORNL-1574. Oak Ridge, Tenn.: Oak Ridge National Laboratory.

Glauz, G., 1974. Private communication.

Golub, G. H., 1973. "Some Uses of the Lanczos Algorithm in Numerical Linear Algebra." In *Topics in Numerical Analysis*, ed. J. J. H. Miller, pp. 173–184. New York: Academic Press.

Golub, G. H.; R. Underwood; and J. H. Wilkinson, 1972. "The Lanczos Algorithm for the Symmetric $Ax = \lambda Bx$ Problem." Technical Report STAN-CS-72-270. Computer Science Dept., Stanford University.

Golub, G. H., and J. H. Welsch, 1969. "Calculation of Gauss Quadrature Rules." *Math. Comp.*, 23:221–230.

Hald, O. H., 1976. "Inverse Eigenvalue Problems for Jacobi Matrices." *Lin. Alg. and its Appl.*, 14:63–85.

______, 1977. "Discrete Inverse Sturm–Liouville Problems." *Num. Math.*, 27:249–256.

Hammerling, S., 1974. "A Note on Modification to the Givens Plane Rotation." *J. Inst. Math. Appl.*, 13:215–218.

Henrici, P., 1958. "On the Speed of Convergence of Cyclic and Quasicyclic Jacobi Methods for Computing Eigenvalues of Hermitian Matrices." *J. Soc. Ind. Appl. Math.*, 6: 144–162.

Hochstadt, H., 1975. "On Inverse Problems Associated with Sturm–Liouville Operators." *Jour. of Diff. Equas.*, 17:220–235.

Hoffman, and B. N. Parlett, 1978. "A New Proof of Global Convergence for the Tridiagonal *QL* Algorithm." *Soc. Ind. Appl. Math. J. Num. Anal.*, 15: 929–937.

Hotelling, H., 1943. "Some New Methods in Matrix Calculation." *Ann. Math. Stat.*, 14:1–34.

Householder, A. S., 1958. "A Class of Methods for Inverting Matrices." *J. Soc. Ind. Appl. Math.*, 6:189–195.

______, 1961. "On Deflating Matrices." *J. Soc. Ind. Appl. Math.*, 9:89–93.

______, 1964. *The Theory of Matrices in Numerical Analysis*. Johnson, Colo.: Blaisdell Pub. Co..

Jacobi, C. G. J., 1846. "Concerning an Easy Process for Solving Equations Occurring in the Theory of Secular Disturbances". *J. Reine Angew. Math.*, 30:51–94.

Jennings, A., 1971. "Accelerating the Convergence of Matrix Iterative Processes." *J. Inst. Math. Appl.*, 7:99–110.

______, 1977. *Matrix Computation for Engineers and Scientists*. New York: John Wiley.

Jennings, A., and T. J. A. Agar, 1978. "Hybrid Sturm Sequence and Simultaneous Iteration Methods." in *Proc. of Symp. on Applic. of Computer Methods in Eng*. Los Angeles: University of Southern California Press.

Jensen, P. S., 1972. "The Solution of Large Symmetric Eigenproblems by Sectioning." *Soc. Ind. Appl. Math. J. Num. Anal.*, 9:534–545.

Kahan, W., 1966. "When to Neglect Off-Diagonal Elements of Symmetric Tridiagonal Matrices." Tech. Report No. CS42. Computer Science Dept., Stanford University.

______, 1967. "Inclusion Theorems for Clusters of Eigenvalues of Hermitian Matrices." Tech Report No. CS42. Computer Science Dept., University of Toronto.

Kahan, W., and B. N. Parlett, 1976. "How Far Should You Go with the Lanczos Algorithm?" In *Sparse Matrix Computations*, eds. J. R. Bunch, and D. J. Rose, pp. 131–144. New York: Academic Press.

Kaniel, S., 1966. "Estimates for Some Computational Techniques in Linear Algebra." *Math. Comp.*, 20:369–378.

Kato, T., 1949. "On the Upper and Lower Bounds of Eigenvalues." *J. Phys. Soc. Japan*, 334–339.

______, 1966. *Perturbation Theory for Linear Operators*. New York: Springer-Verlag.

For general perturbation theory see [Kato, 1966, Chap. 2].

Knuth, D. E., 1969. *The Art of Computer Programming. Fundamental Algorithms*, vol. 1 Reading, Mass.:Addison-Wesley Publ. Co.

Krein, M. G., 1945. "On Self-Adjoint Extensions of Bounded and Semibounded Hermitian Transforms." *Dokl. Akad. Nauk. S.S.R.*, 48:303–306.

Krylov, A. N., 1931. "On the Numerical Solution of Equations which in Technical Questions are Determined by the Frequency of Small Vibrations of Material Systems." *Izv. Akad. Nauk. S.S.S.R. Otd. Mat. Estest.*, 1:491–539.

Kublanovskaya, V. N., 1961. "On Some Algorithms for the Solution of the Complete Eigenvalue Problem." *Zh. Vych. Mat.*, 1:555–570.

Lanczos, C., 1950. "An Iteration Method for the Solution of the Eigenvalue Problem of Linear Differential and Integral Operators." *J. Res. Nat. Bur. Standards*, Sect. B 45:225–280.

Lehmann, N. J., 1949. "Calculation of Eigenvalue Bounds in Linear Problems." *Arch. Math.*, 2:139–147.

______, 1963. "Optimale Eigenwerteinschiessungen." *Num. Math.*, 5: 246–272.

______, 1966. "On Optimal Eigenvalue Localization in the Solution of Symmetric Matrix Problems." *Num. Math.*, 8:42–55.

McCormick, S. F., and T. Noe, 1977. "Simultaneous Iteration for the Matrix Eigenvalue Problem." *J. Lin. Alg. Appl.*, 16:43–56.

Moler, C. B., and G. W. Stewart, 1973. "An Algorithm for Generalized Matrix Eigenvalue Problems." *Soc. Ind. Appl. Math. J. Num. Anal.*, 10:241–256.

Ortega, J. M., and H. F. Kaiser, 1963. "The LL^T and QR Methods for Symmetric Tridiagonal Matrices." *Numer. Math.*, 5:211–225.

Ostrowski, A. M., 1958, 1959. "On the Convergence of the Rayleigh Quotient Iteration for the Computation of Characteristic Roots and Vectors, I and II." *Arch. Rational Mech. Anal.*, 1:233–241; 2:423–428.

Paige, C. C., 1971. *The Computation of Eigenvalues and Eigenvectors of Very Large Sparse Matrices*. Ph.D. thesis, Univ. of London.

______, 1972. "Computational Variants of the Lanczos Method for the

Eigenproblem." *J. Inst. Math. Appl.*, 10:373–381.

______, 1976. "Error Analysis of the Lanczos Algorithm for Tridiagonalizing a Symmetric Matrix." *J. Inst. Math. Appl.*, 18:341–349.

Parlett, B. N., 1964. "The Origin and Development of Methods of LR Type." *S.I.A.M. Rev.* 6:275–295.

______, 1971. "Analysis of Algorithms for Reflections in Bisectors." *Soc. Ind. Appl. Math. Rev.*, 13: 197–208.

______, 1974. "The Rayleigh Quotient Iteration and Some Generalizations for Nonnormal Matrices." *Math. Comp.*, 28:679–693.

Parlett, B. N., and W. Kahan, 1969. "On the Convergence of a Practical QR Algorithm." Information Processing 68 (Proc. IFIP Congress, Edinburgh, 1968). *Mathematical Software*, sect. 1, pp. 114–118. Amsterdam: North-Holland.

Parlett, B. N., and W. G. Poole, 1973. "A Geometric Convergence Theory for the QR, LU, and Power Iterations." *Soc. Ind. Appl. Math. J. Num. Anal.*, 10:389–412.

Parlett, B. N., and D. S. Scott, 1979. "The Lanczos Algorithm with Selective Orthogonalization." *Math. Comp.*, 33:217–238.

Parrott, S., 1978. "On a Quotient Norm and the Sz.-Nagy-Foias Lifting Theorem." *J. of Functional Analysis*, 30:311–328.

Rayleigh, Lord (J. W. Strutt), 1899. "On the Calculation of the Frequency of Vibration of a System in its Gravest Mode, with an Example from Hydrodynamics." *Philos. Mag.*, 47:556–572.

Reinsch, C. H., 1971. "A Stable Rational QR Algorithm for the Computation of the Eigenvalues of an Hermitian, Tridiagonal Matrix." *Num. Math.*, 25:591–597.

Ritz, W., 1909. "Über eine neue Method zur Lösung Gewisser Variationsprobleme der Mathematischen Physik." *J. Rein. Angew. Math.*, 135:1–61.

Ruhe, A., 1975. "Iterative Eigenvalue Algorithms Based on Convergent Splittings." *J. Comp. Phys.*, 19:110–120.

______, 1977. "Computation of Eigenvalues and Eigenvectors." In *Sparse Matrix Techniques, Copenhagen, 1976, LNM,* pp.130–184. New York: Springer-Verlag.

______, 1979. "Implementation Aspects of Band Lanczos Algorithm for Computation of Eigenvalues of Large Sparse Matrices." *Math. Comp.*, 33:680–687.

Ruhe, A., and T. Ericsson, 1980. "The Spectral Transformation Lanczos Method in the Numerical Solution of Large, Sparse, Generalized, Symmetric Eigenvalue Problems." *Math. Comp.,* 34:000–000.

Rutishauser, H., 1969. "Computational Aspects of F. L. Bauer's Simultaneous Iteration Method." *Numer. Math.*, 13:4–13.

______, 1971. "The Jacobi Method for Real Symmetric Matrices." In *Handbook for Automatic Computation (Linear Algebra)*, eds. J. H. Wilkinson, and C. H. Reinsch, pp. 202–211. New York: Springer-Verlag.

______,1971. "Simultaneous Iteration Method for Symmetric Matrices." In *Handbook for Automatic Computation (Linear Algebra)*, eds. J. H. Wilkinson, and C. H. Reinsch, pp. 284–302. New York: Springer-Verlag.

Saad, Y., 1974. "Shifts of Origin for the QR Algorithm." Toronto: Pro. IFIP Congress.

______, 1980. "Error Bounds on the Interior Rayleigh-Ritz Approximations from Krylov Subspaces." *Soc. Ind. Appl. Math. J. Num. Anal.*, 17:000–000.

Sack, R. A., 1972. "A Fully Stable Rational Version of the QR Algorithm for Tridiagonal Matrices." *Numer. Math.*, 18:432–441.

Schönhage, A., 1961. "On the Convergence of the Jacobi Process." *Num. Math.*, 3:374–380.

Schwarz, H. R., 1970. "The Method of Conjugate Gradients in Least Squares Fitting." *Z. Ver.*, 95:130–140.

______, 1974. "The Eigenvalue Problem $(A - \lambda B)x = 0$ for Symmetric Matrices of High Order." *Comp. Meth. in Appl. Mech. Eng.*, 3:11–28.

______, 1977. "More Results on $A - \lambda B$." *Comp. Meth in Appl. Mech. Eng.*, 12:181–199.

Scott, D., 1978. *Analysis of the Symmetric Lanczos Algorithm*. Ph.D. dissertation, Dept. of Mathematics, University of California, Berkeley.

Shavitt, I.; C. F. Bender; and A. Pipano, 1973. "The Iterative Calculation of Several of the Lowest or Highest Eigenvalues and Corresponding Eigenvectors of Very Large Symmetric Matrices." *Jour. Comp. Phys.*, 11:90–108.

Stewart, G. W., 1970. "Incorporating Origin Shifts into the QR Algorithm for Symmetric Tridiagonal Matrices." *Comm. Assoc. Comp. Mach.*, 13:365–1367.

______, 1973. *Introduction to Matrix Computation*. New York: Academic Press.

______, 1976. "A Bibliographic Tour of the Large, Sparse Generalized Eigenvalue Problem." In *Sparse Matrix Computations*, eds. J. R. Bunch and D. J. Rose. New York: Academic Press.

______, 1976. "The Economic Storage of Plane Rotations." *Numer. Math.*, 25:137–138.

______, 1979. "Perturbation Bounds for the Definite Generalized Eigenvalue Problem." *Lin. Alg. and its Appl.*, 23:69–85.

Strang, G., 1976. *Linear Algebra and its Applications*. New York: Academic Press.

Szegö, G., 1939. *Orthogonal Polynomials*. No. 23. New York: Am. Math. Soc. Colloq. Publ.

Takahasi, H., and M. Natori, 1971–1972. "Eigenvalue Problem of Large Sparse Matrices." Report of the Computer Centre, University of Tokyo, 4:129–148.

Temple, G., 1933. "The Computation of Characteristic Numbers and Characteristic Functions." *Proc. London Math. Soc.*, 2:257–280.

______, 1952. "The Accuracy of Rayleigh's Method of Calculating the Natural Frequencies of Vibrating Systems." *Proc. Roy. Soc. London*, Ser. A. 211:204–224.

Temple, G., and W. G. Bickley, 1933. *Rayleigh's Principle and its Applications to Engineering*. London: Constable.

Thompson, R. C., and P. McEnteggert, 1968. "Principal Submatrices II, the Upper and Lower Quadratic Inequalities." *Lin. Alg. and its Appl.*, 1:211–243.

Traub, J., 1964. *Iterative Methods for the Solution of Equations*. Englewood Cliffs, N.J.: Prentice-Hall.

Uhlig, F., 1979. "A Recurring Theorem About Pairs of Quadratic Forms and Extensions: A Survey." *J. Lin. Alg. Appls.*, 25:219–238.

Underwood, R., 1975. *An Iterative Block Lanczos Method for the Solution of Large Sparse Symmetric Eigenproblems*. Ph.D. dissertation, STAN-CS-75-496, Stanford University.

van Kats, J. M., and H. A. van der Vorst, 1976. "Numerical Results of the Paige-Style Lanczos Method for the Computation of Extreme Eigenvalues of Large Sparse Matrices." Tech. Report 3. Utrecht, Netherlands: Academic Computer Center.

Weaver, W., and D. M. Yoshida, 1971. "The Eigenvalue Problem for Banded Matrices." *Computers and Structures*, 1:651–664.

Weinberger, H. F., 1974. *Variational Methods for Eigenvalue Approximation*. Philadelphia: Soc. Ind. Appl. Math.

Weinstein, A., 1935. "Sur la Stabilité des Plaques Encastrées." *C. R. Acad. Sci. Paris*, 200:107–109.

Weyl, H., 1912, submitted for publication in 1911. "The Laws of Asymptotic Distribution of the Eigenvalues of Linear Partial Differential Equations." *Math. Ann.*, 71:441–479.

Whitehead, R. R., 1972. "A Numerical Approach to Nuclear Shell-Model Calculations." *Nuclear Physics*, A182:290–300.

Wiberg, T., 1973. "A Combined Lanczos and Conjugate Gradient Method for the Eigenvalue Problem of Large Sparse Matrices." Tech. Report UMIN-4273. Umea, Sweden: Dept. of Information Processing, S-90187.

Wielandt, H., 1967. *Topics in the Theory of Matrices*. (Lecture notes prepared by R. R. Meyer). Madison: University of Wisconsin Press.

Wilkinson, J. H., 1960. "Householder's Method for the Solution of the Algebraic Eigenproblem." *Comp. Jour.*, 3:23–27.

______, 1962. "Note on the Quadratic Convergence of the Cyclic Jacobi Process." *Num. Math.*, 4:296–300.

______, 1964. *Rounding Errors in Algebraic Process*. Englewood Cliffs, N.J.: Prentice-Hall.

______, 1965. *The Algebraic Eigenvalue Problem*. New York: Oxford Univ. Press.

Wilkinson, J. H., and C. H. Reinsch, 1971. *Handbook for Automatic Computation. Linear Algebra*, vol. 2. New York: Springer-Verlag.

Wilson, E. 1978. Private communication.

Zeeman, E. C., 1976. "Catastrophe Theory." *Sci. Amer.*, 234:65–83.

Zimmerman, K., 1969. "On the Convergence of a Jacobi Process for Ordinary and Generalized Eigenvalue Problems." Dissertation No. 4305. Zurich: Eidgenossische Technische Hochschule.

Annotated Bibliography

MATRIX THEORY

Noble, B., and J. W. Daniel, *Applied Linear Algebra*. 2nd ed. Englewood Cliffs, N.J.: Prentice-Hall. 1977.

Popular, broad in scope, not too demanding of the reader. The new edition brings several improvements and removes several errors.

Franklin, J. N., *Matrix Theory*. Englewood Cliffs, N.J.: Prentice-Hall, 1968.

Clear, attractive to mathematicians, brings out the use of matrix theory in the study of differential equations. The Jordan form is treated thoroughly and intelligibly.

Strang, G., *Linear Algebra and its Applications*. New York: Academic Press, 1976.

My favorite. Compact, stylish, every example and exercise brings out a worthwhile point. The thrust is towards applications but here mathematics is used to enhance understanding, not build an elaborate theory.

MATRIX COMPUTATIONS (Introduction)

Fox, L., *An Introduction to Numerical Linear Algebra*. New York: Oxford Univ. Press, 1964.

A very clear exposition at a fairly elementary level. Despite its age it still makes a very good text for a first course in the subject.

Stewart, G. W., *Introduction to Matrix Computations*. New York: Academic Press, 1973.

Comprehensive, authoritative, and readable. More demanding than Fox. Programs are included.

Schwarz, H. R.; H. Rutishauser; and E. Stiefel, *Numerical Analysis of Symmetric Matrices*. Englewood Cliffs, N.J.: Prentice-Hall, 1973.

The flavor is quite different from both Fox, and Stewart. The presentation of classical, background material is masterly but some of the numerical methods described with such care are surprisingly out of date. Algol programs are used to illustrate the methods.

Gourlay, A. R., and G. A. Watson, *Computational Methods for Matrix Eigenproblems*. New York: John Wiley, 1973.

A beautiful little book. Presents the essentials simply and directly. No programs.

MATRIX COMPUTATIONS (Specialized)

Householder, A. S., *The Theory of Matrices in Numerical Analysis*. Blaisdell, 1964.

Not surprisingly, the numerical methods covered are quite out of date. However, the first three chapters present valuable background topics from matrix theory: projections, factorizations, norms and convex bodies, the Perron-Frobenius theory, and inequalities involving eigenvalues. This material, and the extensive problems and exercises retain their value. The book is densely written, as for professional mathematicians, and was very influential in the field.

Wilkinson, J. H., *The Algebraic Eigenvalue Problem*. New York: Oxford University Press, 1965.

Those parts of matrix theory and perturbation theory which bear on numerical methods are presented first. Next comes roundoff error analysis. The major part of the book is a masterly and detailed account of all the important eigenvalue techniques. The analysis covers both exact arithmetic and noisy arithmetic. Several important examples illustrate the often surprising effects of round-off error. Effective use is made of "backward" error analyses, for which Wilkinson is justly renowned, to explain in a simple and compelling way the behavior of each numerical method.

This book is "the bible" for those who work with matrix computations.

Wilkinson, J. H., and C. Reinsch, *Handbook for Automatic Computation, Volume II—Linear Algebra*. New York:Springer-Verlag, 1971.

This is a collection of some 82 programs in the Algol language, and, in a sense, synthesizes 20 years' work aimed at relegating the basic analysis of matrices to the computer. Along with each program is valuable supporting documentation. Not all the contributions are by the two authors but each program bears the stamp of their influence and advice.

Many of these algorithms are available in the EISPACK collection. I have chosen to point to the appropriate, readily available, well-tested programs in the Handbook rather than present versions of my own.

Jennings, A., *Matrix Computation for Engineers and Scientists*. New York: John Wiley, 1977.

Most methods in current use for matrix problems, and some obsolete techniques, are described in good detail. The book is strong on examples and experiments, and less concerned with providing a theory to relate and explain the behavior of various methods. The language is aimed at structural engineers.

Author and Subject Index

D

E

F

G

H

I

J

K